STATISTICS
for math haters

Under the editorship of Wayne H. Holtzman

STATISTICS
for math haters

ELIJAH P. LOVEJOY
University of California, Santa Barbara

HARPER & ROW, Publishers
New York, Evanston, San Francisco, London

TO CHANTAL

Sponsoring Editor: George A. Middendorf
Project Editor: Cynthia Hausdorff
Designer: Frances Torbert Tilley
Production Supervisor: Will C. Jomarrón

STATISTICS FOR MATH HATERS

Copyright © 1975 by Elijah P. Lovejoy

All rights reserved. Printed in the United States of America. No part of this book may be used or reproduced in any manner whatsoever without written permission except in the case of brief quotations embodied in critical articles and reviews. For informaton address Harper & Row, Publshers, Inc., 10 East 53rd Street, New York, N.Y. 10022.

Library of Congress Cataloging in Publication Data
Lovejoy, Elijah P   1940-
    Statistics for math haters.

    1. Statistics.   I.   Title.
HA29.L8494   519.5   74-12279
ISBN 0-06-044069-4

Contents

A Note to the Student ix

A Statement of Thanks xi

PART I: THE SIGN TEST, AND THE LOGIC OF STATISTICAL INFERENCE 1

1. Introduction: An ESP Experiment 3
2. Studying the Random Guesser with a Marble Box 9
3. A Fundamental Question: Decision Rules 16
4. Picking a Decision Rule: A Statistical Test 21
5. Beyond the Marble Box 29
6. A Full Analysis of the Ten-Coin-Toss Model for the ESP Experiment 38
7. Review and Conclusion: ESP Experiment 45
8. Some Practice Problems, with Solutions and Discussion; A Formal Presentation of the Sign Test 51

PART II: THE NORMAL DISTRIBUTION AND THE Z TEST 59

9. Some Descriptive Statistics 61
10. Variance; Populations and Samples; Z Scores 71

11 The Sampling Distribution of the Mean; Probability Density Functions   78
12 The Addition Rule for Probabilities; A Self-Test for Review   91
13 The Normal Distribution; The Central Limit Theorem   102
14 The Central Limit Theorem, Continued   110
15 The Z Test, or Normal Test   118
16 The Z Test for a Single Observation; The Binomial Distribution   128
17 The Normal Approximation to the Binomial; The Normal Distribution Table   137

PART III: THE t TEST, CORRELATION, AND CHI-SQUARE   149

18 The t Test for the Mean of a Sample   151
19 The t Test for a Single Mean, Concluded; The t Test for the Mean of a Set of Difference Scores   161
20 Correlation   170
21 Correlation, Concluded   182
22 The Chi-Square (χ^2) Test to Help Decide Whether Two Variables Are Related   194
23 The Chi-Square (χ^2) Test, Concluded   205
24 The t Test for Means of Uncorrelated Samples   212
25 Two-Tailed Tests and One-Tailed Tests   220
26 Reviewing for Final Exams   225

APPENDIX   237

Z Tables: Areas of the Normal Probability Curve   239

Table of Critical Values for the t Statistic   240

r Table: Critical Values of the Correlation Coefficient   241

Table of Critical Values for the X^2 Statistic   242

Squares, Square Roots, and Reciprocals   243

Table of Squares, Square Roots, and Reciprocals of Numbers from 1 to 1000   245

Table of Decimal Equivalents of Various Fractions Used for Sign Tests   251

FLASH CARDS   253

INDEX   271

A Note to the Student

Statistics courses are all too often a source of considerable anguish to the students taking them. Many students in the social sciences need (or are required to have) some knowledge of statistical inference. Most of them take statistics courses. And too many of those who do never really understand the basic logic of statistical inference. This is especially sad, since it is unnecessary. This book is written with several assumptions. Among them are these:

1. Many students who take statistics are out of practice in mathematics and have trouble with operations that are almost second nature to many teachers of statistics.
2. Consequently, it is better to spell arguments out in detail, even at the risk of boring some of the brighter students.
3. Redundancy is probably helpful to most students in their first statistics course.
4. Having a good many specific, concrete examples simplifies the task of learning the new kinds of abstract thinking required for inferential statistics.
5. A first course of this sort is fundamentally a course in *thinking*.
6. If you ask a student to refer to a table 30 pages earlier, he *may*. So if there's something you really want him to see, you'd better present it again.
7. There is no reason why 95 percent of the students taking a statistics course should not master 95 percent of the material in it.
8. Learning to think statistically is, in the jargon of the day, mind-expanding. It enriches your perception and understanding of the world.
9. If you're going to really acquire the skills, you have to work lots of problems.

Some students have the good fortune to do mathematics problems rather easily. If you are one of these students, you will find the core material in this book very straightforward. To make the book usable for the large range of students who will be reading it, I have had to go a little more slowly than you might have liked. As a small compensation, I have included a number of problems that are specially difficult or interesting. They are marked by a dagger and appear at the end of the regular homework problems. I hope you accept the challenge and find the questions stimulating. Students who prefer to skip these problems will not miss material essential to the course.

I have done everything I could think of to maximize the probability that a student using this book will master the material in it. You can help. When a question is posed in the text, *please* stop reading and try to answer it. Whether you succeed or not doesn't matter. If you try to answer it, then you will probably get full benefit from the subsequent discussion. If you don't, you probably won't.

Statistics requires a certain minimum amount of pure memorization. If you prefer not to learn the 20 or 30 new concepts and words which are required, you might as well stop here. The book is never going to make sense. The number of such concepts and words is small. A set of flash cards is provided at the back of the book to help you memorize new terms. I have tried to include enough redundancy in the text so that the memorizing will be as painless as possible. If you take the very small amount of time required to learn these terms as they are presented in the text, you will move through the chapters easily. Please take the time. We're talking about a list comparable to half of one week's vocabulary for a foreign language course.

There is no reason why *you* should not master everything in this book. A great many students have done so before you; and many of them were confirmed math haters. A few bore the scars of earlier statistics courses. It is not a trivially easy matter, but it is not traumatically difficult, either. You just might even enjoy it. That would please me enormously.

ELIJAH P. LOVEJOY

A Statement of Thanks

This book has been developed over a series of quarters, in conjunction with Psychology 5, Data Analysis in Psychology, at the University of California, Santa Barbara. I am very much indebted to the hundreds of students who have used various approximations to this book, and who have helped me eliminate rough spots. Without the comments of these students—many of them confirmed math haters—this book would be vastly less useful.

I owe a special debt to two colleagues: Arthur J. Sandusky and Francis T. Campos. Each of these two skilled teachers of statistics has spent many hours helping to improve the following pages. I am most grateful. I would also like to thank a third colleague, John Cotton, who took time from a very busy schedule to read and comment on an early version of this book.

Next, I would like to mention several people who, as undergraduates in the course, were good enough to take the time to provide me with written comments on earlier versions of the text. David M. Fogel was outstanding in this group, and he worked through the entire text a second time after completing the course. I have also profited from the comments of Gail A. Valentine, Steger Johnson, Louise Gilbert, Neil Lakin, Claire Noonan, Lynn Morris, Amye Leong, Brett Enders, Becky Calhoun, Cathi Hamre, Douglas Moe, Ron Snyder, V. Georgiou, Gale Sutton, and Marion Kinney.

The "marble box" described in Chapter 2, and referred to frequently in the book, was built by Mr. Akira Saruwatari, of the psychology department at the University of California at Santa Barbara. As often happens in preparing a course demonstration, Mr. Saruwatari was asked to meet an unreasonably short deadline. He not only met it, but he made a fine piece of equipment, which has served a dozen courses since then. It is still going strong.

There has been a prodigious amount of difficult, detailed typing and manuscript preparation, in order to produce the several versions of these chapters, and culminating in the final book. Without the precise and cheerful help of Marilyn Abel, Rosemary Delagrave, JeNeal Bradford, Lois Koepnick, Becci Wamsley, Molly McGinnis, and Melanie McLees this book would simply never have happened. I hope they have not all become more dedicated math haters as a result of the experience. And I earnestly thank them for all their assistance.

My dealings with George A. Middendorf, at Harper & Row, and with Dr. Wayne Holtzman, the editor of this series, have been extraordinarily pleasant and helpful. At their request, several expert referees have gone over earlier versions of this book and helped eliminate many errors, rough spots, and incongruities. Although these referees' names are unknown to me, I shall always be indebted to them. I owe a double gratitude to Dr. Holtzman, who not only provided critical comments but also gave me a critical evaluation of the comments of the other referees. I very much appreciate this wide-ranging and expert criticism.

I am indebted to the Literary Executor of the late Sir Ronald A. Fisher, F. R. S., to Dr. Frank Yates, F. R. S., and to Oliver & Boyd, Edinburgh, for permission to reprint Tables II_1, III, IV, and VII from their book *Statistical Tables for Biological, Agricultural and Medical Research*.

The Sign Test, and the Logic of Statistical Inference

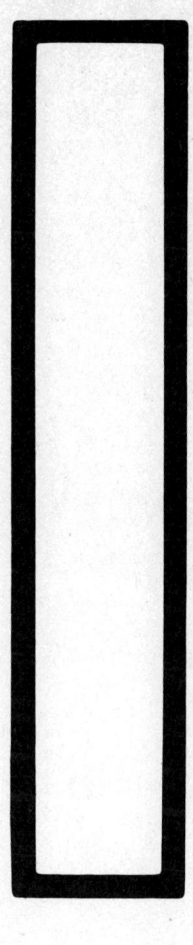

Introduction: An ESP Experiment

1

INTRODUCTION[1]

The word *statistics* is charged with emotion. Many students enroll in statistics courses only because they are required to, and they anticipate the course in a state of mind somewhere between mild displeasure and acute dread. A central purpose of this book is to convince you that statistics is not a black art, and that statistical techniques represent a refinement and elaboration of certain fundamental and important parts of basic reasoning or thinking. This book should sharpen your skills in statistical thinking and make you a better decision maker. At the same time, it should arm you with the ability to spot misuses of "statistics" accidentally or deliberately used to support unsound arguments.

An enormous body of knowledge exists in the general area of statistics. At some universities statistics departments exist independently of mathematics departments. Plainly, you are not going to learn a very large fraction of that whole body of knowledge in a single undergraduate course. Of all the people who take statistics courses as part of an undergraduate social science major, only a relatively small number will ever regularly use statistical tests. Most people using a text such as this only rarely will be called upon to carry out detailed statistical analyses; those who do expect to use statistics regularly will usually take advanced courses. Accordingly, this course is aimed at making you a more accurate and critical thinker. It is intended to help you understand *completely* the basic logic involved when someone conducts a statistical "test." Understanding this logic will enable you to spot many mistakes made by others. And you will also be able to understand and appreciate the role of statistics in the many situations where it is appropriately used.

[1] If you have not yet read "A Note to the Student" a few pages back, please do so before continuing. Thank you.

Statistics is a set of tools that you can use to help you make good decisions on the basis of limited and variable information or data.

This definition is probably quite different from the one you would have given before starting this book. Since it is very important, it wouldn't be a bad idea to memorize it, so that you can see how material to be presented in subsequent chapters relates to this simple, fundamental idea. At the end of this course, you should be a more exact thinker and better decision maker than you were when you started.

You are Already a Statistician

Suppose you watch two boys flipping pennies: One boy flips a penny and calls it in mid-air. If he correctly predicts how it will fall, the other boy gives him a penny. Now suppose that after you start watching, the boy who is flipping wins twice in a row. Would you be surprised? Would you think that he was cheating? Probably not. But suppose that, as you watch, the same boy continues to win, over and over. You keep track and note that he wins 20 times in a row! At this point would you be surprised? Would you be suspicious? Indeed, would you be convinced that he was cheating somehow?

These are statistical questions. Most people would agree that when someone wins twice in a row, they would not conclude that he's cheating. It's the sort of thing that happens very often when coins are tossed. But 20 wins in a row is another story. You probably said that you would find it very hard to believe after 20 consecutive wins that the victor was playing fairly. And I certainly agree with you. But what would you say if you observed an intermediate case: say, 5 wins in a row. Then what would you conclude?

If you decided that 2 wins would not be surprising but that 20 would be, then you are already a statistician, and a skilled one at that. Not all cases are so clear-cut, however, and that is why statistics is studied and why this course is being taught. Intermediate cases are more difficult; in the coming chapters you will learn a systematic set of tools for dealing with them.

We defined statistics above as a set of tools to help you make good decisions on the basis of limited and variable information. To illustrate how broad a range of decisions is included, four situations will now be given in which statistical tools would help you make decisions:

Example 1. A book called *Your Baby's Sex: Now You Can Choose* (David Rorvik with Landrum B. Shettles, M.D., Dodd, Mead, New York, 1970) was recently published. In this book a set of procedures is described for couples who wish to have a female child, and it is reported that 19 out of 22 couples who followed these procedures were successful. Does this prove that the procedures work? How confident can you be that this isn't just a lucky result of chance variability?

Example 2. Two young plants of the same size were used in an experiment. One was sprinkled with powdered moon rock; the other was not. The plant sprinkled with the moon rock grew noticeably faster than the other plant. Does this prove that moon rock increases plant growth? Probably not, since it is widely known that plants grow at different rates, even when you try to treat them in exactly the same way. But if one pair of plants can't convince you, how many such pairs would you want to study before believing that moon rock influences plant growth? Suppose you and a friend give different answers to this question? He said you'd need at least 100 plants, and you said 20 would suffice. Is there any way you could resolve the matter?

Example 3. A new kind of psychotherapy is tried on 1 "hopeless" case in a state mental hospital. Six months after the start of therapy the patient leaves the hospital, finds a job, and appears cured. What conclusions can be drawn? Suppose that a sample of 10 more "incurable" patients is treated and 3 of them are able to leave the hospital within a year. What conclusions can you make?

Example 4. In a large lower-division course like introductory psychology the content of the course is defined to be the factual knowledge necessary to correctly answer 2000 multiple-choice questions. Suppose that on the final exam 100 of these questions are presented and 1 student answers correctly only 32 out of 100, where 25 out of 100 correct would be expected by pure chance. The student later goes to the teaching assistant and notes that this was a very unfair test because, in fact, he knew about 1500 of the 2000 "facts," and the final exam simply happened to choose questions in the remaining 500 to which he did not know the answer. Should we believe the student's assertion that he knew 3 out of every 4 questions and was unlucky on the final exam?

In each of these examples, a decision must be made on the basis of limited and variable information. After reading a few more chapters you will be in a position to make such decisions much more accurately than you can now. These are examples of situations where statistical thinking helps.

AN ESP EXPERIMENT

At this point I would like you to imagine an experiment in extrasensory perception (ESP), which can be conducted as a class demonstration in statistics. During this demonstration a number of very important points will be raised — points that we shall have occasion to return to repeatedly in the coming chapters.

Ten opaque envelopes are placed at the front of the class. Each contains a piece of paper on which is marked either an X or an O. A student will be asked to come forward and make the best possible guess about which envelopes contain Xs and which contain Os. Before he does this, however, several important points are raised and a brief class poll is taken.

1. We can anticipate in advance all the possible ways the experiment could turn out. The student may guess 10 out of 10 envelopes correctly; or he may guess 9 out of 10 correctly; or he may guess 8 out of 10 correctly; or 7 out of 10; or 6 out of 10; or 5 out of 10; or 4 out of 10; or 3 out of 10; or 2 out of 10; or 1 out of 10; or 0 out of 10. There are thus eleven possible results of the experiment,[2] and it is certain that one of these eleven will occur. Together, they cover all the possibilities.
2. Certain results would be taken to be very "surprising" if there were no ESP here nor collusion between the student and me nor other such chicanery. For example, if a person guessed all 10 envelopes correctly, this would suggest to many observers that the guesses were not purely random. Based on a survey taken in one class with 100 students, Table 1.1 shows for each possible experimental result how many students said they would think that something funny was going on.

TABLE 1.1 NUMBER OF STUDENTS WHO WOULD BE SURPRISED BY VARIOUS RESULTS

Number in class who would believe we had "an effect"	73	73	30	7	1	1	0	0	38	69	75
Number of correct guesses in ESP experiment	10/10	9/10	8/10	7/10	6/10	5/10	4/10	3/10	2/10	1/10	0/10

To read the table, read vertical pairs of numbers. For instance, 30 of the students said they'd be surprised if the guesser was correct 8 times in 10.

[2] Count them.

3. Note that if the student guessed all 10 envelopes *wrong*, that result would be just about as surprising as if he guessed them all *right*! (Please refer to the table!) This illustrates a very important point in statistics – one that we shall return to repeatedly during the course.
4. Even if the student in question had no ESP powers at all, it is still possible that he would be lucky and guess all 10 envelopes correctly. That is, it is possible that we would be led to believe there is an effect, when in fact all we are observing is random variation or chance fluctuation. This is a possible source of error.
5. It is also possible that a student might have a small amount of ESP. Suppose that overall he guessed correctly 60 percent of the time. Even if we observed 6 correct choices out of 10, most people would conclude that this would not prove he had ESP. This is another possible source of error – that is, we might decide "He has no ESP" when in fact he has some.
6. Prior to the experiment, different people might be inclined to make different hypotheses about extrasensory perception:
   a. Some might assume that if the subject has ESP, we would expect to see 10 correct choices out of 10. If ESP exists, there should be a big, visible effect.
   b. Some people would not be quite so demanding but would expect the subject with ESP to do quite a lot better than chance.
   c. Still others might feel that with a weaker form of ESP the guesser would usually do better than chance, but that on some occasions he would not, thus hypothesizing that ESP is a chancy business and is only visible when you look at the results of a long series of experiments.
   d. A final hypothesis is that there is no such thing as ESP and that we would expect the overall results to be the same as completely random guessing. In other words, we expect that a person guessing in an experiment such as this would do no better and no worse than he would do by flipping a coin in front of each envelope and saying "X" when the coin came up heads and "O" when the coin came up tails. This is a very testable hypothesis since we could compare our ESP guesser with a coin-tossing guesser for as many experiments as we wish and thus decide whether the ESP guesser is doing any better.
7. If we wanted to be more sure of our conclusion about the presence or absence of ESP in this situation, we could repeat the experiment over and over, say ten times, and see how the student did with 100 guesses. Alternatively, we could enlarge the experiment by having a much larger set of envelopes about which to make guesses.

If you understand the above, you understand some of the most important points that will be made in this course. We shall elaborate on them of course. And it will be useful to introduce some new vocabulary. But the basic logic of the situation won't change. In subsequent chapters, when things begin to get a bit confusing, think back to how simple they were here, when you were just getting started.

To conclude this chapter let's go back over some of these points and introduce some vocabulary which will be important as we continue. Remember: These are new words to describe simple, old ideas.

First of all, look back at Table 1.1, which shows how many people in a class of 100 said they would be surprised by the various possible results of the experiment. Some people said they would consider 7 out of 10 correct guesses surprising and would interpret that to mean that the student had ESP, or something akin to it. And 73 people said that they would find 9 out of 10 correct guesses convincing proof of ESP. In other words, there was a difference in opinion about what to conclude with an observation of 7/10 or 8/10 or even 10/10. Disagreement about when to conclude "that result couldn't be due to chance" is common. We see many examples of it here. *A major goal of statistics is to devise public, open, decision making procedures so that people can agree with each other as much as possible about what conclusions to make in situations where there is uncertainty.* When a survey of this kind is conducted later in the course, there is much greater agreement about how to interpret the results.

It is worth noting once again that seeing 0/10 correct guesses is just about as surprising as seeing 10/10 correct guesses.

In point 4 above the possibility was mentioned that even without any special powers a person might get lucky and guess all 10/10 correct. Although in this case most observers would be inclined to conclude that we have evidence for ESP, the subject was in fact guessing randomly. This type of error has a special name in statistics: it is called a *Type One error*. This is the error you make when you say "that couldn't possibly be due to chance variation" — and you are wrong. To help you remember this term, think "There's *one* chance in a thousand to get this answer as a result of random variation." When that one-in-a-thousand chance happens as a result of random variation you are lured into making a Type One error.

In point 5 we note that we might fail to be impressed by a small amount of ESP. Although a person with weak ESP might get 6/10 correct, that would not be sufficient evidence for most of us to conclude that we have a case of ESP. This is another important kind of error in statistics. The error we make when we conclude "That looks like the result of chance variability," when we are actually dealing with a nonrandom effect, is called a *Type Two error*. A mnemonic which may help you here is that you make a Type *Two* error when the effect is *too* small to see. (Or, if you wish, *two* small to see.)

It is interesting and important to note that it is impossible to avoid all chance of error. Whenever we decide "That couldn't be the result of chance fluctuations," we run the risk of making a Type One error. We could be wrong. Similarly, whenever we conclude that the result we have observed is probably just the result of random fluctuations, we run the risk of a Type Two error. There might have been a real effect, but one that was too small to see.

In point 6d above we mentioned a very precise hypothesis — that there is no such thing as ESP. This was the hypothesis that the guesser would do no better and no worse than he would do by flipping a coin in front of each envelope, saying "X" if the coin came up heads and "O" if it came up tails. Such a very precise hypothesis — that is, that we are seeing the results of chance variation — is an example of what is called a *null hypothesis*. A null hypothesis is always stated in such a way that it is very precise and so can be tested comparatively easily. The symbol H_0 is often used, in statistics, to refer to a null hypothesis.

Finally, in point 7 above we mentioned that if we wanted to be more confident of our conclusion, we could increase the size of the experiment. An experiment that is likely to answer the question for which it was designed can be called a *powerful* experiment. We shall have occasions in the future to refer to the *power* of an experimental or statistical test.

Summary

Statistics is a set of tools to help you make good decisions on the basis of limited or variable information or data. It can be thought of as a systematic extension of careful thinking and inference. You are already a statistician, and you make statistical decisions often. This book will help you make better decisions. Four examples were presented of situations where statistical techniques might be helpful. An ESP experiment was described, and several important ideas and concepts were introduced in reference to the experiment. A *Type One error* is like a false alarm: you decide there is something interesting happening, when in fact the result is due only to random variation. A *Type Two error* occurs when you fail to detect a real phenomenon: for example, when the effect is *too* small to detect.[3] A null hypothesis (abbreviated H_0) is a precise statement that one is seeing the results of chance variation. It is a theory about the situation, and you are interested in deciding whether or not it is a tenable theory. An experiment or statistical test is *powerful* if it is likely to correctly discover a real effect when, in fact, there is one.

[3] To help yourself learn the definitions of Type One and Type Two errors, see the flash card supplement at the back of the book.

Problems

1. Try to write down from memory the definition of statistics. Then check the definition on page 4.
2. Please write a short paragraph on each of the four examples discussed in this chapter, giving your opinions of the questions asked and trying to indicate what sort of decisions you think would be appropriate.
3. Suppose that an ESP experiment like the one described in the chapter were carried out with 50 envelopes, and a person guessing whether each envelope contained an X or an O. How many correct guesses out of 50 would you require before you would decide "That many correct couldn't be the result of chance, I think the guesser has ESP or some special knowledge"?
4. Suppose that in the experiment above the person guessed only 7 out of 50 envelopes correctly. What would you conclude? Is it possible that a person with no special extrasensory powers and no special information could still guess all 50 envelopes correctly?
5. Suppose that a person had very slight ESP and guessed right on the average 11 times out of 20; 10 times out of 20, of course, would be pure chance. Is there any way that you could convince a skeptic that this person has ESP? What would it take to convince you?
6. Under what conditions do you make a Type One error in the ESP experiment described in this chapter? (See page 7.) Under what circumstances do you make a Type Two error? (See page 7.)
7. Suppose that in a different ESP experiment a card is selected at random from a well-shuffled deck. Before it is exposed, the subject tries to guess whether it is a club, a spade, a heart, or a diamond. State a null hypothesis appropriate for a test of whether or not the person has ESP. Review page 7 if you're not sure how to do this. An answer is given below.

Answer to problem 7. Remember that a null hypothesis is a *hypothesis that we are seeing the results of chance variation*. In the case of the man trying to guess the suit of the card chosen at random, the null hypothesis would state that his guessing is unrelated to the card actually drawn. Suppose that on a given try a person guesses clubs. Now if you choose a card at random from the deck, there's 1 chance in 4 that it will be a club. Similarly there's 1 chance in 4 that it will be a heart, 1 chance in 4 that it will be a diamond, and 1 chance in 4 that it will be a spade. On those trials when the guesser says clubs there's 1 chance in 4 that the card drawn will agree with his guess. The same is true if he guesses hearts, spades, or diamonds. So, regardless of what the subject guesses, there is 1 chance in 4 of his being correct. A precise null hypothesis, then, would be:

H_0 *We assume that the subject has no extrasensory perception, and that he has 1 chance in 4 of being correct on any one trial.*

Some people find the following equation helpful to aid them in remembering that the null hypothesis is a precise statement that all we are observing is *chance* fluctuations:

H_0 = **Chance**

Studying the Random Guesser with a Marble Box

2

AN EXAMPLE OF DUBIOUS STATISTICAL REASONING

An important reason for learning about statistics is that it represents an improvement in general reasoning powers. The good statistical thinker will be better equipped to reach accurate conclusions than will someone ignorant of statistics.

A politician once made the following statement about marijuana: "I know there is a great deal of controversy surrounding the use of marijuana. But marijuana is an illegal substance. Most medical authorities have stated that it is at least a hallucinogen with no known medical purpose. If used excessively, it can be dangerous. Furthermore, current statistics indicate that many hard narcotics users started off on marijuana." The implication is clearly intended that we should crack down on marijuana in order to reduce hard drug use. But does this implication necessarily follow from the observation "many hard narcotics users started off on marijuana"? Before reading on, you might find it a useful challenge to try to think of two reasons why the implication does *not* necessarily follow.

First, the fact that marijuana smoking generally precedes hard drug use does not mean that it is a cause. Most murderers drink milk as children; but this does not mean that milk drinking is related to their subsequent behavior. There is no relationship here, since non-murderers also drink milk as children. So one weakness in the argument is that the politician didn't say that hard drug users are more likely to have a past history of marijuana smoking than non-hard drug users. But suppose the latter were true, as I suspect it is. Then could we conclude that marijuana use *causes* hard drug use and that by eliminating marijuana use we could reduce or eliminate hard drug use? We could not.

Let me make this clear with another example. Swallows often fly low before a storm. This relationship has been known for centuries by sailors, and they have used it to predict bad weather: When you see the swallows flying low, prepare for bad weather. "Swallows flying low" often precedes "a storm." Does this mean that the swallows flying low cause the storm? That seems unlikely. It's hard to believe that by killing the swallows, or somehow keeping them from flying low, one could change the weather pattern. Rather it seems that both the first observation (low-flying birds) and the second (storm) result from another, prior event: low barometric pressure. When the pressure is low, insects fly low, and the swallows follow them while hunting. Also, low barometric pressure often precedes a storm.

It may be the case that in a similar way, a person who is likely to use hard drugs is also a person who is likely to use marijuana, and that for various reasons such a person will often smoke marijuana before using hard drugs.

Before reading on please answer the following question: Does the previous set of arguments demonstrate that pot does not lead to hard drug use?

If you answered "Yes," then you need the rest of this book. There is room for improvement in your statistical reasoning ability. If you answered "No," then you have successfully avoided a very common statistical pitfall: When a person tries to prove a proposition and fails, this does not mean that the proposition is false. It may be false, or it may be true. All we know is that one particular argument is invalid. In this example it may indeed be the case that marijuana smoking leads to hard drug use, at least part of the time. But the observation that many hard narcotics users started off on marijuana doesn't mean that there is a casual relationship. We just don't know without further information.

THE ESP EXPERIMENT: A GUESSER WITH NO ESP

Let's return to the ESP experiment. Remember in Chapter 1 it was mentioned that different people have different opinions about how well a person would have to do in the ESP experiment before they would be convinced that he wasn't just guessing. An important consideration here is how likely various scores are in a situation where a person guesses randomly. For example, if 8/10 happens pretty often even when a person is guessing randomly, then seeing someone guess 8/10 wouldn't convince you that he had ESP, just as seeing someone win 2 coin tosses in a row wouldn't convince you that he was cheating.

For the rest of this chapter, we shall look at the following question: How often would we see each of the eleven possible scores in an ESP experiment if the guesser had no ESP and guessed randomly for each of the ten envelopes? Before reading further, please take a couple of minutes to try to figure out how you would go about answering this question if it were up to you to find a good answer. There is a way.

The basic way of finding out what would happen if a person guessed randomly is to take a person with no ESP and run the experiment over and over again. Set up the ten envelopes, record ten guesses, and count the number correct. Rearrange the envelopes, get ten more guesses, and compute a new score. Continue this procedure for many, many repetitions of the experiment. After running it, say, 1000 times, you would have a pretty good idea of how likely each of the possible outcomes is. This is exactly the way gamblers in the sixteenth century found out about odds. Suppose a gambler wanted to answer the following question: Which is more likely when two dice are tossed — a *double 4* or a *total of 3*? He could toss dice over and over and keep track of the results. At the end of an evening he could look over his records and find out how often each of the two events occurred. Wealthy gamblers hired other people to do this dirty work for them; they called them "statisticians." Modern statistical theory grew out of the search for an easier way to make such predictions.

The method of "try the experiment over and over" is a perfectly good way to answer our question, assuming that we can find someone who, we are sure, has no ESP. The problem is that it is painfully slow. There must be a faster solution, and there is.

Making a Model

The essence of the solution is to make believe you are running the experiment by building a model for it. We said some time ago that if a person has no ESP, then there's a 50 : 50 chance that he'll guess correctly about the contents of any single envelope. When you toss a coin, there's a 50 : 50 chance that it will come up heads. So maybe we can use coin tossing to imitate the behavior of the guesser, and say "Correct guess" each time it lands heads, and "Incorrect guess" each time it lands tails. We say that we are using a coin-tossing game as a *model* for single guesses in the ESP experiment.

This use of the word model is very like using miniature wooden airplane *models* in the wind tunnels to find out how a real airplane will behave in the wind. We make a model that resembles the finished plane in important respects, and then we study the model in a wind tunnel rather than having to build the full-size plane and a big wind tunnel in order to test the effect of various wind velocities and angles. There is a close correspondence between the parts of the model and the real airplane. If the airplane has two wings, so does the model. If the airplane has a landing gear, so does the model. If the plane has a radio antenna, so does the model, and so on.

The use of models to help understand the real world is very common. Here are some other examples:

1. In considering whether to build a new building in a city, it is common to make a scale model of that area of the city, and to place a scale model of the proposed building in the appropriate location. This gives a vivid impression of how the finished structure will look. Note that the scale model of the building contains many features of the real building. But there are many other features, such as interior structure, plumbing, and wiring, that are not represented in the model.
2. In deciding how to arrange the furniture in a living room, one procedure is to simply start hefting it around, piece by piece, until a pleasing result is obtained. That is tiring. However, a useful model of the living room can be drawn on a piece of graph paper. Then a piece of paper the shape of each piece of furniture, and cut to scale, can be quickly produced. It is much easier to move the model chairs and sofas around the model room than to make the corresponding changes in the real room.
3. In personal financial planning most of us make use of a simple mathematical model. We represent anticipated income and expenditures by positive and negative numbers. An alternative way to find out "What would happen if I bought the phonograph, paying for it in three months, and sold my bicycle during the second month?" would be to get out a pile of cash and start moving it around on a table. The number system provides a simple and elegant *model* for the actual transactions, while retaining the essential features.

In our study of the ESP experiment, the coin-toss model is very simple, and the relationships between the real situation (the ESP experiment) and the model are quite clear:

Real World (Actual Experiment)	Model World (Imitation or Simulated World)
Envelope contains X or O	Envelope not represented
Person guesses X or O	Person's guess of X or O not represented
Decide "right" or "wrong" guess	Toss coin: heads corresponds to "right"; tails corresponds to "wrong"
We assume that without ESP there's a 50 : 50 chance of guessing right for each try	There's a 50 : 50 chance of heads
Ten guesses make an experiment	Ten coin tosses are used to model an experiment

Why do we use a model rather than carrying out the experiment over and over? Because it's a lot more convenient.

How do we know we'll get the same results by studying the coin-tossing model as by running the experiment? For some people this is reasonable, since they've worked with lots of models and always found that studying the model gives the same sort of information as studying the real world. For others this is an act of faith, based on the visible similarity between the model and the real world situation. Think about it.

Now, in order to try to find out how things would go if we ran the ESP experiment over and over using a guesser who had no ESP, we run the model experiment over and over.

Suppose a coin is flipped 10 times, and the results recorded H for heads, and T for tails:

HTTTHHTHHT   5 heads

This run corresponds to a complete experiment in which the subject guessed exactly 5 right. Let's try again:

THTTHTTHTH   4 heads

This run corresponds to a complete experiment in which the volunteer guessed exactly 4 right.

By repeating this 10-toss process over and over, we could find out what proportion of the time we'd expect 0/10 correct (zero heads), how often we'd expect 1/10 correct (1 head), and so on, for the other possible results. This would be substantially simpler than running the actual guessing experiment over and over.

The coin-toss model is still time consuming. We can further accelerate our investigation by introducing yet another model: In class I use a large wooden box containing 600 marbles, 300 red and 300 green. I have a specially made sampling paddle which randomly removes ten marbles from the box at a time. The paddle contains ten holes, each just big enough for one marble. By putting the paddle into the box and shaking it, I can take a sample of 10 marbles out of the box. Then I can use the set of ten marbles as a *model* to represent a complete 10-trial ESP experiment with the set of correspondences as follows:

Real World (ESP Experiment)	**Model (Marble box)**
If the guesser has no ESP, there's a 50 : 50 chance he'll be correct	There are an equal number of red and green marbles, so there's a 50 : 50 chance that any one hole in the paddle will be filled by a green marble
Correct guess	Green marble
Wrong guess	Red marble
Entire 10-guess experiment	Ten marbles in paddle
Number of correct guesses	Number of green marbles

In the table you can see the correspondences between parts of the model and parts of the real experiment. For example, drawing a *green marble* in the model corresponds to a *correct guess* in the real experiment. Please think through the other four correspondences in the table.

Now, using the marble box, we can easily simulate the results of running the experiment over and over without any ESP. Each time we take a sample of 10 marbles out of the box, we simulate a complete run of the experiment. After taking a sample, the marbles are returned to the box before taking the next sample. (This is called sampling *with* replacement.)

In a recent class we went through the procedure 50 times. Each time we recorded the number of green marbles in the paddle (0/10, 1/10, 2/10, 3/10, etc.). Then we found the total number of times that each result occurred. These totals are shown in Figure 2.1.

This kind of a table is known as a *frequency distribution table*. We can also draw a picture or graph of the same information; this is done in Figure 2.2.

STUDYING THE RANDOM GUESSER WITH A MARBLE BOX

Number of green marbles in paddle	0/10	1/10	2/10	3/10	4/10	5/10	6/10	7/10	8/10	9/10	10/10
Frequency (number of times out of 50 that we observed this number of green marbles)	0	0	0	10	9	12	9	8	1	1	0

For example, we observed 3/10 green marbles exactly 10 times

Figure 2.1 How often each total was observed in a 50-trial ESP experiment.

Several interesting features about this frequency distribution are worth noting now. The most frequent result, which occurred 12 times out of 50, is 5/10 green marbles.[1] That seems reasonable. But it is interesting that 4/10, 6/10, 3/10, and 7/10 turned up quite frequently, too. Each of those results occurred at least eight times. The other possible results are much less common. Four of them were not seen at all in our series of 50 samples.

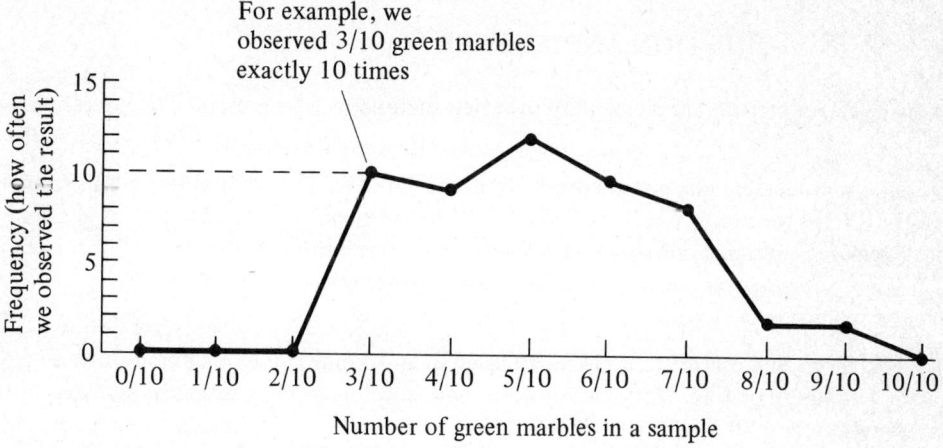

Figure 2.2 Frequency distribution (or frequency polygon) showing how often each of 11 possible totals was seen in a 50-trial marble box experiment. Note that frequencies of zero are also shown (for example, for 0/10, 1/10, and 2/10 green).

Now, if this marble box model is a good model for the ESP experiment when the guesser has absolutely no ESP, this suggests that even without ESP it wouldn't be surprising to see someone guess 7/10 correctly. But it seems as if he shouldn't do much better than that, most of the time, if he has no ESP.

---

[1] Please refer to the graph (Figure 2.2) to see that this is indeed the case. You will save a lot of time in the long run if you get in the habit of *checking carefully* any assertion like this. By doing this you make sure that you are following the argument. You may also find some errors in the text.

At this point, if you have understood all that has been said, you have understood some of the most important ideas in this book. Now let's take advantage of this example to introduce some vocabulary which we'll be seeing increasingly.

The set of 600 marbles in the box is a *population* of marbles, from which we took lots of *samples*. Each sample contained ten marbles: It was a *sample of size 10.*

For each sample we found a single number — the proportion of green marbles in the sample (0/10, 1/10, 2/10, etc.). This score is an example of a *statistic*: a number calculated on the basis of the sample.

When we had taken 50 samples and looked at the frequencies with which each of the eleven possible results occurred, we had a *frequency distribution*, which can be presented as a table or as a graph. The graph is sometimes called a *frequency polygon.*

To prepare for important future developments, it is useful to introduce one more technical term here. The frequency distribution in Figures 2.1 and 2.2 is the result of *sampling* marbles from a box. It is also helpful to think of it more abstractly as the result of sampling *scores* (0/10, 1/10, 2/10, etc., up to 10/10) from a population in which each of these scores occurs with a characteristic probability. Suppose you had a box containing thousands of index cards, and each card had a single score, like *7/10 correct,* marked on it. By drawing one card at a time, recording the observation, and returning the card to the box, you could similarly sample scores from a population of scores.

In this example, the population of scores is abstract: it contains all the scores or experimental results that could possibly be observed. By taking 50 samples, we in effect sample 50 scores from that infinite population. The figures, then, are an approximation to *the sampling distribution of the score.* We shall see several further examples of sampling distributions as the book continues. This is a central concept.

QUESTION AND ANSWER

From time to time in this book, interesting supplementary or review material will be presented in a question and answer format.

Question: In this chapter there is considerable talk about models. In particular, a big box of marbles, and a system for taking samples of size 10 from the box, are proposed as a model for a person with no ESP making guesses about the contents of ten opaque envelopes. How can we tell whether this is a *good* or *useful* model for that situation? More generally, how do you ever decide if a model is a good one? (*Dear reader*: PLEASE try to answer the question before reading on.)

Answer: A model is like a theory. You find out if it is accurate by using it, and seeing if the predictions it leads you to make are accurate. Suppose that John Wilton is known to have absolutely no extrasensory perception and a lot of free time. We set up the ESP experiment time and time again, and John makes a set of guesses. We keep track of how often he gets 0 out of 10 correct; 1/10; 2/10; etc. After 500 runs through the experiment, we let him go home and get out the marble box. Then we take 500 samples using the marble box, and see how many times we observe 0 out of 10 green; 1/10; 2/10; etc. Then we compare the two sets of numbers. Did John have 0/10 correct guesses about as often as the marble box produced 0/10 green marbles? Likewise for 1/10 correct guesses and 1/10 green marbles? The other events? If so, the model is a useful one, for it is certainly easier to take a sample of 10 marbles than it is to run the ESP experiment once. Not to mention the problem of finding someone *known* to have absolutely no ESP!

Summary

An example of inaccurate statistical reasoning was presented. Hard drug users often have used marijuana before they tried heroin. The common conclusion that marijuana *causes* hard drug use was discussed.

Returning to the ESP experiment, we considered this question: How likely would various possible scores be if the person had *no* ESP? A marble box model was proposed to help answer the question. We presented the results of a 50-trial sampling experiment with the marble box. These results were summarized into a *frequency distribution table,* and then displayed as a *frequency distribution graph,* or *frequency polygon.* The terms *population* and *sample* were introduced. It was noted that the frequency distribution of results can be described as an estimate of the *sampling distribution of the score,* where the score gives the proportion of green marbles for the marble box.

Problems

1. Suppose that an ESP experiment is conducted by rolling a single six-sided die and having a blindfolded subject try to guess which face is on top. State a null hypothesis that would be appropriate for analyzing this situation. (An answer is given following these problems.)

2. Suppose that in a dice ESP experiment, like the one just discussed, a subject makes 30 guesses. Design a procedure like the marble box experiment that could be used to find out experimentally how likely various possible results or *events* would be in this situation. The possible results or events are: 0/30, 1/30, 2/30, 3/30, etc., 30/30 correct guesses. How could you find out experimentally the likelihood of each of these possible events, assuming as we did before that the null hypothesis is true: The subject has just 1 chance in 6 of guessing correctly on any single try?

3. Given the statement, "Furthermore, current statistics indicate that many hard narcotic users started off on marijuana," many people would deduce that marijuana causes hard narcotic use, and that by stopping marijuana smoking we could stop or reduce hard drug use. Give two reasons why this deduction is not necessarily logical. Just because the deduction is illogical, does that mean that it is false? (Turn back to the beginning of the chapter if you need help.)

4. Give four examples, in addition to those mentioned in the text, in which we use a *model* to find out the properties of some part of the real world. Try to find examples that are very different from those given in the chapter. What do you think of this statement: All thinking and understanding is the result of models.

5. Look at the frequency distribution graphed in Figure 2.2. Suppose that the marble box experiment had been carried out 200 times instead of 50 times. Describe the differences between the frequency distribution you think would result from that exercise and the distribution found on page 13.

6. Rolling a die, I observed the following outcomes in order: 5, 6, 2, 3, 6, 6, 2, 6, 3, 5, 4, 2, 6, 4, 5, 3, 2, 3, 2, 5, 3, 5, 2, 1. Compile a frequency distribution like the one shown in Figure 2.1 based on these data. Draw a graph of the resulting frequency distribution like the one shown in Figure 2.2.

†7.[2] There is a slight difference between the marble box model and the ten-coin-toss model for the ESP experiment. This difference, for example, will result in the events 10/10 and 0/10 being slightly less likely in the marble box situation than in the coin-toss situation. Can you identify the difference that causes this effect? How could one minimize the importance of the effect?

Answer to Problem 1. Remember that a null hypothesis is a hypothesis that we are seeing the results of *chance variation*. For this situation, the null hypothesis is that the person's guesses are unrelated to the face of the die which actually happens to be on top. Suppose that a person guesses "3." Since there's no relationship between the die and the guess, there should be just 1 chance in 6 that the die actually is showing a 3 on this trial. So a null hypothesis of the following sort is appropriate: The subject has no ESP, and so there is exactly 1 chance in 6 that he will guess correctly on any one trial.

---

[2] The questions marked by a dagger (†) in the book are not essential to the core material. They raise interesting but often somewhat difficult points relating to the material. They are intended as enrichment for the especially well-prepared student who finds the regular homework straightforward.

A Fundamental Question: Decision Rules

AN EXAMPLE IN WHICH YOU MAY BELIEVE MORE THAN YOU'VE BEEN TOLD

In a recent advertising campaign, it was proudly announced that 90 percent of the Volvos sold in the United States in the last eleven years were still on the road. This means that Volvos last a long time, right? Not proven!

Suppose that the ad said that 90 percent of the Vegas sold in the last eleven years were still on the road. That wouldn't mean much, since the Vega was first sold in 1970. None were sold before that. But consider the case of the Volvo: Sales have probably been much greater in the past few years than they were ten or eleven years ago. Indeed, it might well be the case that 90 percent of the Volvos sold in the last eleven years were sold in the last three or four years.

If we knew that 90 percent of the Volvos sold in the United States eleven years ago were still on the road, that would be much more impressive.

One last point is worth making: The statement in the ad was presumably accurate as printed. If the reader believes that the statement demonstrates that Volvos last a long time, that's his error! All the more reason to be an accurate statistical thinker.

A FUNDAMENTAL QUESTION IN STATISTICS

We have already had occasion to consider several questions of this sort: *Is the observation the result of a real underlying process, or is it just the result of random variation?* As this book continues, we shall return repeatedly to this basic question. In many of the chapters we will present techniques to help answer this question in a variety of situations. The examples in Table 3.1 are intended to make it clear that this question comes up very often.

TABLE 3.1

Observation	Question
1. Coin-toss gambler wins five times in a row.	1. Is he cheating, or is he just lucky? Is the observation the result of a real underlying process (cheating), or is it just the result of random variation?
2. After 22 couples try a procedure which is supposed to help them have a female child 19 are successful.	2. Did the procedure really influence the sex of the child, or was the observation just the result of chance variation?
3. Two plants start out the same size. One is sprinkled with moon rocks. It grows faster.	3. Did the moon rocks really influence growth, or was this just a chance difference?
4. Acme Rain Service has tried on 12 occasions to produce rain by seeding clouds. It was successful 3 times.	4. Did the cloud seeding really increase the chance of rain, or was the Acme man lucky 3 times in 12?
5. One hundred students learning math with teaching machines average 74.2 percent correct on the final exam. One hundred students learning math in a traditional way average 75.7 percent correct.	5. Is there really a difference in the effectiveness of the two systems, or is the observed difference just the result of random variation?
6. In a 10-envelope ESP experiment a person guesses correctly 7 times.	6. Is the observation the result of random variation, or is there really an effect (ESP, or collusion, or cheating, or something like that)?

DECISION RULES

A complete statistical analysis of an experimental situation, made before an experiment is run, can lead to a *decision rule* — a rule that will describe how you'll answer the fundamental question for each possible result you might observe. For example, the following is a plausible decision rule for the ESP experiment: If the subject guesses 10/10 correct, or 9/10 correct, I'll be convinced that there's a real underlying process (like ESP) which is helping him guess correctly. But if he gets 8 or fewer correct, I'll conclude that this is just the result of random variation. Note that this decision rule specifies what decision you'll make for every possible result of the experiment. We could also summarize this decision rule in a table like Table 3.2.

What would be the "best" decision rule for our ESP experiment? There are many different decision rules we could consider. Here are some of the more plausible ones:

1. Say "He has ESP" if you observe 10/10 correct. Otherwise say "He is just guessing."
2. Say "He has ESP" if you observe 10/10 or 9/10 correct. Otherwise say "Just guessing."
3. Say "He has ESP" if you observe 10/10 or 0/10 correct. Otherwise say "Just guessing."
4. Say "He has ESP" if you observe 10/10 or 9/10 or 0/10 or 1/10 correct. Otherwise say "Just guessing."

It would be a good exercise for you to try to add to this list until it contains all the rules that *you* would possibly consider. We shall present a more extensive list of plausible decision rules below.

How do you choose the best decision rule? An ideal decision rule would have the following properties: Whenever the subject had ESP it would lead to the decision "He has ESP," and whenever he had no ESP it would lead to the decision "just guessing." Unfortunately, as will become painfully clear, we can almost never find so perfect a decision rule. We never know for sure whether or not the subject has ESP. All we know is how many times he guessed correctly.

18 THE SIGN TEST, AND THE LOGIC OF STATISTICAL INFERENCE

TABLE 3.2   A SINGLE DECISION RULE

This is a *single* decision rule. For each possible observation, it specifies a decision.

Result of ESP experiment (observation)	Decision	
If you observe 10/10 correct,	decide	"He has ESP"
If you observe 9/10 correct,	decide	"He has ESP"
If you observe 8/10 correct,	decide	"He was just guessing" (and this was a random fluctuation)
If you observe 7/10 correct,	decide	"He was just guessing"
If you observe 6/10 correct,	decide	"He was just guessing"
If you observe 5/10 correct,	decide	"He was just guessing"
If you observe 4/10 correct,	decide	"He was just guessing"
If you observe 3/10 correct,	decide	"He was just guessing"
If you observe 2/10 correct,	decide	"He was just guessing"
If you observe 1/10 correct,	decide	"He was just guessing"
If you observe 0/10 correct,	decide	"He was just guessing"

On the basis of this score we must make a decision, using a rule like those listed above. Recall that we are trying to answer the fundamental question: Is the observation the result of a real underlying process, or is it just the result of random variation? There are two possibilities: either the subject has ESP (there is a real underlying process), or he is just guessing. Similarly, there are two possible decisions we could make after an experiment: we can decide "He has ESP," or we can decide "He was just guessing." We can summarize the various possibilities in a two by two table like Table 3.3.

TABLE 3.3   A 2-BY-2 DISPLAY OF DECISIONS AND WORLD STATES

		Actual state of the world	
		Subject has ESP	*Subject is just guessing*
Decision	We decide "He has ESP."	This cell of the table corresponds to the case when the subject has ESP and we correctly say that he does. A correct decision.	This cell of the table corresponds to the case when we make a severe error: Although in fact we are seeing only the effects of random variation, we conclude that the subject has ESP. This is called a *Type One error*, or *false alarm*.
	We decide "He was just guessing."	This cell of the table is also an error: Although the subject actually has ESP, we decide that he is just guessing. We attribute the results to chance: in fact there is a real underlying process. This is called a *Type Two error*, or a *miss*.	This cell of the table corresponds to the case when we correctly decide that we're seeing nothing but chance variation. A correct decision.

Now we can start to compare different decision rules. Statisticians are especially concerned with the error called a Type One error, or a false alarm. You make this kind of error when you decide that the results reflect a real underlying process when in fact they arise from chance variation alone. This seems reasonable. Since chance variation is ubiquitous, we have to deal with random fluctuations in all kinds of situations. Sometimes there are systematic relationships at the same time, and we would like to identify them. But if we make many Type One errors, that is rather like seeing lots of ghosts where there are none. It is difficult to accurately understand something if you make the mistake of hypothesizing that there are real underlying processes when in fact there are none. Accordingly, we try to control the risk of this kind of error.

A question to consider: In the ESP experiment we have been discussing, what would be a Type One error (or false alarm)?

As may be seen from the previous two-by-two table, this is the error of deciding that the subject has extrasensory perception when in fact he has none. Now we can ask a question that lies at the heart of our statistical analysis of the ESP situation: How do various possible decision rules compare in terms of the risk of a Type One error (false alarm)? That is, how do they compare in terms of the risk of our deciding that a subject has ESP when in fact he has none? We shall use the results of our marble box experiment to begin to answer this question in the next chapter. If you have understood most of what has been said so far, you should be able to do the following exercise. This will greatly aid your comprehension and memory for Chapter 4.

Exercise: One plausible decision rule for the ESP situation is the following: If the guesser is correct 10/10 or 9/10 or 8/10 or 7/10 times, I'll decide he has ESP. Otherwise, I'll decide he was just guessing. Try to figure out: *If we use this decision rule, how big a risk do we run of a Type One error, or false alarm?* That is, suppose that we use this decision rule and that the subject has no ESP but guesses randomly. What is the chance that we will erroneously conclude "He has ESP"? Try to answer this question before going on to the next chapter. You will probably want to look back at the results of the marble box experiment, which are found in the frequency distribution in Figure 2.1.

Here is a hint: How many envelopes will the subject have to guess correctly if we are to be led to make this kind of error? What is the chance that he will do sufficiently well?

Summary

We discussed a car advertisement that takes advantage of the fact that most people are not very skillful at thinking statistically. A *fundamental question,* which is often of interest when we use statistical tools, was raised: Is the observation the result of a real underlying process, or is it just the result of random variation?[1] Several examples were presented to show that this kind of question is common. The concept of a *decision rule* was presented: A decision rule is a rule that specifies how you will answer the fundamental question for each possible result of your experiment. It is the *rule* you use in making *decisions* on the basis of an experiment. An ideal decision rule will always decide "There's a real effect" when in fact there is: It will never make a Type Two error. At the same time, it will never decide "There's a real effect" when in fact the result is due to random variation: It will never make a Type One error. Alas, such decision rules are possible only in ideal worlds. In general, we have to accept some risk of making errors.

At the end of the chapter, a very important exercise was presented. The exercise involved figuring out how likely a Type One error (false alarm) would be for one particular decision rule for the ESP experiment.

Problems

1. Complete the exercise at the end of the chapter.
2. Review the list of four plausible decision rules in the text. Write down as many additional plausible decision rules as you can. These should be rules you or someone else might consider good ones.

---

[1] To help yourself remember this *fundamental question,* see the flash card supplement in the back of the book.

20 THE SIGN TEST, AND THE LOGIC OF STATISTICAL INFERENCE

3. Without looking back at the text, try to write down as much as you can of the 2-by-2 table presented in this chapter, showing the possible results of two different states of the world, coupled with two different decisions.

4. Suppose that for a certain car manufacturer, car sales for the past eleven years are as follows (in thousands of cars):

10 years ago	0
9 years ago	1
8 years ago	1
7 years ago	2
6 years ago	1
5 years ago	3
4 years ago	7
3 years ago	16
2 years ago	27
1 year ago	24
This year	33

   Would it be surprising if 90 percent of the cars sold in the past eleven years were still on the road? What is the most "representative" age in this group of cars, the age such that half of the cars are newer than it and half of the cars are older? (As we will see in Chapter 9 this is called the *median* age.)

5. Remember the fundamental question in statistics, which comes up in a great many different forms: Is the observation the result of a real underlying process, or is it just the result of random variation? Find three examples from your own life in the past few days where you've asked a question of this sort. Write them down.

6. Why do you suppose a Type One error is sometimes called a false alarm?

7. What is your risk of a Type One error, or false alarm, with the following decision rule: No matter how well the subject performs, I will not decide that he has ESP. For every possible result, I'll decide that he is just guessing. Is this a perfect decision rule? Why (or why not)?

8. Mrs. Westphal said to her neighbor, "I'm going to switch to the new supermarket. Last year it cost me $27.40 per week, on the average, for groceries; this week, at the new market, I spent only $26.00. I figure that I'll save $72.00 this year." Is she justified in making this confident decision? Is there any chance she might be making a mistake? Type One or Type Two?

Picking a Decision Rule: A Statistical Test

THE CHANCE OF A TYPE ONE ERROR

At the end of the last chapter the following exercise was presented: One plausible decision rule for the ESP situation is the following: If the guesser is correct 10/10 or 9/10 or 8/10 or 7/10 times, I'll decide he has ESP. Otherwise, I'll decide he was just guessing. Try to figure out: *If we use this decision rule, how big a risk do we run of a Type One error, or false alarm?* That is, suppose that we use this rule, and that the subject has no ESP but guesses randomly. What is the chance that we will wrongly conclude "He has ESP"? If you have not tried to do this exercise, you will do yourself a favor, and maybe even enjoy it, if you try to do so now. You will probably want to use the results of the marble box experiment, which are reprinted in Table 4.1.

TABLE 4.1 THE RESULTS OF THE MARBLE BOX EXPERIMENT

Number of green marbles in paddle	0/10	1/10	2/10	3/10	4/10	5/10	6/10	7/10	8/10	9/10	10/10
Frequency (number of times out of 50 samples that we observed this number of green marbles)	0	0	0	10	9	12	9	8	1	1	0

Here's the logic of the situation. We want to find out how often someone would guess 7/10 or 8/10 or 9/10 or 10/10 correct, if he had no ESP whatsoever. Remember that the marble box is a model for the way someone would guess if he had no ESP whatsoever. It has been argued that a person without ESP should guess 7/10 correct just about as often as we observe 7/10 green

22 THE SIGN TEST, AND THE LOGIC OF STATISTICAL INFERENCE

marbles with the marble box. Similarly, a person without ESP should guess 8/10 correct just about as often as we observed 8/10 correct in the marble box. And so on. Now we must ask how often we observed 7/10 or 8/10 or 9/10 or 10/10 correct in the marble box experiment. The answer is 8 plus 1 plus 1 plus 0, which is equal to 10 times, in 50 samples (see Table 4.1). By analogy, we reason that a person without ESP would get one of these four scores just about as often, or about 10 times in 50. Since any one of these scores would, according to our decision rule, lead us to decide "He has ESP," this means that about 10 times in 50 we would be led to say this when in fact he has none. About 10 times in 50, we would make a Type One error. Associated with this particular decision rule, then, we have deduced that there is a chance of a Type One error of about 10 in 50, or 0.2 (since 10/50 = 20/100 = 0.2). This is the solution to the problem in the exercise.

CHOOSING A DECISION RULE

We have just finished an analysis of one particular decision rule. We can now return to a problem that was raised in the last chapter: How do we choose the best decision rule? In order to answer that question, we will go through the following steps:

1. Write down all the plausible decision rules we can.
2. Figure out the *chance of a Type One error* associated with each.
3. On the basis of these calculations, pick a decision rule.

First we will write down all the plausible decision rules we can:

1. We might say "He has ESP" only if he guesses all 10/10 correct, and otherwise decide he has none. This is a conservative decision rule, since it takes a very superior performance to convince us that the subject has ESP.
2. We might say "He has ESP" if he guesses all 10 correct (10/10), or if he does the opposite and guesses all 10 wrong (0/10 correct). The reasoning here is that if a person is just guessing, it would be as surprising for him to guess all 10 wrong as for him to guess all 10 right. Either event would suggest ESP.
3. We might say "He has ESP" if he guesses 10/10 or 9/10 correct, and otherwise say "He is just guessing." This is quite like rule 1 above, but somewhat more lenient.
4. We might say "He has ESP" if he guesses 10/10 or 9/10 or 1/10 or 0/10 correct. This is a rule like rule 2 above, but somewhat more lenient.
5. We might say "He has ESP" if he guesses 10/10 or 9/10 or 8/10 correct, and otherwise say "He is just guessing."
6. We might say "He has ESP" if he guesses 10/10 or 9/10 or 8/10 or 2/10 or 1/10 or 0/10 correct, and otherwise say "He is just guessing."
7. *Et cetera.*

Now for each of these decision rules, and any other of interest, we can perform a calculation like the one at the start of this chapter: We can find out how likely we would be to make a Type One error with that particular decision rule. For example, consider rule 3. We want to know how likely it is that, using this rule, we would be led to decide "He has ESP" when in fact the subject was guessing randomly. As before, we use the marble box. We say that the chance of a random guesser getting 9/10 or 10/10 correct is just the same as the chance of the marble box producing 9/10 or 10/10 green marbles. Referring back to the results presented in Table 4.1, we see that 9/10 occurred 1 time and 10/10 occurred 0 times out of 50. The chance of one of these results, then, is just 1 plus 0 times in 50, or 1/50. If you have followed the argument up to this point, you should have no difficulty making up a table which shows the estimated chance of a Type One error for each of the possible decision rules. You will have a chance to prove this in the problems. The results are in Table 4.2.

TABLE 4.2 DECISION RULES AND THEIR CONSEQUENCES

Decision rule: Say "There's ESP" if you observe any of these results	Estimate of the chance of being led to claim ESP when there is none, based on the marble box model (estimated chance of Type One error)
10/10 correct	$\frac{0}{50}$ (since the event 10/10 never occurred in marble box)
10/10 or 0/10 correct	$\frac{0}{50}$ (since neither event ever occurred)
9/10 or 10/10	$\frac{1+0}{50}$ (since 9/10 occurred once)
9/10, 10/10, 0/10, or 1/10	$\frac{1+0+0+0}{50} = \frac{1}{50} = .02$
8/10, 9/10, or 10/10	$\frac{1+1+0}{50} = \frac{2}{50} = .04$
7/10, 8/10, 9/10, or 10/10	$\frac{8+1+1+0}{50} = \frac{10}{50}$ (see example at start of unit)
7/10, 8/10, 9/10, 10/10, 3/10, 2/10, 1/10, or 0/10	$\frac{8+1+1+0+10+0+0+0}{50} = \frac{20}{50}$
6/10, 7/10, 8/10, 9/10, 10/10	$\frac{9+8+1+1+0}{50} = \frac{19}{50}$
6/10, 7/10, 8/10, 9/10, 10/10, 4/10, 3/10, 2/10, 1/10, or 0/10	$\frac{9+8+1+1+0+9+10+0+0+0}{50} = \frac{38}{50}$
Any result	$\frac{50}{50}$

Understanding a sample entry of this table: If the decision rule is to say "There is ESP" if we observe 8/10 or 9/10 or 10/10 correct guesses, then the estimated chance of a Type One error associated with this decision rule is 2/50. In order for a Type One error to occur, a person without ESP has to guess correctly 8/10 or 9/10 or 10/10 times. We estimate that the chance of this happening is the same as the chance of observing 8/10 or 9/10 or 10/10 green marbles in the marble box experiment. Referring to the results in Table 4.1, you can see that 8/10 occurred once; 9/10 occurred once; and 10/10 occurred 0 times out of 50. In all, then, we observed these events 1 plus 1 plus 0 times, or 2 times out of 50.

We have listed a series of decision rules and calculated the risk of Type One error associated with each one. Now how do we go about picking one we think is best? You will get the most out of the following discussion if you try to answer this question before reading on.

Note that the various possible decision rules which were included in the left-hand column of Table 4.2 are arranged in a sort of order: the most conservative rule is on top, and the least conservative rule is at the bottom. People adopting the first decision rule (only if the person gets all 10 correct will we conclude he has ESP) run a very small risk of a Type One error. On the basis of our experiment, we are sure that it is very unlikely that 10/10 would occur by pure chance. On the other hand, consider a less conservative person who agrees to the decision rule which says that 7/10 or 8/10 or 9/10 or 10/10 is sufficient evidence to be convincing: this person runs a higher risk of a Type One error. According to the right-hand column, one of these results will happen about 10 times in 50, or 20 percent of the time, *even if the guessing is purely random.* So in this respect, this decision rule is more risky than the 10/10 decision rule.

Why, then, would anyone use any but the most conservative decision rule? Before reading on, you might like to try to stop here to see if you can anticipate the next paragraph, which answers this question.

Consider this decision rule: No matter what the result of the experiment, you can't convince me there's any ESP or other nonrandom effect going on here. That's the most conservative possible rule. And with it, there is absolutely no chance of a Type One error, right? But what is wrong with it? No matter how much ESP the subject has, a person with this decision rule will never recognize it. He has no chance of correctly identifying ESP when there is some present. And, looking over the various rules on the left side of Table 4.2, you can see that those at the bottom have the greatest chance of correctly identifying ESP when there is, in fact, some ESP present. In different words, the decision rules at the bottom of the list have the smallest chance of a Type Two error. (Remember that a Type Two error is the error of failing to find ESP when it is in fact present.)

At this point you should be able to see the trade-off that exists between the two kinds of error. The decision rules at the top of the table have a low chance of Type One error: with them, there's very little chance you'll decide "There's ESP" when in fact there is none, and the person is guessing randomly. But in keeping your chance of Type One error low, you run a high risk of Type Two error: failing to find ESP when it's really there! For the decision rules at the bottom of the table, the opposite situation prevails.

At this point in the analysis of the ESP experiment, we must reconsider the basic question: Which decision rule shall I use? A traditional way of answering this question is to take the best rule available which does not have too large a chance of Type One error. It is quite common for someone to decide that he is willing to take a 1-in-20 chance of making a Type One error. The logic in this situation goes something like this: If this result, which looks like a case of ESP, is so strong that there's only 1 chance in 20 that we'd see this strong a result when a person guesses randomly, we'll be convinced that he has ESP. This number, the maximum acceptable risk of a Type One error, is called a *significance level.* Here, our significance level is 1/20, or 0.05.

In our particular case, looking at the right-hand column in Table 4.2 we can see that any one of the top six decision rules meets this general criterion: the chance of a Type One error is less than 1 in 20. (One chance in 20 is the same as 5 in 100, or 0.05.) Accordingly, suppose we limit our choices to the first six decision rules. Which one of these shall we select? The sixth is best in one important respect: it has the greatest possible chance of finding ESP when it is in fact present. It has the smallest chance of Type Two error of any of the six rules.

This rule says that we will conclude we have a case of ESP, or some other special effect, if the guesser gets 8/10, 9/10, 10/10, 0/10, 1/10, or 2/10 correct. Otherwise we'll decide he is guessing randomly.

Note that this entire analysis, including the use of the marble box to draw up Table 4.2, could be done before you even run the ESP experiment. This is important: You can pick your decision rule so as to control your errors of the two kinds before the experiment takes place. This is a common feature of statistical analysis. A decision rule should be chosen before an experiment is carried out.

An ideal decision rule would result in correct decisions in every case. In the real world, we can't do that. We have to accept some chance of each kind of error, and try to choose with knowledge of how likely each error is.

A Special Decision Rule: *One possible decision rule, which doesn't appear in the left side of the table, is the decision that no matter how the experiment turns out you will assume that the subject was just guessing. This means that whenever he is just guessing, you'll be correct. But if he ever had ESP or special information, you would be doomed to a Type Two error.*

This isn't a totally implausible decision rule for this experiment. Suppose that the experiment contained 1000 envelopes, and the person guessed correctly every single time. Then even the most skeptical person one could imagine would conclude that he wasn't guessing randomly. That would be convincing. But it may be the case that for some of you the present experiment is just too small. No matter what happens, you can't be convinced that something nonrandom has happened. This is a situation that occurs from time to time in statistical problems. When it does, it is usually a good idea to run a larger, more powerful experiment.

What we have just gone through, with reference to the marble box as a model for the ESP experiment, is essentially the development of a *statistical test* for the presence or absence of ESP (or other special information) in the experiment. As we go on to discuss other tests later in the book, you will see many steps which are completely parallel to those we have taken in this case. Remember the definition that was offered in Chapter 1:

Statistics is a set of tools to help you make good decisions on the basis of limited and variable information or data.

In this case, we have used statistics to study a variety of plausible decision rules which you might use for a 10-envelope ESP experiment. Now that our analysis is finished, you are fully informed about the properties of the various rules and can make a more "rational" decision than would otherwise be possible.

SIX STEPS

The following six steps were involved in building a statistical test for the presence of ESP in the ten-envelope experiment:

1. Specify all the possible events that could occur in the experiment: in this case, the results were 0/10, 1/10, 2/10, 3/10, etc., 10/10 correct guesses.
2. Make a formal statement, called a *null hypothesis* (abbreviated H_0), about the state of the world when all we are observing is random variation. This was written: The guesser's choices are indistinguishable from those that would result from flipping a coin before each envelope and saying "X" if the coin came up heads and "O" if it came up tails. The null hypothesis is a precise statement about how things would be if there were no ESP.
3. We studied the null hypothesis: We set up a model for the world when the null hypothesis is true, and found out how likely each event was under those conditions. In our case this model was the marble box, and our probability estimates were based on the number of times we observed 0, 1, 2, etc., 9, and 10 green marbles out of 10.
4. Using a table, we looked at each of a variety of possible decision rules. For each decision rule we found the associated chance of a Type One error. At this point in a decision making situation it is necessary to decide how conservative you are going to be: How big a risk of a Type One error are you willing to take? Once you have made that decision, it is clear which decision rule to use. The maximum acceptable risk of a Type One error is called the *significance level*.
5. At this point, after choosing your chance of Type One error and the corresponding decision rule, you can make a list of those results which, if observed, would convince you that the null hypothesis is *not* true. That is, you list those results which will prompt you to *reject* the null hypothesis. (This set of results is called the *rejection region*.)
6. Now, after completing your statistical study, you are ready to run the experiment, observe the result, consult your list (point 5), and announce your decision.

Each of you must have gone through something like the above steps in deciding what conclusions you would draw on the basis of each possible result in the experiment. What statistics does is to formalize your intuition, extend it to those cases where you really don't know what to expect, and help you reduce your chances of making mistakes.

If you have understood everything up to here, you are in a very good position to understand the chapters to come!

QUESTION AND ANSWER

Question: Much of this chapter was devoted to figuring out the chance of a Type One error for each of several decision rules. Could we, in a similar manner, calculate or estimate the chance of a Type Two error? (Please try to answer this question before reading on.)

Answer: The chance of a Type One error is just the chance that we will wrongly decide "He has ESP" when in fact the guessing is purely random. We were able to estimate this chance using the marble box. Recall that a Type Two error occurs when a person actually *has* ESP, but we decide that he has none. Can we figure out how likely we are to make this kind of error? It turns out that we cannot, at least not without some additional information. Suppose a person has just a tiny bit of ESP: he usually does better than chance, but not by much. Maybe he averages 6 correct guesses out of 10. For this person there is quite a large chance that we'll fail to detect his ESP. On the other hand, suppose that a person has very powerful ESP and almost always guesses all 10 envelopes correctly: for him, with any of the decision rules we've listed in Table 4.2, we would correctly detect his ESP. Thus there would be almost no chance of a Type Two error. The important point, then, is that we can't calculate the chance of a Type Two error unless we know exactly how much ESP our subject has. And since we wouldn't be running the experiment if we knew that, we're stuck. This is another reason why we pay more attention to the Type One error: We know more about it.

Question: I've been able to follow most of what's been presented so far, but I'm having trouble seeing how this is going to help in making decisions in the face of uncertainty. Could you provide an example?

Answer: Here's an example, which itself can be divided into a question and an answer. Ten students decide to test the effect of meditation on learning rate. Each student engages in two standard learning situations one after an hour's meditation and the other without such preparation. Then they are tested to see how much they learned in the two conditions. Seven learned *more* after meditation, while three learned more without meditation. Do these results prove that meditation helps, or do you think the difference could be the result of random variation?

An analysis of this case can be made in a way very similar to the ESP experiment. We can follow the six steps outlined in the Summary at the end of this chapter.

1. Specify all the possible results of the experiment. In this case, the results *could have been* 0 out of 10 students doing better after meditation; or 1/10 students doing better after meditation; or 2/10; or 3/10; up to 10/10 students doing better after meditation.
2. Make a formal statement, called a *null hypothesis* (H_0), about the state of the world when all we are observing is chance variation: For any one person, there's a 50 : 50 chance that he'll do better after meditation, and a 50 : 50 chance that he'll do better without meditation.
3. Study the null hypothesis. As you have probably guessed by now, the marble box experiment can be used as a model for this situation, just as it was as a model for the ESP situation. For each of the 10 marbles in each sample in that experiment, there was a 50 : 50 chance that it would be green. And if meditation has no effect, as stated in the null hypothesis, then there will be a 50 : 50 chance that a person will do better after meditation than without it. The entire sample of 10 marbles represents the entire set of 10 students. And, when the null hypothesis is true, we expect to see 10/10 students doing better after meditation just about as often as we see 10/10 green marbles. This holds for 9/10 students and 9/10 green marbles; 8/10; 7/10; etc. Thus we can use the results of the 50-sample marble box experiment, which were presented at the start of this chapter, to give us an idea of how likely various results would be if the meditation had no influence.
4. Look at various decision rules, find the associated chance of Type One error, and decide how big a risk you are willing to take of a Type One error. The analysis in Table 4.2 works for this situation, too. Looking back

at the Table, we can review the risk of Type One error associated with each of the decision rules. If we follow the same convention as before and say that we will not tolerate a risk of Type One error which is greater than 1 chance in 20, or 0.05, then we decide on the sixth decision rule: Say that the meditation had an effect on learning performance if we observe any of these results: 8/10, 9/10, 10/10, 2/10, 1/10, or 0/10 students doing better after meditation. Otherwise conclude that we haven't proven that meditation influences learning rate.

5. List the set of results which, if observed, will lead you to conclude that the null hypothesis is not true. Any of these results will lead you to *reject* the null hypothesis. For the decision rule we choose, these results are 8/10, 9/10, 10/10, 2/10, 1/10, and 0/10. These events constitute the *rejection region* for this statistical analysis.
6. Consider the actual results of the experiment, and see if the observed data lead you to reject the null hypothesis. We reported that 7 out of 10 students did better after meditation. The result 7/10 doesn't lie in the rejection region of step 5. So we conclude that we have no conclusive proof of a relationship. We are not convinced that meditation influences learning rate.

Summary

A variety of possible decision rules was considered, and the chance of making a Type One error with each rule was estimated. The reasoning involved was this: The chance of a Type One error is the same as the chance of seeing a convincing result if the subject is just guessing. The chance of seeing, say, 10/10 correct if the subject is just guessing is the same as the chance of seeing 10/10 green marbles in the marble box. Thus by using the marble box results, we can estimate the chance of seeing by pure chance a result so good that it prompts us to say "There is ESP" when in fact there is none.

In choosing a decision rule, we first decided how big a risk of a Type One error we were willing to tolerate. Then we looked at all the rules which were acceptable in this respect, and chose one which had the best chance of correctly finding ESP when in fact the person has some (the rule which minimized the Type Two error probability).

The six steps which underlie many statistical tests were presented:

1. List all possible results of the experiment.
2. Make a formal statement, called a null hypothesis, about how things would be if all we were observing was random variation.
3. Study the null hypothesis to find out how likely various results would be if the null hypothesis were true.
4. Decide how big a risk we are willing to take of a Type One error. This number, the maximum acceptable risk of a Type One error, is called the *significance level*. Look at various decision rules and find the associated risk of a Type One error.
5. Specify a *rejection region*: those results which, if observed, will prompt us to reject the null hypothesis and decide that there's really an effect.
6. Run the experiment, find the result, and see if it lies in the rejection region.

Problems

1. Suppose a person were going to use the following, very bizarre, decision rule: If the guesser gets 7/10 or 8/10 correct, I'll say he has ESP. If he gets *any other number* correct, I'll decide he is just guessing. For this crazy decision rule, what do you estimate to be the chance of a Type One error?
   (*Hint*: In other words, what is the chance that a person using this decision rule would make the mistake of saying "He has ESP" *when in fact* the subject is just guessing randomly?)
2. Why is this a crazy decision rule?
3. Suppose that I continue to take samples from the marble box until I have repeated the process 1000 times. I observe the eleven different possible events with the following frequencies:

Number of green marbles in paddle	0/10	1/10	2/10	3/10	4/10	5/10	6/10	7/10	8/10	9/10	10/10
Frequency (number of times out of 1000 samples that I observed this number of green marbles)	1	10	45	117	205	247	245	120	44	9	1

a. On the basis of these results, what do you estimate to be the chance that a person guessing randomly would get 0/10 or 10/10 correct?

b. What would be the probability of a Type One error associated with the following decision rule: Say "He has ESP" if he guesses all 10 right or all 10 wrong; otherwise say "He is just guessing"?

c. What would be the probability of a Type One error associated with the following decision rule: Say "He has ESP" if he guesses correctly 8 times or more, out of 10; otherwise say "He is just guessing"?

†4. Criticize the following statistical reasoning: A friend of mine was upset when he heard that there was 1 chance in 1000 that someone would put a bomb on any plane he was riding. Having studied a bit of probability, however, he deduced that the chance of *two bombs* on the same plane was 1/1000 x 1/1000 or 1 in 1 million. He now carries a bomb with him whenever he flies, and is happy knowing that he has improved the odds terrifically.

Beyond the Marble Box

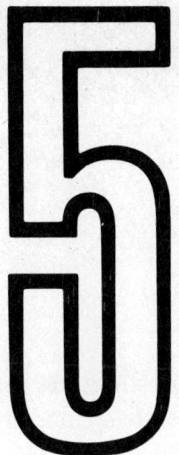

The marble box provided a model for the ESP experiment when the null hypothesis was true: when the person was just guessing, and had no special powers. But it was cumbersome and variable. For example, we happened to see 9/10 green on 1 sample out of 50. But we certainly might have seen 9/10 zero times, or 2 times, or maybe even 3. How are we to deal with the variability? This chapter is intended to develop a new, less noisy, and more accurate model which we can use to study how things would be if a person were just guessing randomly. We are going to introduce a new way to *study the null hypothesis*.

The marble box is made of wood, plastic, and glass. And it is tedious, noisy, and inaccurate. We shall now move toward a *symbolic* model, made up of simple mathematics. It is simple, silent, and very accurate.

We used the marble box to estimate how likely each of the eleven possible experimental results would be if a person were guessing at random. And it provided us with estimates. Now we shall ask: Can we figure out by pure reasoning power how likely each possible result is, when the null hypothesis is true and a person is guessing randomly? We can.

A COIN-TOSS MODEL

First we shall set up a slightly new model for the null hypothesis. Imagine that you tossed 10 coins, one after the other, and wrote down the sequence of results: HTTHTHHHTT (H for heads, and T for tails). This is like the marble box situation where we generated "strings" of 10 marbles, such as GRRGRGGGRR where G stands for green, and R for red. And it is like the guessing situation, where we could see strings like CWWCWCCCWW where C stands for correct, and W for wrong. Now we are going to study the ten-coin-toss model to find out how likely it is that we would see each of eleven possible *events*: 0/10 heads, 1/10 heads, 2/10 heads, 3/10 heads, etc., when you flip 10 coins one after another.

How do you figure that out? Those of you who enjoy puzzles could stop here and try to figure it out before reading further.

If you didn't figure it out before reading further, you might be able to do so with a hint: Start with some very simple cases and work up. That is how we shall proceed.

The single-coin situation: Suppose you flip one coin. This might be a model for an ESP experiment with just one envelope. There are two possible outcomes: heads and tails. And there are two possible scores or *events*: 0 heads, and 1 head. The event "1 head" will happen, we all know, half the time. We say it will happen with probability 1/2. Similarly, the event "0 heads," which corresponds to a tail, will happen with probability 1/2. We can make a simple table of these results, which is not necessary here but will pave the way for more complicated situations.

Possible Outcomes	Possible Events	Probabilities
H	1 Head	½
T	0 Heads	½

Before we take up the case of a two-coin situation, let's consider an important side argument which we will need. The following two possible strings may be seen in the marble box situation:

GGGGGGGGGG   (string A)
RGGRGRRRGG   (string B)

I recently asked a class of 100 students how many thought that string B was more likely to happen than string A, and 56 thought it was. I then asked how many thought that string A was more likely to happen than string B: nobody thought so. Forty-four people were right. The exact string RGGRGRRRGG is just as likely as the string GGGGGGGGGG. This is not to say that you are as likely to see 5 greens as to see 10 greens — we are all convinced that that is wrong. But it turns out that every exactly specified string is just as likely to happen as every other such string. This is quite a surprising result, and an important one. We shall have more to say about this later.

The two-coin situation: Suppose you flip two coins, one after the other. And suppose, to keep things straight, that the first coin is a dime, and the second is a quarter. Now let's look at the various possible *outcomes* (or arrangements) that could result. Then let's look at the various possible scores, or *events*. If we represent the dime by a small circle and the quarter by a large one, the following are the possible *outcomes*:

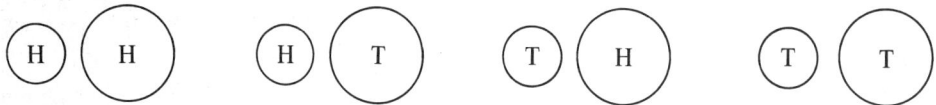

It is important to recognize that the second and third *outcomes* are different, even though each contains one head and one tail.

Figure 5.1 translates the above to a "tree." On the left side are the four possible arrangements or outcomes; the middle is the *tree*. A tree like this shows all the possible results of an experiment in a single picture. Each outcome corresponds to one path from the left to the right, through the tree. From Start, you go to the dime. There are now two possibilities: either the dime comes up heads, or it comes up tails. Now consider the upper part of the diagram, corresponding to heads on the dime. After the dime comes up heads, there are two ways the quarter can come up: heads and tails. And the possibilities are similar for the bottom part of the tree. At the very right of this

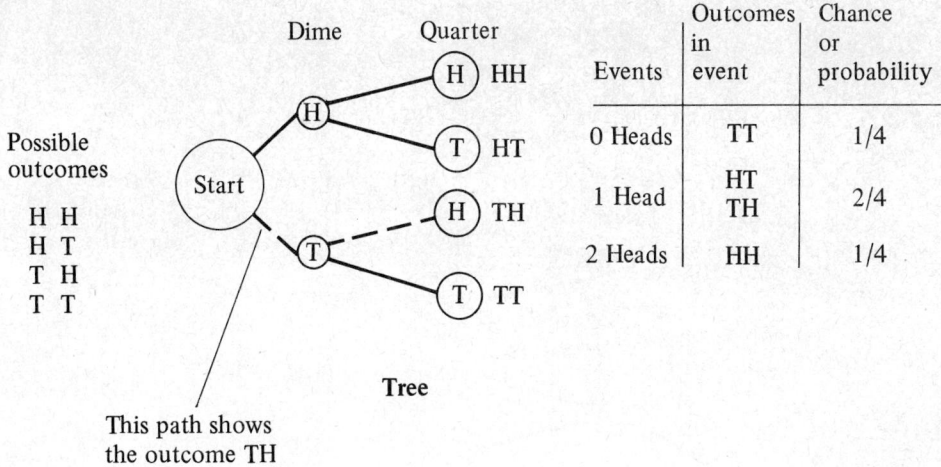

Figure 5.1 The two-coin-toss situation.

tree are labels for the four possible outcomes: HH, HT, TH, and TT. We'll be looking at more trees in this and subsequent chapters.

On the right-hand side of the picture the three possible scores, or *events* are listed: 0 heads, 1 head, and 2 heads. If the two-coin experiment were a model for a two-envelope ESP experiment, these would correspond to the results 0 correct, 1 correct, and 2 correct. The chance, or probability, of each event at the far right of the picture is based on the number of possible outcomes corresponding to the event. There are four possible outcomes; one of these (TT) corresponds to the event "0 heads." Since each of the four outcomes is just as likely as each of the others, there is just 1 chance in 4 that we'll see the event "0 heads." Similarly, there is just one outcome corresponding to the event "2 heads." So the chance of that event is just 1/4. But there are two different outcomes which contain 1 head, and so correspond to the event "1 head" these outcomes are HT and TH. So the chance of seeing one of these two outcomes is 2/4. There is a probability of 2/4 that you will see exactly 1 head when you flip two coins. We have figured out the probabilities of each of the three possible events, or scores. We have finished the analysis of a two-coin situation.

The three-coin situation: At this point it would be a very good exercise, dear reader, if you would try to complete a similar analysis for the three-coin situation before reading on. List all the possible outcomes and events; draw a tree; find the probability of each *event*. If you have fully understood what has been said so far, you should be able to do this. The result you should obtain is shown in Figure 5.2.

When you examine Figure 5.2 you should be starting to see a pattern for the total number of different outcomes possible in a situation like this. With one coin, there are two different outcomes. With two coins, there are 2 x 2 = 4 different outcomes possible. With three coins, there are 2 x 2 x 2 = 8, or 2^3 different possible outcomes. What do you suppose the number is for the four-coin situation?

How many possible different *events* are there here? Put differently, how many different possible numbers of heads could you see? You could see 0 heads, 1 head, 2 heads, or 3 heads. The *outcomes* corresponding to each of these *events* are given in Table 5.1.

At this point you should be starting to get the picture. Suppose the ESP experiment had contained three envelopes. We could have done a statistical analysis, using a three-hole paddle and the marble box, to find out how likely 0/3, 1/3, 2/3, and 3/3 green would be. But, by means of

32 THE SIGN TEST, AND THE LOGIC OF STATISTICAL INFERENCE

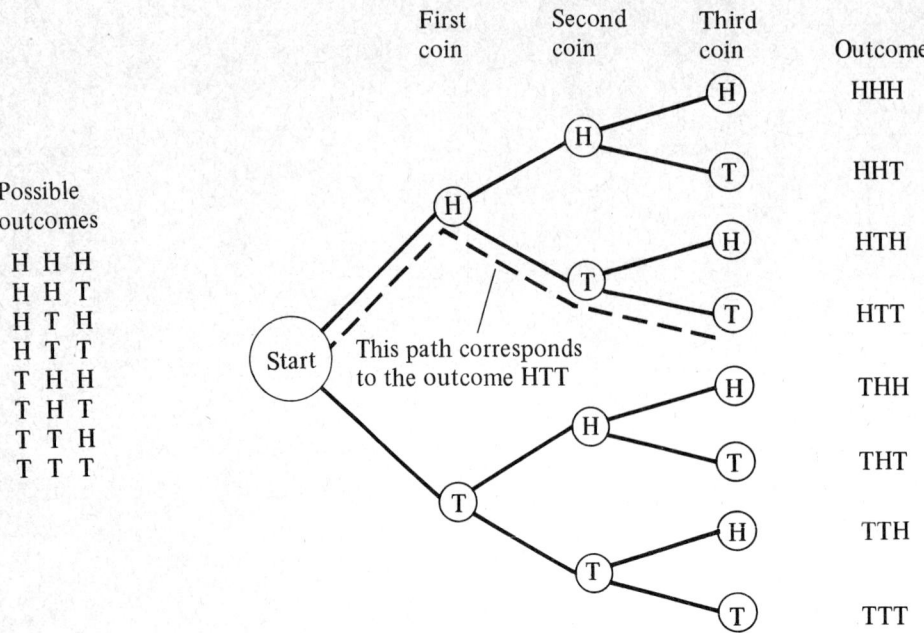

Figure 5.2 The three-coin-toss situation.

the analysis just completed, we have arrived at a theoretical answer to the question: How likely is each possible event? After this analysis, we would know enough not to be very impressed if a subject guessed correctly on 3/3 envelopes, for we know that he has 1 chance in 8 of doing that well completely by chance, even if he has no ESP at all. We would probably be intrigued by 3/3 correct, but not willing to bet much money on the person's extrasensory powers.

You should see that there is no reason, in principle, why one couldn't conduct an analysis of

TABLE 5.1   THREE-COIN-TOSS SITUATION

Event	Corresponding outcomes*	No. of outcomes	Chance of event
0 heads	TTT	1	1/8
1 head	HTT, THT, TTH	3	3/8
2 heads	HHT, HTH, THH	3	3/8
3 heads	HHH	1	1/8
	Total number of outcomes	8	

*Note that each of the eight possible outcomes appears *exactly once* in this column

just this sort for the four-coin situation, the five-coin situation, and so on, up to the ten-coin situation. You would thus arrive at a theoretical answer to the question: How likely is each of the possible results in the ten-envelope ESP experiment, if the guesser has no ESP at all?

The problem is that you would run out of paper!

There's a better way. First we have to know the rule for figuring out how many different possible outcomes there are for each different situation:

1 coin means   2 outcomes: $2^1 = 2$
2 coins mean $2 \times 2 = 4$ outcomes: $2^2 = 4$
3 coins mean $2 \times 2 \times 2 = 8$ outcomes: $2^3 = 8$
4 coins mean $2 \times 2 \times 2 \times 2 = 16$ outcomes: $2^4 = 16$
5 coins mean $2 \times 2 \times 2 \times 2 \times 2 = 32$ outcomes: $2^5 = 32$
etc.
n coins means $\underbrace{2 \times 2 \times 2 \times \text{etc.} \times 2}_{n \text{ times}}$ outcomes: 2^n

In case this isn't completely clear, you can see that every outcome in the one-coin case leads to two outcomes in the two-coin case: H leads to HH and HT; T leads to TH and TT. Similarly, for every outcome in the two-coin case, there is a pair of outcomes in the three-coin case, as seen in Table 5.2.

TABLE 5.2   MAKING UP THREE-COIN PATTERNS FROM TWO-COIN PATTERNS

Two-coin case	Corresponding outcomes in the three-coin case
HH	HH H, HH T
HT	HT H, HT T
TH	TH H, TH T
TT	TT H, TT T

A similar argument can be made each time you add a coin: You double the number of possible outcomes each time you add a coin.

Now in this manner we could have deduced, without listing the eight possible outcomes, that there are exactly $2 \times 2 \times 2 = 8$ possible outcomes in the three-coin case. If we also knew how many outcomes had 0 heads, how many had 1 head, how many had 2 heads, and how many had 3 heads, we could figure out the probabilities of the events 0, 1, 2, and 3 without making the list of outcomes.

How many possible arrangements can you make with three coins that have exactly 0 heads? The answer is one: TTT. How many possible arrangements can you make that have exactly 1 head? That head can be the first toss (HTT), or the second (THT), or the third (TTH). There are three different possible outcomes. So the chance of the *event* 1 head must be 3 out of 8, or 3/8. Similarly, in a symmetric argument, there must be three different outcomes with exactly 2 heads: an arrangement with 2 heads must have 1 tail. And there are exactly three different places where that tail could be. So the chance of seeing 2 heads must be 3 out of 8, or 3/8, as well. Finally, there is just one way of arranging three coins so that there are 3 heads: HHH. So the chance of seeing 3 heads is just 1 out of 8, or 1/8.

The four-coin situation: Now let's work through a similar argument for the case where there are four coins.

First question: How many different arrangements or outcomes are there? $2 \times 2 \times 2 \times 2 = 16$ different outcomes: $2^4 = 16$.

Next: How many different arrangements can you make that have exactly 0 heads? Just one: TTTT. So the chance of the event "0 heads" is just 1/16.

Next: How many different arrangements can you make that have exactly 1 head? There are four different places where the head can go: HTTT, THTT, TTHT, and TTTH. So there are four ways of having exactly 1 head, and the chance of exactly 1 head must be 4/16.

Now things get more troublesome: How many ways are there of having exactly 2 heads out of four coins?

It turns out that there are 4 times 3 divided by 2 different ways. The following argument is intended to convince you of that. We shall consider a close analogy: The number of ways you can arrange four coins with exactly one of them showing heads is the same as the number of ways you can put one ball into four pots (Figure 5.3).

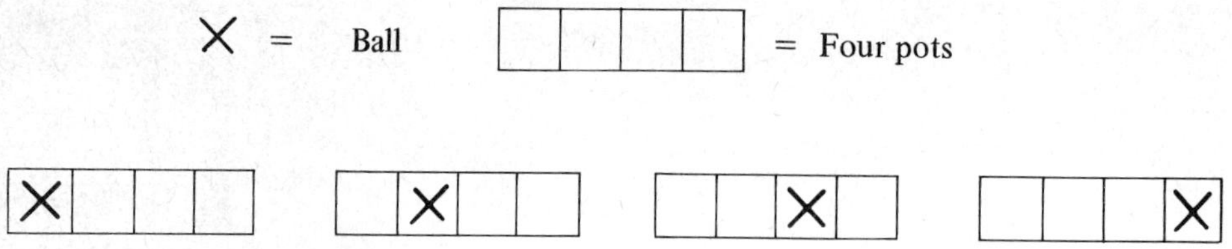

$\times$ = Ball   ☐☐☐☐ = Four pots

☒☐☐☐   ☐☒☐☐   ☐☐☒☐   ☐☐☐☒

Figure 5.3 Four ways to place one ball in four pots.

There are four ways of putting one ball into four pots. Now the question is: How many ways can you have exactly two heads in four tosses? This is the same as asking, how many ways can you put two balls into four pots? As Figure 5.4 illustrates, for each arrangement with one ball in four pots, you can create three new arrangements with two balls in four pots since there are three places left for putting the second ball. In the process, however, you make each pattern twice, so the desired number is only half of 4 times 3.

By means of similar arguments, we can fill out Table 5.3:

TABLE 5.3   CALCULATIONS FOR THE FOUR-COIN CASE

Event	No. of outcomes corresponding to this event		Probability of event
0 heads		1	1/16
1 head		4	4/16
2 heads	$\frac{4}{1} \times \frac{3}{2} = \frac{12}{2}$	= 6	6/16
3 heads	$\frac{4}{1} \times \frac{3}{2} \times \frac{2}{3} = \frac{24}{6}$	= 4	4/16
4 heads	$\frac{4}{1} \times \frac{3}{2} \times \frac{2}{3} \times \frac{1}{4} = \frac{24}{24} = 1$		1/16

TABLE 5.4   THE TEN-COIN CASE

Event	No. of outcomes corresponding to this event		Probability
0 heads		1	$1/1024 \cong .001$
1 head		10	$10/1024 \cong .01$
2 heads	$\frac{10}{1} \times \frac{9}{2} = 45$		$45/1024 \cong .04$

Ten-coin case: Here the number of possible outcomes = 2 x 2 x 2 x 2 x 2 x 2 x 2 x 2 x 2 x 2 = 2^{10} = 1024

This is slightly easier than listing all the possible outcomes for this case. But the economy of effort is even more marked for the ten-coin case. The first three lines of that table are shown in Table 5.4.

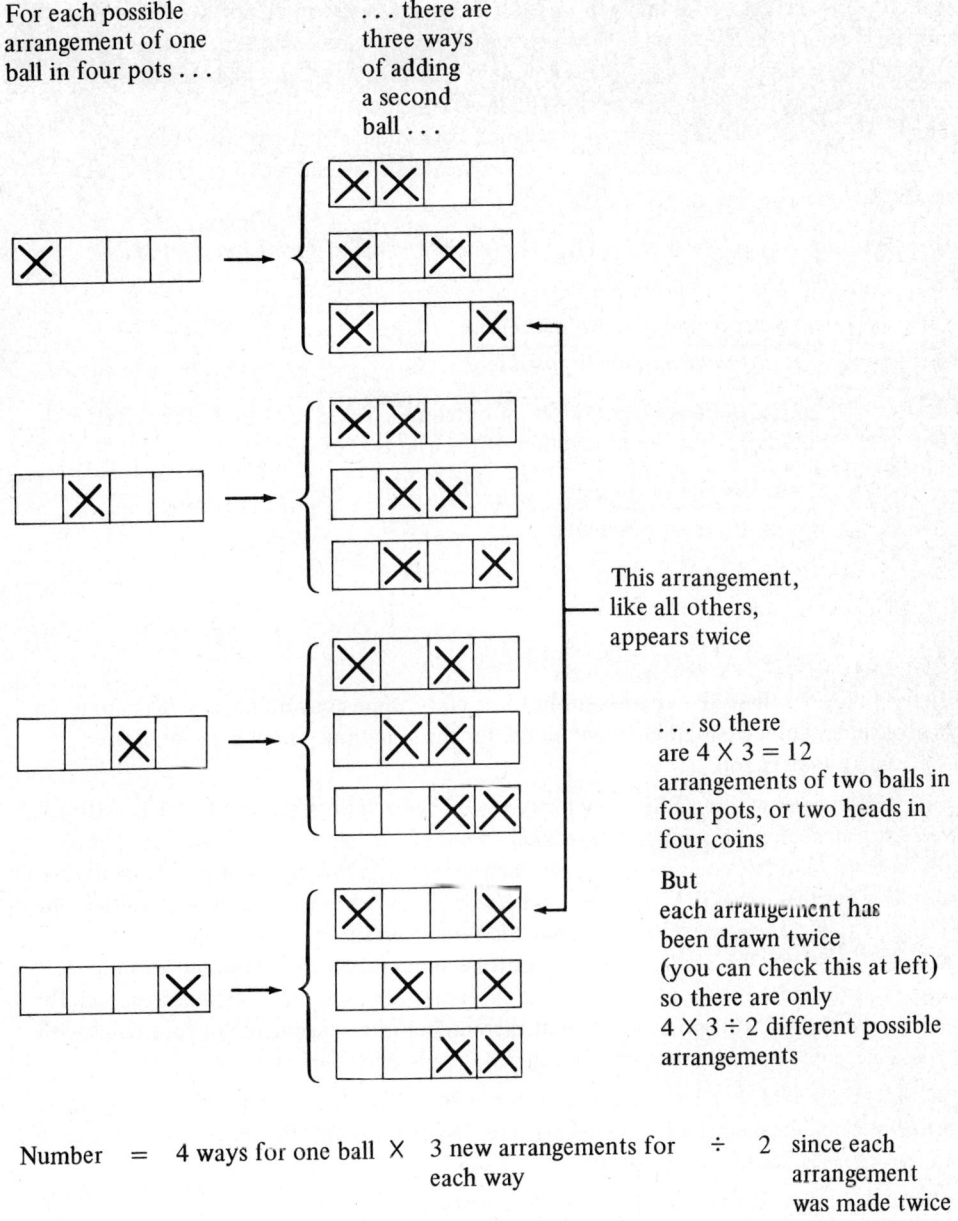

Number = 4 ways for one ball × 3 new arrangements for each way ÷ 2 since each arrangement was made twice

Figure 5.4 Six ways to put two balls into four pots. Each way is shown twice.

QUESTION AND ANSWER

Question: By counting all the possible different outcomes (arrangements of Hs and Ts) in various coin-tossing situations, and then looking at how many of these outcomes correspond to different *events* (numbers of heads), you have come up with numbers which give the probabilities of these events. What rule were you using to do this? Can you always use it?

Answer: This problem will be discussed more fully in Chapter 12. Here's some advance information to hold you until then. It was asserted that the probability of GGGGGGGGGG was exactly the same as the probability of the precise arrangement RGGRGRRRGG in the marble box. Similarly, the probability of TTTTTTTTTT is exactly the same as the probability of HTTTHTHHTT, or any other specific arrangement of ten coin-toss results. And a similar statement can be made for any number of coins. You're asked to take this statement on faith, for the moment; we'll come back to it later. In this sort of situation, every one of the different arrangements, or outcomes, is equally likely. It is as if we wrote each arrangement on a slip of paper, put all the slips into a big hat, mixed thoroughly, and drew one at random. The chance of any one arrangement would be the same as the chance of any other. Now suppose that we consider the arrangements in a three-coin-toss situation. There are $2 \times 2 \times 2 = 8$ different arrangements. Each is equally likely. But we're not interested in the chance of an arrangement, or outcome. We're interested in the chance of an *event*, like "2 heads." To do this, we use the following rule:

When all the separate arrangements or outcomes are equally likely, the chance of seeing one particular event is given by

$$\text{Probability of event} = \frac{\text{Number of outcomes corresponding to that event}}{\text{Total number of different outcomes possible}}$$

For example, in the three-coin situation there are eight different outcomes. Three of them (HHT, HTH, and THH) correspond to the event "2 heads." Thus the probability of the event "2 heads" is

$$\text{Probability of "2 heads"} = \frac{\text{Number of outcomes with 2 heads}}{\text{Total number of different outcomes}} = 3/8$$

Summary

The marble box is a useful model for the ESP experiment, but it is noisy, time consuming, and inaccurate. In this unit we make progress toward developing a mathematical model for the situation, which is faster, more accurate, and only slightly harder to understand.

Our goal is to figure out, for a guesser with no ESP, how likely would be each possible event: 0/10; 1/10; 2/10; etc., up to 10/10. We attack an analogous problem: How likely would 0/10 heads, 1/10 heads, etc., up through 10/10 heads be, if you tossed ten fair coins in a row? We then work up to this problem by first studying the one-coin situation: how likely are the events 0/1 heads and 1/1 heads? Then we study the two-coin situation: How likely are the events 0/2, 1/2, and 2/2 heads? Then we study the three-coin situation, and find the probability of the events 0/3, 1/3, 2/3, and 3/3 heads. At this point it becomes useful to extend our techniques by developing some counting rules to find out how many different outcomes can be found which correspond, for example, to the event "2 heads in 4 tosses." It is pointed out that the number of arrangements of four coins with two of them heads is just the same as the number of ways you can put 2 balls into four pots.

The counterintuitive result is presented that every specific arrangement of 10 Hs and Ts is just as likely as every other specific arrangement. Thus the chance of seeing HHHHHHHHHH is exactly the same as the chance of seeing the precise order HTTHTTTHTH, or any other specific order.

Throughout this chapter, the word *outcome* is used to refer to a particular arrangement: for example, HTTT is one possible outcome of the four-coin-toss experiment. The word *event* is used to refer to a *score*: for example, "1/4 heads" is one possible event in the four-coin-toss situation. That event contains four outcomes: HTTT, THTT, TTHT, and TTTH. These words will be defined more formally below.

Problems

1. It was argued, on page 35, that there are exactly $4 \times 3 \div 2$, or 6 different ways to put two balls into four pots, and that there are similarly six different patterns of four coin tosses which contain exactly 2 heads. Prove this by writing down all six.

2. Make a tree, like the one in Figure 5.2, for the four-coin situation. Show each of the 16 different possible outcomes, or branches of the tree. Then list the five possible *events* (0, 1, 2, 3, or 4 heads). How many outcomes are there which correspond to each of these events? What do you think is the probability of seeing exactly 4 heads when you toss four fair coins?

3. Critically evaluate the following example of questionable statistical reasoning: An Englishman was recently ordered by the local constable to stop practicing boomerang throwing in public parks, because of the possible risk to innocent bystanders. The gentleman was furious, and pointed out that it was far more likely that someone would be hurt while crossing a street than that someone would be hurt by a boomerang. "Show me a single case of someone being hurt by a boomerang," he said, "and I will be pleased to stop." Was his point well taken?

4. Describe how we could confirm our calculations for the three-coin situation using a marble box. What should be the proportion of red marbles in the box? How many holes should there be in the paddle? How many samples should you take?

A Full Analysis of the Ten-Coin-Toss Model for the ESP Experiment

THE TEN-COIN MODEL

At the end of Chapter 5, we started to derive the probabilities of the eleven possible different scores, or events, in the ten-coin situation. We shall now finish that job. The first step in this derivation is to figure out how many different outcomes, or arrangements, could result when you toss ten coins, and write down heads or tails for each, in the order obtained. Recall that for one coin there are two possible outcomes: H and T. For two coins, there are $2 \times 2 = 4$ or 2^2 different outcomes: HH, HT, TH, and TT. For three coins there are $2^3 = 8$ different outcomes, or patterns: HHH, HHT, HTH, HTT, THH, THT, TTH, TTT. You should be able to see that proceeding in the same way, you arrive at the conclusion that if you toss ten coins, there are $2^{10} = 1024$ different possible outcomes or patterns or arrangements. What remains to be done is to figure out how many of these outcomes contain 0 heads, how many contain exactly 1 head, how many contain exactly 2 heads, and so on, up to 10 heads.

The first case is simple: How many different patterns of heads and tails can you create which contain exactly *zero* heads? There is only one such pattern, or outcome: TTTTTTTTTT. So there is just 1 chance in 1024 that we will observe the event "0 heads." The probability of the event 0/10 is just 1/1024.

The next case is only slightly more tricky: How many different outcomes are there which contain exactly *one* head? Phrased differently, how many different patterns can you make which contain exactly 1 head and 9 tails? If you stop to think, you'll see that there are ten different places where the single head can be: the first, or the second, or the third, etc., up to the tenth coin. The patterns are HTTTTTTTTT, THTTTTTTTT, TTHTTTTTTT, TTTHTTTTTT, TTTTHTTTTT, TTTTTHTTTT, TTTTTTHTTT, TTTTTTTHTT, TTTTTTTTHT, and TTTTTTTTTH. There is no other pattern that contains exactly 1 H and 9 Ts.

Now things get more challenging. If you can master the next step, you will have no trouble for the rest of this chapter. How many different patterns can you make which contain exactly *two* heads? It turns out that there are 10 x 9 ÷ 2 = 45 different patterns. Here's how to see that this is so, without having to write them all out: The question, "How many patterns can you make with 1 H and 9 Ts?" is exactly analogous to the question, "How many different patterns can you make by putting one ball into ten boxes?" The answer, it should be clear, is that there are ten different ways to put one ball into ten boxes. These are shown in Figure 6.1.

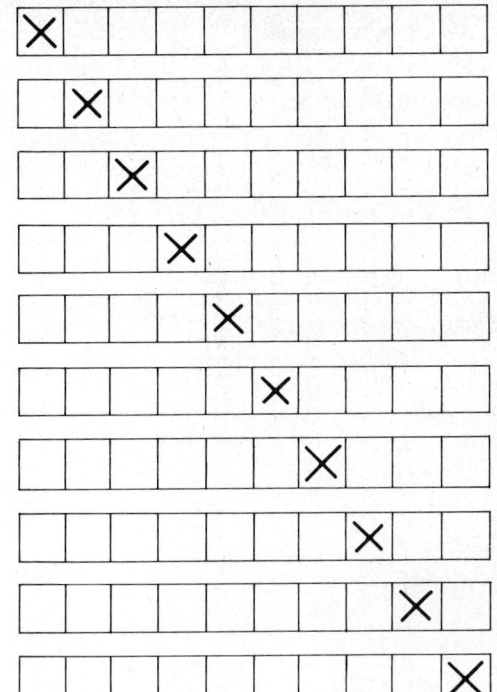

Figure 6.1 Ten possible ways of putting one ball (shown with an X) into ten pots.

For each of the ten patterns in Figure 6.1, there are nine empty boxes. That means that there are nine places where you could put a second ball. You can thus create nine different new patterns by adding a second ball to the top pattern in Figure 6.1

You can similarly make nine different new patterns by adding one ball to the second display in Figure 6.1.

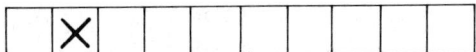

This can be repeated for each of the ten patterns (in Figure 6.1) with one ball. This makes a total of 10 x 9 = 90 different new patterns. But are they all different? If you think for a minute, you will see that by making up all 10 x 9 = 90 patterns, you will actually have made each pattern twice. For example, the pattern below will result once when you add a second ball to the top pattern of Figure 6.1 and once when you add a second ball to the bottom pattern of Figure 6.1.

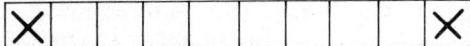

There is no other way that this pattern could be obtained! Similarly, each other pattern will have been generated twice when we make up the 90 new patterns. So the total number of distinct, different patterns is not 10 times 9, but rather it is 10 × 9 ÷ 2 = 45. So we conclude that there are 45 different possible ways to put two balls into ten boxes. And, by direct analogy, in a ten-coin experiment there are 45 different possible outcomes that have exactly *two heads*.

A very similar argument can be used to deduce the number of different patterns which have exactly *three* heads: For each of the 45 different two-headed patterns, we would have eight places to put the third head, making 45 × 8 = 360 patterns with exactly three heads. But again, they are not all different, for we would have created three copies of each pattern. So the correct number of different patterns is 45 × 8 ÷ 3, or 10/1 × 9/2 × 8/3 outcomes.

By similar reasoning we can now make up an entire table for all the different possible *events* (or scores) in a ten-coin situation. This, you may still remember, was the model we were working up to for the ten-envelope ESP experiment. Using the results we've already obtained, and the others that follow in exactly the same way, we end up with Table 6.1.

TABLE 6.1   CALCULATIONS FOR THE TEN-COIN MODEL

Event or score or number of heads in ten tosses	No. of different outcomes which correspond to that event, or which contain that number of heads	Probability of event
0/10 heads	1	1/1024 = .001
1/10 heads	10	10/1024 = .010
2/10 heads	$\frac{10}{1} \times \frac{9}{2} = 45$	45/1024 = .044
3/10 heads	$\frac{10}{1} \times \frac{9}{2} \times \frac{8}{3} = \frac{360}{3} = 120$	120/1024 = .118
4/10 heads	$\frac{10}{1} \times \frac{9}{2} \times \frac{8}{3} \times \frac{7}{4} = 210$	210/1024 = .205
5/10 heads	$\frac{10}{1} \times \frac{9}{2} \times \frac{8}{3} \times \frac{7}{4} \times \frac{6}{5} = 252$	252/1024 = .246
6/10 heads	$\frac{10}{1} \times \frac{9}{2} \times \frac{8}{3} \times \frac{7}{4} \times \frac{6}{5} \times \frac{5}{6} = 210$	210/1024 = .205
7/10 heads	$\frac{10}{1} \times \frac{9}{2} \times \frac{8}{3} \times \frac{7}{4} \times \frac{6}{5} \times \frac{5}{6} \times \frac{4}{7} = 120$	120/1024 = .118
8/10 heads	$\frac{10}{1} \times \frac{9}{2} \times \frac{8}{3} \times \frac{7}{4} \times \frac{6}{5} \times \frac{5}{6} \times \frac{4}{7} \times \frac{3}{8} = 45$	45/1024 = .044
9/10 heads	$\frac{10}{1} \times \frac{9}{2} \times \frac{8}{3} \times \frac{7}{4} \times \frac{6}{5} \times \frac{5}{6} \times \frac{4}{7} \times \frac{3}{8} \times \frac{2}{9} = 10$	10/1024 = .010
10/10 heads	$\frac{10}{1} \times \frac{9}{2} \times \frac{8}{3} \times \frac{7}{4} \times \frac{6}{5} \times \frac{5}{6} \times \frac{4}{7} \times \frac{3}{8} \times \frac{2}{9} \times \frac{1}{10} = 1$	1/1024 = .001

Several points about this table are worth noting. First, the number of outcomes with exactly 4 heads is 210, the same as the number of outcomes with exactly 6 heads; similarly, the number with 3 heads is the same as the number with 7 heads, or 120; and so on. This makes sense, since if there are 120 patterns you can make with 3 heads and 7 tails, by symmetry there should be 120 patterns you can make with 7 heads and 3 tails.

Another feature of this table is interesting: If you add up the numbers in the center column, 1, 10, 45, 120, 210, 252, 210, 120, 45, 10, and 1, what total do you get? Does this make sense? Why?

At this point, we have just about finished our analysis of the ten-coin experiment. Let's consider the implications for the ESP experiment. According to Table 6.1, there is just 1 chance in 1024 that you will get 10/10 heads in this situation. This means that if a person had no special information or extrasensory perception, he would have only 1 chance in 1024 of getting all ten envelopes correctly. So if anyone did this well in the ESP experiment, that would be pretty convincing proof that something strange was happening. Even the event 9/10 is quite improbable: it would happen only 1 time in 100, when the null hypothesis was true and the person was just guessing. We'll return to this later.

DRAWING A PROBABILITY DISTRIBUTION

It is easier to see how a collection of numbers like those in Table 6.1 interrelate if you draw a graph of them. Figure 6.2 shows what this looks like:

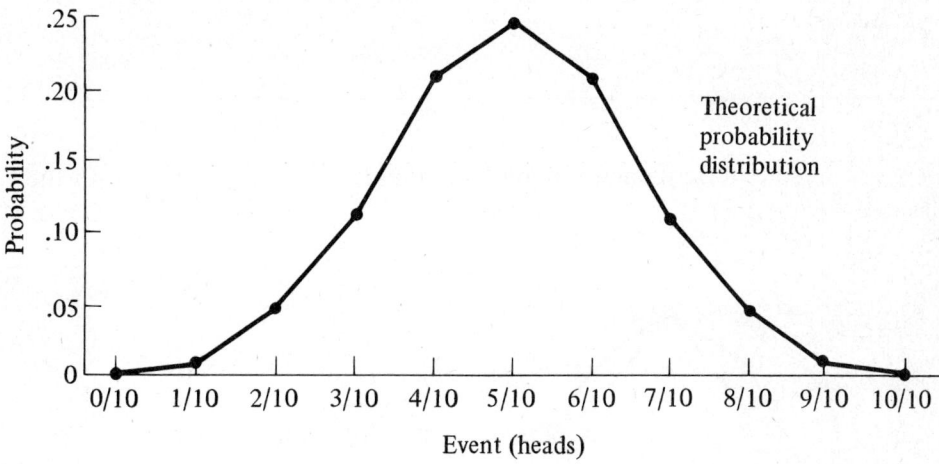

Figure 6.2. Theoretical probability distribution for the ten-coin model for the ESP experiment. Based on Table 6.1.

Recall that this theoretical analysis is intended to do the same job that the marble box did, but to do it more accurately and more simply. As a check on what has happened up to here, let's see if there is any similarity between the theoretical probability distribution graphed in Figure 6.2 and the results of the marble box investigation. Remember (see Figures 2.1 and 2.2) the data obtained, which are listed again in Table 6.2

TABLE 6.2 THE DATA FROM THE MARBLE BOX EXPERIMENT

Number of green marbles	0/10	1/10	2/10	3/10	4/10	5/10	6/10	7/10	8/10	9/10	10/10
Frequency with which we observed this result	0	0	0	10	9	12	9	8	1	1	0

In order to compare this with our theoretical probability distribution, we have to change something; the probabilities range from almost zero to a maximum of 0.246. The frequencies are much larger numbers. To make them comparable, we can look at *relative frequencies*: the *proportion* of the time when we observed each possible number of green marbles. Since we made 50 observations, we can obtain overall proportions, or relative frequencies, if we divide each of the frequencies by 50, the total number of observations. This gives Table 6.3.

TABLE 6.3 CHANGING THE FREQUENCIES IN TABLE 6.2 INTO RELATIVE FREQUENCIES

Number of green marbles	0/10	1/10	2/10	3/10	4/10	5/10	6/10	7/10	8/10	9/10	10/10
Frequency with which we observed result	0	0	0	10	9	12	9	8	1	1	0
Relative frequency	0/50	0/50	0/50	10/50	9/50	12/50	9/50	8/50	1/50	1/50	0/50
Relative frequency expressed as a decimal	0	0	0	0.20	0.18	0.24	0.18	0.16	0.02	0.02	0

Now we can compare these relative frequencies by plotting them on the same graph with the theoretical probabilities (see Figure 6.3).

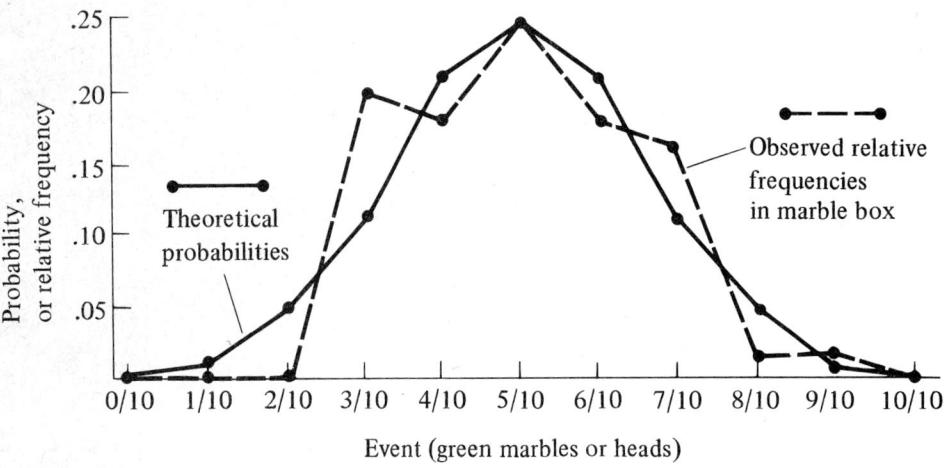

Figure 6.3 Theoretical probability distribution for the ESP experiment compared with the results of the marble box experiment.

Is there any similarity between the relative frequencies from the marble box experiment and the theoretical probabilities we have calculated? The two graphs are certainly not identical; it would be suspicious if they were, since we know that the marble box results are subject to random variation. But there is a strong similarity between the two graphs. Both show a maximum likelihood of the event 5/10. And both show almost no likelihood of the events 0/10 and 10/10. We do not now know any statistical tools that will help us decide whether these two are different in important respects, though we will know some before Chapter 23. But at this stage it certainly seems plausible that our theoretical and our marble box models have very similar properties.

EVENTS AND OUTCOMES

During the past few chapters, the words "event" and "outcome" have been used in a very precise way, although we have waited until now to state the exact relationship. These two everyday words are given a precise and special meaning in statistics. In defining them, we shall introduce one more term — sample space — which will be useful in coming chapters.

A sample space is the set of all the possible outcomes from a conceptual experiment.

Here are some examples of sample spaces: In a three-coin-toss experiment, the sample space includes eight outcomes: HHH, HHT, HTH, HTT, THH, THT, TTH, TTT.

In the marble box experiment, there were $2^{10} = 1024$ different outcomes in the sample space: RRRRRRRRRR, GRRRRRRRRR, RGRRRRRRRR, RRGRRRRRRR, and 1020 others.

In a two-envelope ESP experiment, the sample space includes four outcomes: CC, CW, WC, and WW, where the C indicates a correct guess and W indicates a wrong guess.

An outcome is one specific result of an experiment.

Exactly one outcome is obtained each time the experiment is done. In the examples we have been discussing, order is important in differentiating various outcomes. The outcome CW is different from the outcome WC.

An event is a collection of outcomes in a sample space.

Here are some examples of events: The event "two heads" in the three-coin-toss experiment consists of three different outcomes: HHT, HTH, and THH.

The event "1 green" in the marble box experiment consists of ten different outcomes: the outcome with a green marble in the first hole, and the rest red; the outcome with a green marble in the second hole, and the rest red; and so on.

Note: As we have seen, some events consist of only a single outcome. In the two-coin toss experiment, the event "0 heads" consists of the single outcome TT.

It is possible to imagine an event that has no outcomes; for example, the event "12 green marbles" has no corresponding outcomes in the marble box experiment.

Summary

Counting rules were developed for finding the number of outcomes corresponding to each of the eleven possible events in the ten-coin-toss situation, which is analogous to the marble box and the ESP experiment.

The number of outcomes with exactly 0 heads is 1

The number of outcomes with exactly 1 head is 10

The number of outcomes with exactly 2 heads is $\frac{10}{1} \times \frac{9}{2}$

The number of outcomes with exactly 3 heads is $\frac{10}{1} \times \frac{9}{2} \times \frac{8}{3}$

The number of outcomes with exactly 4 heads is $\frac{10}{1} \times \frac{9}{2} \times \frac{8}{3} \times \frac{7}{4}$

and so on

There is a symmetry here: The number of outcomes with exactly 2 heads is the same as the number with exactly 2 tails (or 8 heads); and likewise for 3 heads and 3 tails (7 heads); etc.

On the basis of the numbers resulting from this counting rule, we can compute the exact probabilities of the

44 THE SIGN TEST, AND THE LOGIC OF STATISTICAL INFERENCE

eleven different events. It turns out that the exact probabilities are similar to the observed relative frequencies with which we observed the eleven possible events in the marble box experiment.

Relative frequency is defined as the frequency with which we observed a given event, divided by the total number of samples (50, in our case).

An *outcome* is any specific result of an experiment, with the order of various components retained, for example, HTTHTHHTTH in a ten-coin-toss experiment.

A *sample space* is the set of all the possible outcomes from a conceptual experiment.

An *event* is a collection of outcomes in a sample space. For example, all those outcomes that contain exactly 3 heads in a ten-coin-toss experiment are included in the event "3 Heads."

1. Make a tree for the five-coin-toss situation, showing each of the 32 possible branches. Find out how many outcomes correspond to each of the six possible events (0, 1, 2, 3, 4, and 5 heads). Then compute the probability of each of these six events, using the rule

$$\text{Probability of event} = \frac{\text{Number of outcomes corresponding to event}}{\text{Total number of outcomes in sample space}}$$

2. Now figure out the same probabilities using the same kind of counting procedure that was used in this chapter. Proceed as follows:

   a. How many different possible outcomes are there in the five-coin-toss situation?

   b. How many of these have exactly *zero* heads?

   c. How many of these have exactly *one* head?

   d. How many have exactly *two* heads? (*Hint*: For each of the outcomes from step c, there is a way to put one ball into five boxes. For each of these ways, there are _____ places where you can put a second ball. In this fashion, you can generate _____ x _____ different patterns. But you have made two copies of each pattern, so the final number of outcomes with exactly two heads is one-half of this number, or
   _____ x _____ ÷ 2.)

   e. How many outcomes have exactly *three* heads? For each pattern of two balls in five boxes, there are three places where you could put a third ball. There are thus 3 x _____[1] different patterns you can generate in this manner. But in so doing, you have made each distinctive pattern three different times.
   For example,

   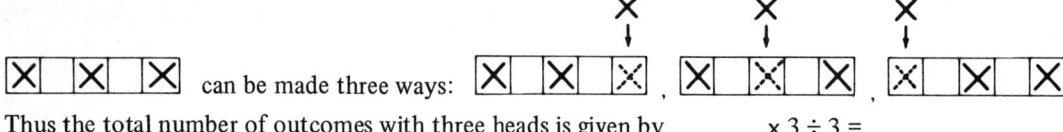

   Thus the total number of outcomes with three heads is given by _____ x 3 ÷ 3 = _____.

   Note: This is the same answer as for part d. Does this make sense? Should the number of patterns with *three heads* be the same as the number of patterns with *three tails* (and thus two heads)?

   f. How many outcomes have exactly *four* heads? How can you find this on the basis of results already obtained?

   g. How many patterns have exactly *five* heads?

   h. Now find the probabilities of each event, on the basis of the counting rules. Do you get the same results you got with the tree?

3. Look back at problem 3 in Chapter 4. Draw a frequency distribution for these observations.

4. What conclusion can be drawn from the similarity of the two curves in Figure 6.3.

---
[1] Insert the answer from part d here.

Review and Conclusion: ESP Experiment

We are now ready to put together the results of the past chapters to arrive at a complete statistical analysis of the ESP experiment. This is the sort of analysis you should do before running a ten-guess experiment, to help make up your mind what decision you will make about ESP for each possible result of the experiment.

In Chapter 6 we computed the probability of each of eleven different possible events in the ten-coin situation. These are reproduced here, to save having to look back and forth.

TABLE 7.1

Event (Number of heads)	0/10	1/10	2/10	3/10	4/10	5/10	6/10	7/10	8/10	9/10	10/10
Probability	0.001	0.010	0.044	0.117	0.205	0.246	0.205	0.117	0.044	0.010	0.001

With these numbers in mind, let's go through the basic steps for a statistical test, following the list in Chapter 4.

A STATISTICAL TEST

Step 1. Specify all possible results of this experiment. As before, the possible results are 0/10 correct, 1/10 correct, 2/10 correct, etc., up to 10/10 correct. (Note that when we say "results" we mean *events*.)

Step 2. Make a formal statement, called a null hypothesis, about the way things would be if we were observing only random variation. In this case, that would be something like this: The guesser has no ESP and should have the same chance of getting 0 correct as the chance of seeing 0 heads if a perfectly fair coin were tossed 10 times in a row. And this would hold for 1, 2, etc., up to 10 correct. (A "perfectly fair" coin is one in which the chance of heads is exactly 1/2 and the chance of tails is exactly 1/2.)

Step 3. Study the null hypothesis. This is a fairly involved process, and results in the table above. When the null hypothesis is true (and the guesser is choosing randomly) the chance of each possible result is shown in Table 7.1.

Step 4. Study various possible decision rules, to find the associated probability of a Type One error for each. *Choose a significance level* which is the maximum tolerable risk of Type One error. Using the numbers in Table 7.1, we can easily find the chance of Type One error for any decision rule we wish to consider. For example, suppose our rule is: Decide there's ESP if you see 10/10 correct; otherwise, don't. We know that the chance of a Type One error is 0.001 — the chance of seeing 10/10 when the null hypothesis is correct. If our decision rule is to conclude there's ESP (reject the null hypothesis) if we observe 10/10 or 0/10, the chance of a Type One error is 0.001 + 0.001 = 0.002. Similarly, using the numbers in Table 7.1, we can fill in Table 7.2.

TABLE 7.2

Decision rule: Decide "There's ESP" if you see any of the following results (reject H_0 if:)	Chance of any of these events when the null hypothesis is true, or probability of Type One error
10/10	0.001
10/10 or 0/10	0.001 + 0.001 = 0.002
10/10 or 9/10	0.001 + 0.010 = 0.011
10/10, 9/10, 0/10, or 1/10	0.001 + 0.010 + 0.001 + 0.010 = 0.022
10/10, 9/10, or 8/10	0.001 + 0.010 + 0.044 = 0.055
10/10, 9/10, 8/10, 0/10, 1/10, 2/10	0.001 + 0.010 + 0.044 + 0.001 + 0.010 + 0.044 = 0.110

It should, I hope, be clear to the reader how to add to this table if one wished to consider even less conservative decision rules.

We must still choose a significance level. That is, we must decide how big a chance of a Type One error we are willing to tolerate. Remember that a Type One error happens when you mistakenly decide "There is ESP" when in fact the subject is just guessing. Your significance level is directly related to the decision rule you finally choose. Your decision rule specifies under what circumstances you will decide "There is ESP." From Table 7.2 you can figure out how likely you are to say so when in fact there is none. That is the same as the significance level, or chance of a Type One error.

How big a chance should one accept? There is no single right answer for this question. A very conservative rule is to keep the chance of a Type One error less than 1 in 1000, or 0.001. The symbol for significance level generally used in statistics is the Greek letter alpha, α. This significance level then would be denoted $\alpha = 0.001$. A significance level of 1 in 100, or $\alpha = 0.01$, is still quite strict. Many people are satisfied with a 0.05 chance of making a Type One error: $\alpha = 0.05$. Let's use that choice here. It is quite common in the social sciences.

Step 5. Choose a decision rule appropriate for the significance level just specified and identify the rejection region which goes with it.

A question to consider: Given that we have committed ourselves (in step 4) to a 0.05 significance level, what decision rule should we pick? We must pick one which involves a chance of Type One error, which is 0.05 or smaller. Any of the first four rules in Table 7.2

would meet this criterion. (Please check!) The fifth is a narrow miss, with a 0.055 chance of Type One error. Now, if we are to pick one of the first four rules in the table, which one shall we choose? The fourth one has the advantage of maximizing our chance of correctly detecting ESP when it is really present. That is the best one for us. Picking any of the other rules would really be like picking a more conservative (smaller) significance level.

Therefore, our rejection region is that specified by the fourth decision rule. We shall reject the null hypothesis if we observe 10/10, 9/10, 0/10, or 1/10 correct guesses in the ESP experiment. These four events lie in our rejection region.

Step 6. Do the experiment and see whether the result lies in the rejection region.

This completes the statistical analysis of this situation, using a mathematical model to replace the marble box model. Our next task is to review quickly four fundamental assumptions which were parts of that mathematical model: four assumptions we needed to make in order to find the numbers in Table 7.1.

Four Assumptions

The following four assumptions were essential to our argument. If you are willing to accept the general argument given, that means in particular that you accepted these assumptions. Think about it.

Assumption 1. In the ten-coin-toss situation, there are $2^{10} = 1024$ different patterns or outcomes which could result. I tried to make this plausible, and most of you should be able to appreciate that each time you add another coin you double the number of different patterns possible. You could check this by drawing a giant "tree."

Assumption 2. Each of these 1024 different outcomes is exactly as likely to occur as each of the others. For example, HHHHHHHHHH is just as likely as HTTHTTTHTH. This is somewhat counterintuitive. (Recall that half of the members of a large class voted that the second was more likely than the first.) An attempt to make this assumption more plausible follows just below.

Assumption 3. The numbers of different outcomes (patterns) corresponding to each of the eleven possible different events (0/10, 1/10, etc.) are as given in Table 6.1 on page 40.

Assumption 4. The probability of an event occurring is the number of outcomes corresponding to that event, divided by 1024, which is the total number of different outcomes. We shall have occasion to return to this assumption later, in Chapter 12.

Plausibility Argument in Support of Assumption 2: Before reading further, please write down one specific pattern of ten tosses, such as HTTHTHHTHT, TTTTTTTTHH, HHHHHHHHHH, THTHTHTHTH, or any other you wish, in the following space:

____ ____ ____ ____ ____ ____ ____ ____ ____ ____

Now, suppose that I flip a coin once. What is the chance that it will come up so as to agree with what you put in the first blank? The chance is 1/2, regardless of what you put in the blank! Now suppose I flip it a second time: What is the chance that it will agree with what you wrote in the second blank? Again, the chance is just 1/2, regardless of what you wrote there! And this is the result all the way through. No matter what specific pattern you wrote down, the chance is just the same that if I toss a coin ten times I will match it exactly:

$1/2 \times 1/2 \times 1/2 \times 1/2 \times 1/2 \times 1/2 \times 1/2 \times 1/2 \times 1/2 \times 1/2 = 1/2^{10} = 1/1024$

48 THE SIGN TEST, AND THE LOGIC OF STATISTICAL INFERENCE

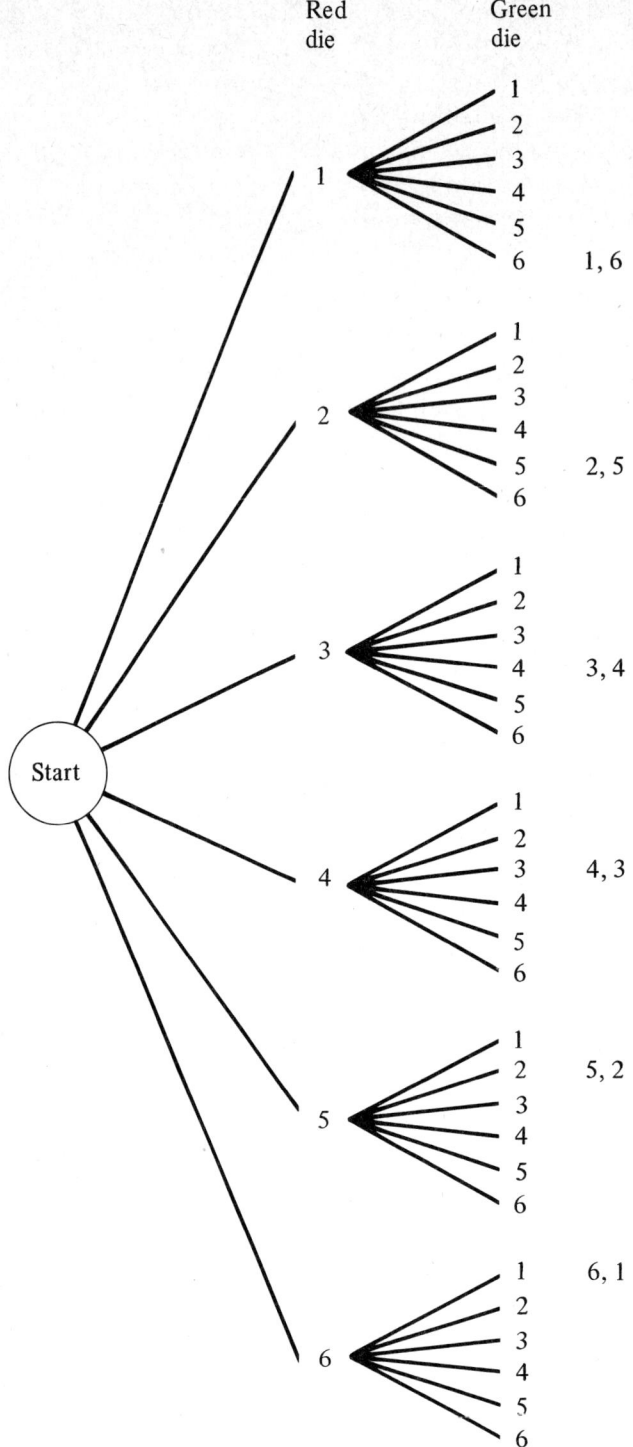

Figure 7.1 A tree showing every possible outcome in an experiment in which two dice are tossed and the outcomes recorded in order.

Review and Illustration of the Notions *Sample Space*, *Event*, and *Outcome*

Recall that a *sample space* is the set of all the possible outcomes of a conceptual experiment. And an *event* is a collection or combination of outcomes in a sample space. The following example is intended to provide another specific case for using these ideas.

Example: Suppose you toss two dice. Suppose further, to make it clear, that the first is red and the second is green. What is the sample space corresponding to this experiment? What outcomes are in the sample space? What are a few interesting events? (You will learn more if you take a few minutes to try to answer this before reading on.)

Suppose we represent an outcome when the first die was 3 and the second 2 as follows: 3, 2. Using this notation, we can list the possible outcomes quite easily:

1, 1; 1, 2; 1, 3; 1, 4; 1, 5; 1, 6;
2, 1; 2, 2; 2, 3; 2, 4; 2, 5; 2, 6;
3, 1; 3, 2; 3, 3; 3, 4; 3, 5; 3, 6;
4, 1; 4, 2; 4, 3; 4, 4; 4, 5; 4, 6;
5, 1; 5, 2; 5, 3; 5, 4; 5, 5; 5, 6;
6, 1; 6, 2; 6, 3; 6, 4; 6, 5; and 6, 6.

These 36 outcomes make up the sample space. A simple way of showing them all, without a nasty list, is to make a tree like the one in Figure 7.1. All six of the outcomes corresponding to the event "total = 7" are indicated on the tree. Since there are six such outcomes, the overall probability of seeing a total of 7 when you roll two dice is 6 out of the 36 different outcomes, or 6/36 (1/6).

The *event* "total = 4" would contain only three *outcomes*: 1, 3; 2, 2; and 3, 1. So the probability of this *event* is just 3/36 or 1/12.

One could define all kinds of different events. For example, the event "Those outcomes with the second die showing twice the number of spots on the first die" would contain the following outcomes: 1, 2; 2, 4; and 3, 6. So the probability of that event would be 3/36, too.

In all of the above computations, we have been using the rule which was first presented in the Question and Answer section for Chapter 5: *Whenever the sample space is made up of a number of equally likely outcomes, the probability of any event is given by*

$$\text{Probability of an event} = \frac{\text{Number of outcomes corresponding to that event}}{\text{Total number of outcomes in the sample space}}$$

Summary

In Chapter 4 we carried out a statistical analysis of the ESP experiment, using the data from the marble box experiment to compute estimates of the chance of Type One error associated with various decision rules. In the past two chapters we have developed a way to use mathematics instead of marbles for finding probabilities. Using the probabilities of the various events which we computed in Chapter 6, we carried out a new statistical analysis in this chapter. As a result of this analysis, we concluded that for a given significance level of 0.05, our best decision rule was to conclude we had evidence of ESP if our subject guessed 10/10 or 9/10 or 0/10 or 1/10 correct, and otherwise to conclude we could not be sure he wasn't just guessing.

Four assumptions underlying the mathematical model for the ESP experiment were listed and discussed.

The important concepts of event, outcome, and sample space were reviewed and illustrated in the context of a two-die experiment.

Problems

1. Construct two additional lines of Table 7.2 to cover the following two decision rules:
   a. Decide "There's ESP" if you see 10/10, 9/10, 8/10, or 7/10 correct. (What is the associated chance of a Type One error?)

b. Decide "There's ESP" if you observe 10/10, 9/10, 8/10, 7/10, 3/10, 2/10, 1/10, or 0/10 correct. Otherwise conclude subject could just be guessing. (What is the associated chance of a Type One error?)

2. In the two-die experiment diagramed in Figure 7.1, there were 36 possible outcomes. How many outcomes would there be in a similar experiment which used three dice?

3. Four aces are placed in a hat: hearts, spades, diamonds, and clubs. One is drawn at random, the result is recorded, and the card is replaced. Then a second draw is made, and the result is recorded. Draw a tree to represent this experiment. How many individual outcomes are there? Which outcomes are included in the event "At least one heart"? Which outcomes are included in the event "Both cards are red"? What are the probabilities of these two events?

4. Define each of the following as accurately as you can. At this point, you should be able to do this quite easily, and without errors.

   a. Significance level.
   b. Sample space.
   c. Rejection region.
   d. Event.
   e. Type Two error.
   f. Null hypothesis.
   g. Outcome
   h. Type One error.
   i. Decision rule.

5. Write down, from memory, the six steps involved in the statistical analyses we have carried out. After you have done so, you can check your answer by looking back at page 27.

Some Practice Problems, with Solutions and Discussion; A Formal Presentation of the Sign Test

At this point you have learned many of the most important ideas connected with statistical testing. To consolidate your knowledge, you should now try to do some statistical tests. People learn much more by active problem solving than by passive reading. *After* you have made an honest effort to do the two problems below, read the solutions. If you jump directly to the solutions, you will be cheating yourself.

PRACTICE PROBLEMS

1. In an English composition class, pretest and posttest scores are available for seven people. A dean and the teacher involved got into an argument whether the differences are convincing proof that the students improved in the course. The scores are given below:

Pretest score	Posttest score	Difference (post − pre)
46	63	+17
57	59	+ 2
72	80	+ 8
71	64	− 7
48	57	+ 9
55	73	+18
67	78	+11

    a. Carry out a statistical analysis to help decide whether these differences are the result of random variability in testing results, or whether they reflect a real change in performance between the tests. Use a significance level of $0.05 (\alpha = 0.05)$.

(*Hint*: If the course had no effect, we'd expect half the students to have negative difference scores and half the students to have positive difference scores, on the average.)

b. What is a Type One error in this situation?

c. What is a Type Two error?

2. *Null hypothesis practice*: for each of the following situations, write down clearly and completely *in words* the null hypothesis you think appropriate for a statistical test of the results. Then, if you wish, write it down in symbols.

a. Are prices in ghetto neighborhoods different from prices in suburbs? Prices are obtained on 20 items available in markets in both places. For each item you record a G if an item is more expensive in the ghetto and an S if it is more expensive in the suburb. You observe 13 Gs and 7 Ss, and run a statistical test. *State the null hypothesis* and then go on to the next question.

b. Do people prefer good Scotch (at $8.00 per bottle) to cheap Scotch (at $5.00)? Suppose that you give each of five people a pair of glasses, one containing good Scotch and the other containing cheap Scotch. Suppose further that all five say that they prefer the glass containing the good Scotch. In order to carry out a statistical test, first state a null hypothesis.

c. How do TV lectures compare to live ones? Suppose that 40 people see a course, half of which is taught on TV and half of which is taught live. If 25 of the 40 say they prefer live presentation, whereas 15 prefer TV, what can we conclude? State the null hypothesis appropriate for a statistical test.

SOLUTIONS

Solution to Problem 1a

Seven students were tested before and after an English composition class. A difference score with a plus (+) indicated net improvement, and one with a minus (−) indicated a loss. In the seven cases, there were six plusses and one minus. (The scores were +17, +2, +8, −7, +9, +18, +11.) There are many cases like this in which we ask whether we have significant evidence that the course, or other treatment, had any effect on the people involved. Here is one way of answering the question.

Verbal level: Were there so many plus scores that we are convinced that these differences wouldn't have resulted by pure chance, if the class had no effect?

Precise statistical analysis: Follow the six steps from Chapter 4.

Step 1. Specify all the possible results of the experiment. This step causes severe problems for many people, especially at first. There are an enormous number of different possible lists of seven "difference scores" like those shown in the quiz. But for the moment we are deliberately ignoring the size of the differences, and looking only at whether a person improved (+) or deteriorated (−) as a result of the course. This makes the situation much simpler. There were seven students in the course. In such situations, where you obtain seven *difference scores*, each with a plus or minus, how many different number-of-plus scores could you have? You could have 0/7 plusses, 1/7 plusses, 2/7 plusses, etc., up to 7/7 plusses. There are thus only eight different number-of-plus scores possible. In this sense, there are eight different possible results of the experiment.

Along this line of thinking, then, the possible different scores or *events* are 0/7, 1/7, 2/7, 3/7, 4/7, 5/7, 6/7, and 7/7 plusses.

Step 2. Make a formal statement (a null hypothesis) about the state of the world when all we are observing is random variation, and there is no "effect." In our case, a null hypothesis would be that the class has no systematic effect on English composition ability, and so any differences in scores should be the result of chance variation. We can write this in slightly more precise form as follows. *Null hypothesis*: The chance that a person will show a positive (+) difference score is just 1 in 2. There is an equal chance that he will show a negative (−) difference score.

Step 3. Study the null hypothesis, and find out how likely each of the various events would be if the null hypothesis were true. We shall do this in a manner exactly analogous to the way we studied the ESP experiment. We will figure out how many different patterns of seven plusses and minuses we could draw, and then count how many of those patterns have 0, 1, 2, etc., up to 7, plusses. Then we shall compute the probability of each of the eight different events.

We must first consider a preliminary question: How many different patterns can you make with seven symbols, each of which is a plus (+) or a minus (−)? By now, you should realize that there are $2^7 = 2 \times 2 \times 2 \times 2 \times 2 \times 2 \times 2 = 128$ different patterns. If you don't believe it, consider how many patterns you can make of length 1: +, −, or a total of 2; of length 2: + +, + −, − +, and − −, for a total of $2^2 = 4$; etc.

Now we need only figure out how many of these patterns, or *outcomes*, correspond to each possible event. Consider first the event 0/7 plusses. Plainly, only one pattern corresponds to this event: Since only one of the 128 outcomes corresponds to this event, the probability of this event is 1/128 when H_0 is true. How many outcomes correspond to the event 1/7 plusses? There are seven different patterns with exactly 1 plus:

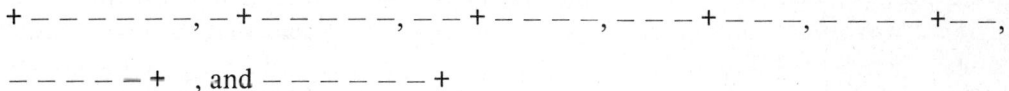

So the chances of seeing exactly 1 plus, when the null hypothesis is true, is just 7/128, or about 0.05. Following an argument exactly analogous to that for the ESP experiment, we can fill in Table 8.1. (Look back at Chapter 6 if you don't quite understand how it all works out.)

This completes our analysis of how likely various events are when the null hypothesis is true: when a plus is just as likely as a minus. As may be seen, the scores 3/7 and 4/7 are most likely; 0/7 and 7/7 are least likely, and each would be expected to occur only about 8 times in 1000.

Step 4. Pick a significance level, α, and study various possible decision rules. In this case you were instructed to tolerate at most a 0.05 chance of a Type One error. That is, we must choose a decision rule such that when the null hypothesis is true, the chance of our saying "The null hypothesis is false, and there's really a difference between pretest and posttest performance" is less than 0.05, or 5 out of 100. (Recall that significance level is defined as how large a chance of a Type One error you are willing to tolerate in a decision.)

We must now look at various possible decision rules, to see how great a risk of Type One error they entail. We want to include as many extreme events as possible without getting too great a chance of Type One error. If we look at the two most extreme events, 0/7 and 7/7, we find that their combined probability is 0.008 + 0.008 = 0.016, when the null hypothesis is true (see Table 8.1). This is still less than 0.05, which is our maximum. What would happen if we added any more events? The total would be too great. Indeed, as may be seen in the table

TABLE 8.1   PROBABILITY OF EACH EVENT IF THE CLASS HAD NO EFFECT

Event or score (plusses)	No. of different outcomes corresponding to that event	Probability of that event
0/7	1 $(-------)$	$1/128 = 0.008$
1/7	7 $+------, -+-----$, etc.	$7/128 = 0.055$
2/7	$\frac{7}{1} \times \frac{6}{2} = 21$	$21/128 = 0.164$
3/7	$\frac{7}{1} \times \frac{6}{2} \times \frac{5}{3} = 35$	$35/128 = 0.273$
4/7	$\frac{7}{1} \times \frac{6}{2} \times \frac{5}{3} \times \frac{4}{4} = 35$	$35/128 = 0.273$
5/7	$\frac{7}{1} \times \frac{6}{2} \times \frac{5}{3} \times \frac{4}{4} \times \frac{3}{5} = 21$	$21/128 = 0.164$
6/7	$\frac{7}{1} \times \frac{6}{2} \times \frac{5}{3} \times \frac{4}{4} \times \frac{3}{5} \times \frac{2}{6} = 7$	$7/128 = 0.055$
7/7	$\frac{7}{1} \times \frac{6}{2} \times \frac{5}{3} \times \frac{4}{4} \times \frac{3}{5} \times \frac{2}{6} \times \frac{1}{7} = 1$	$1/128 = 0.008$

Note: To find decimal equivalents for fractions in calculations like this (for example, to find that $21/28 = 0.164$), you can use the table of Decimal Equivalents on page 251.

either 1/7 or 6/7 by itself will happen slightly more than 5 times out of 100. So neither of these events could be included in a rejection region with this significance level.

Step 5. Specify the rejection region. As a result of the argument just completed, we will reject the null hypothesis if we observe 0/7 or 7/7 plusses. These two events, then, constitute the rejection region. If we observe any other event, we cannot be confident that the results aren't just the result of chance fluctuation. But if we observe 7/7 plusses, we'll conclude the course helped the grades. And if we observe 0/7 plusses (or 7/7 minusses) we'll conclude that the course hurt performance.

Step 6. Run the experiment, and interpret the result. In this case, we know the result: 6/7 of the students showed a plus, and 1/7 showed a minus. *Decision*: Since 6/7 does not lie in the rejection region, we decide not to reject the null hypothesis. That is, we are not convinced that the class has an effect on writing ability. We don't have statistically significant evidence that the class helps.

Solution to Problem 1b

What is a Type One error in this situation? The answer is that a Type One error is the error of rejecting the null hypothesis when you shouldn't. That would happen if you decided that the class has a significant effect on writing scores, when in fact it doesn't.

Solution to Problem 1c

What is a Type Two error here? A Type Two error is the other kind of error: concluding that the class has no influence on writing ability when in fact it does influence writing ability. We may have made a Type Two error in our analysis.

Problem 2: Null hypothesis practice

Solution to part a: Are prices in ghetto neighborhoods different from prices in suburbs? Prices are obtained on 20 items available in markets in both places. For each item you record a G if the item is more expensive in the ghetto, and an S if it is more expensive in the suburb. You observe 13 Gs and 7 Ss, and run a statistical test.

An appropriate null hypothesis is: Prices are no different in ghettos and suburbs. So, for any item differing in price in the two stores, there's a 50 : 50 chance it will be more expensive in the ghetto and a 50 : 50 chance it will be more expensive in the suburb. (Note that here, as in general, the null hypothesis is *a precise statement that there is no difference*.)

Solution to part b: Do people prefer good Scotch to cheap Scotch? Suppose that you give each of five people a pair of glasses, one containing good Scotch and the other containing cheap Scotch. Suppose further that all five say that they prefer the glass containing the good scotch. In order to carry out a statistical test, first you state a null hypothesis.

An appropriate null hypothesis is: When a person makes a choice between a glass of good Scotch and a glass of cheap Scotch, there's a 50 : 50 chance that he will prefer cheap Scotch. People can't tell the difference.

Solution to part c: How do TV lectures compare to live ones? Suppose that 40 people see a course, half of which is taught on TV and half of which is taught live. If 25 of the 40 say they prefer live presentation, whereas 15 prefer TV, what can we conclude?

An appropriate null hypothesis is: When a person in this situation makes a choice between TV and live presentation, there's a 50 : 50 chance that he will prefer the former, and an equal chance that he will prefer the latter. Both kinds of presentation are equally acceptable.

In each of these three situations you know about the result of the experiment before you are asked to state the null hypothesis, but in every case the hypothesis is stated in a way completely unrelated to the actual result. Remember that *the null hypothesis is a precise statement of how likely various results would be if there were no phenomenon except chance or random variation*.

After our analysis of the first problem in this chapter, we decided not to reject the null hypothesis: we are not convinced that the class has an effect on writing ability; we don't have statistically significant evidence that the class helps. Many people in such a situation are tempted to write: We *accept* the null hypothesis; we have proof that the course didn't help. This is an error which you should try to avoid. Failing to reject the null hypothesis is not the same as being sure that it is true. Very often we are simply left unconvinced: It might be true, or it might not. We just don't know.

Thinking back to the ESP experiment may help clarify this. If a person guesses 9/10 correct, we decide to reject the null hypothesis and conclude that he has ESP (using a 0.05 significance level). Suppose that a subject guesses 8/10 correct. This isn't enough evidence to convince us, but it might well make us wonder. It might lead us to design a larger, more powerful experiment, to see if his better-than-chance performance continued. A score of 8/10 wouldn't prompt us to reject the null hypothesis. But it certainly wouldn't lead us to accept the null hypothesis or to be convinced that it were true. We are left in doubt.

THE SIGN TEST

The statistical analysis we carried out in the answer to problem 1a above is called a *sign test*. The analysis we did for the ESP experiment is another example of a *sign test*. The following characteristics are always present in situations where a sign test can be used.

When to Run a Sign Test

1. The situation can be thought of as a series of independent observations, each of which falls into one of two separate classes. In the ESP experiment each observation was a guess for one envelope. Each guess was either *correct* or *wrong*. In the English composition situation each observation was a difference score for one student. Each score was either *positive* or *negative* (it had either a plus or a minus sign before it). In both of these examples the observations are independent. In the ESP experiment one can assume that the second guess (correct or wrong) is independent of the first guess, and so on, for the remaining guesses. In the English composition class the difference score for the second student does not depend on the difference score for the first student, or on the difference score of any other student. The individual difference scores are *independent*.

2. We are interested in answering the following question: Are we confident that the source of the observations has a definite tendency to produce more of one kind than the other?

    In the ESP experiment, we wanted to know if the person over the long run would do significantly better (or worse) than a coin-toss random process. In the English composition example, we were interested in knowing if the teacher in the long run would produce more students who improved (and had a positive difference score) than students who retrogressed between pretest and posttest.

3. When we run a sign test, we are always interested in *testing* a null hypothesis of this form: Each of the two classes of observation is equally likely; there is a 50 : 50 chance that you will see an observation in either class. The logic here gets somewhat contorted. In order to decide whether the guesser has ESP, we test the following null hypothesis: He is guessing purely randomly, with a 50 : 50 chance of being correct on each guess. If we reject this null hypothesis, that amounts to concluding that he has ESP. In order to decide whether the teacher is influencing her class, we test this null hypothesis: For each student going through the course, there's a 50 : 50 chance he'll do better on the posttest than the pretest (and so have a positive difference score) and a 50 : 50 chance he'll do better on the pretest than on the posttest (and so have a negative difference score). If we *reject* this null hypothesis, that amounts to concluding that the teacher has a significant influence on her class.

Students often have considerable difficulty stating an appropriate null hypothesis for a sign test. They say things like, "There's a 50 : 50 chance that the subject has ESP," which is quite different from a precise statement that the subject has *no* ESP. A workable rule, to help write a good null hypothesis for any sign test situation, is to follow this format:

For each _____ *there is a 50 : 50 chance that* _____ *will occur and a 50 : 50 chance that* _____ *will occur.*

By filling in the blanks, one can state good null hypotheses and avoid many errors. The null hypothesis above, it will be noticed, is in this form.

We shall see many more examples of the sign test in chapters to come. If you are able to understand the two you have seen so far, you will be in a very good position for the material that follows. (By the way, as you have perhaps deduced, the *sign test* gets its name from problems like the one at the start of the chapter, where we count plus and minus *signs*.)

QUESTION AND ANSWER

Question: In answering problem 1a above about the difference scores in the English composition course, you looked only at the *sign* of the differences and ignored the size. Shouldn't you consider the size, too? The way you do things, a difference of +1 and a difference of +37 count equally. Is that right?

Answer: The sign test *does* ignore the size of the differences; that is a weakness of the test. Later on in this book we will learn techniques that take the size of the difference into account. These techniques involve more difficult computations than the sign test, and they cannot be used in all the situations where a sign test is applicable. As you will learn in coming chapters, the sign test often can be done in your head, once you understand the logic fully and know some computational shortcuts. It is a very useful test for many purposes but sometimes the fact that it ignores the *size* of differences makes it less useful than other techniques we will be learning about later.

Question: What would have happened if someone in the English composition class had obtained the same score on the pretest and the posttest? Then his difference score would be zero, which is neither positive (+) nor negative (−). Could you still use a sign test? How would you proceed?

Answer: The way a *zero* difference is handled is to delete that subject from the series of observations and analyze the results as if he had never existed. Suppose in a similar class the following year, the teacher had ten students, and the difference scores (posttest minus pretest) were +6, +9, −2, +12, +11, 0, −15, +6, +4, −1. There are six positive (+) differences, three negative differences, and one zero. We restrict our analysis to those students who had a nonzero difference score. We ask, then, whether the observed result that 6 of the 9 nonzero scores were plusses could plausibly have resulted from a teaching system in which plus and minus signs each happen with equal probability. Our sign test analysis would involve looking at all the possible events when *nine* coins are tossed. We want to know how likely 0/9, 1/9, 2/9, 3/9, etc., 9/9 would be, when nine fair coins are tossed.

There is nothing improper about this procedure. It is necessary to remember that any conclusions are limited to those cases with nonzero difference scores. But since in any real situation the zero cases are a small minority, this restriction is usually unimportant.

Question: In step 5 of the solution to problem 1a, you describe the rejection region: 0/7 or 7/7 plusses. Then you say, "If we observe any other event, we cannot be confident that the results aren't just the result of chance fluctuation." Why so many negatives? Why not speak English like everybody else, and say, "If we observe any other event, we *can* be confident that the results *are* just the result of chance fluctuation"?

Answer: Because we can't. Unfortunately, it is not the case that we are always confident about something. We actually observed 6 plusses out of 7: 6/7. This is not sufficiently extreme to *convince* us that the course had an effect. It doesn't definitely convince us to reject the null hypothesis. But it is certainly suggestive. The teacher certainly will be reluctant to *accept* the null hypothesis. And we would probably agree with her: It looks as if the class has an effect but this particular analysis isn't conclusive enough to prove it. You will see similar cases as we proceed.

Summary

Another example of a statistical analysis like the one for the ESP experiment was presented. In this case we looked at a series of seven *difference scores* between a posttest and a pretest in English composition. The scores were categorized into two classes: those with a positive difference (indicating improvement as a result of the course) and those with a negative difference (indicating retrogression). The name of this statistical analysis is the *sign test*.

The following are the basic conditions for a sign test:

1. The situation can be thought of as a series of independent observations, each of which falls into one of two separate classes.
2. We are interested in answering the following question: Are we confident that the source of the observations has a definite tendency to produce more of one kind than the other?
3. When we run a sign test, we are always interested in testing a null hypothesis of this form: Each of the two classes of observation is equally likely. There is a 50 : 50 chance that you will see an observation in either class.

Problems

1. Suppose that for his class of 20 students a professor tries to guess the month in which each student was born. State a null hypothesis appropriate for this situation.

2. Describe how one could carry out a procedure like the marble box experiment to study the above null hypothesis and analyze the situation. How could you estimate experimentally the chance of the professor guessing 0/20, 1/20, etc. up to 20/20 times correctly? If you suggest a marble box, say how many marbles of each color you would use, how many holes in the paddle, and how many samples you would draw.

3. In a study of hemispheric dominance, words are presented to either the left eye or the right eye of a subject. The presentation is very rapid, so the subject doesn't always have time to identify the words. For seven subjects the following data are observed (all subjects are right handed):

Percent correct when watching with right eye (R)	Percent correct when watching with left eye (L)	Difference (R − L)
33	27	+ 6
59	47	+ 12
36	30	+ 6
44	42	+ 2
28	26	+ 2
51	51	0
49	60	− 11

Run a sign test to see whether there is a significant tendency to see better with one eye than the other. Use a *significance level of 0.20*. Show all your steps. (*Hint*: See the second Question and Answer for this Chapter.) Use the table on page 251 to find the decimal equivalents of fractions.

4. A psychologist believes he has noticed that school children are more aggressive in the afternoon than in the morning. Observers record instances of aggressive behavior (pushing, hitting, taking pencils and books from others, etc.) for each of 15 children, for an hour each morning and an hour each afternoon. For each child a summary score of M (for those more aggressive in the morning) or A (for those more aggressive in the afternoon) is assigned. Of the 15 children, 10 are assigned A and 5 are assigned M ratings. Do not run a sign test unless you want to, but state the null hypothesis appropriate to this situation as a first step in a sign test.

5. Add up the numbers in the right-hand column of Table 8.1. Does the total make sense?

6. Look back at the six examples of statistical questions on page 17. For which of these six do you think a sign test could be used to analyze the situation?

The Normal Distribution and the Z Test

Some Descriptive Statistics

DESCRIPTIVE STATISTICS AND INFERENTIAL STATISTICS

Statistics is often characterized as being divided into two parts. On the one hand there is descriptive statistics, which includes those techniques used to organize and make understandable a great deal of data. The frequency distributions you have seen and drawn are examples of descriptive statistics, as are most of the ideas to be presented in this chapter. *Descriptive* statistics is used to help *describe* a sample. On the other hand is inferential statistics, which has been and will be our primary concern in this book. This is the set of tools you use when you observe a limited number of cases and wish to make an inference about the whole population from which you have taken the sample. When you compare the prices of 20 items in ghetto and suburb stores, you are really interested in more than those 20 items: you are interested in the price level generally. You study the sample of 20 items in order to make *inferences* about the prices of all items.

Descriptive statistics is essentially a set of tools for number crunching. These tools have been developed to make your life bearable when you deal with great quantities of data and wish to arrange them into a more understandable and easily organized form. For the rest of this chapter we will be concerned with some very simple descriptive statistics.

MEASURES OF CENTRAL TENDENCY

Consider the following data: 7, 29, 3, 5, 6, 10, 21, 14, 3, 4, 9, 5, 3. If I asked you to tell me a single number which best represents this group of numbers, you might find the question vague. But if I told you that I had observed these numbers when watching a roulette wheel for 13 consecutive spins, and I asked you which number you would care to bet on for the following spins, one number stands out as the best: this is the number 3, which occurs more frequently than any other.

The number which occurs more often than any other number in a set of data is called the mode.

On the other hand, if I told you that these numbers were test scores and that I wanted to use them to choose one most representative student with whom to have a long talk, to find out how the course was going, which score would lead me to the most representative student? One plausible answer to that question is to find the score of the student who is right in the middle of the distribution — a student such that half the people in the class scored lower than he did on the test and half scored higher. To find this student's score, all we need to do is arrange the scores in increasing order, from lowest to highest, and then find the middle one. When we do this we obtain the following list: 3, 3, 3, 4, 5, 5, **6**, 7, 9, 10, 14, 21, 29. Since there are 13 numbers, the middle one is the seventh, indicated by bold type. The middle score here is the score 6, and this is the *median* of this set of data.

The median of a set of numbers is the middle number, when all the numbers are arranged in increasing order by size, from the smallest to the largest. (Median means middle.)

If you are on your toes you will realize that this won't always be quite as easy as it was for the preceding list of numbers. Suppose there is an even number of data entries: then what is the median? Consider the data set 4, 4, 8, 11, 17, 99. What is the median of this set? There are two numbers, 8 and 11, at the middle of this list. The definition generally given for such cases is the halfway point between the two middle numbers. The midpoint between 8 and 11 is 9½. This value, which is the average of the two middle numbers, is the median for this list. A similar procedure is used whenever a list of data contains an even number of entries.

A third way of finding a single number to represent this group is to take the mean.

In statistics, the simple arithmetic average is called the mean.

Suppose that these numbers represented the numbers of masters degrees in psychology given at each of 13 state colleges in California during 1972–1973. If you wished a single number to best represent the production of masters degrees per school, you would take the *mean*: you would add up the numbers and divide by 13. We will use the letter M to represent the mean:

$$M = \frac{7 + 29 + 3 + 5 + 6 + 10 + 21 + 14 + 3 + 4 + 9 + 5 + 3}{13}$$

$$= \frac{119}{13} = 9\frac{2}{13} = 9.15$$

We have now seen three different ways to pick a single number to best represent a whole group of numbers: the mode, the median, and the mean. Each of these measures is valuable in certain circumstances, and they are all simple. You should memorize them all.

BIMODAL DISTRIBUTIONS

Suppose you turn on a brand new light bulb and measure the time until it finally burns out. Call this time the *life* of the light bulb. Now if you repeat this operation a great many times — say 1000 times — and record the frequency with which each possible *life-length* occurs, you will see a curve something like the one in Figure 9.1.

This frequency distribution has two peaks: a small one very near 0 hours of life, resulting from

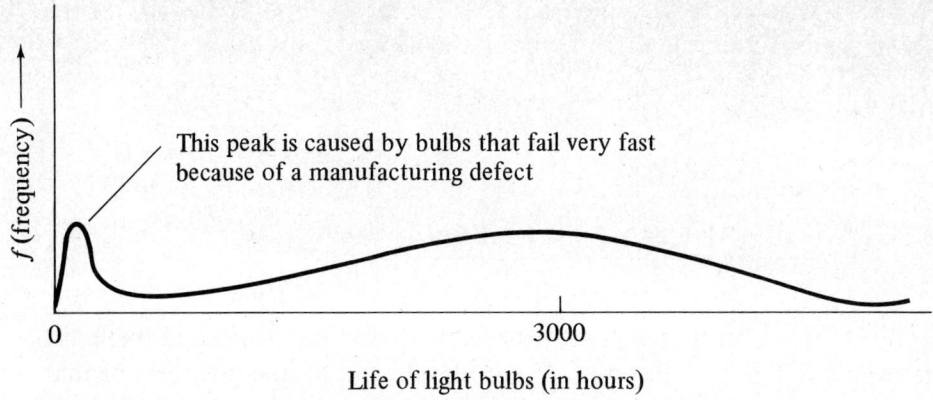

Figure 9.1 An example of a *bimodal distribution*

bulbs which failed very quickly as a result of a manufacturing defect, and a broad one at around 3000 hours of life, representing the nondefective bulbs. A distribution of this sort, with two distinct and separate peaks, is called a *bimodal* distribution. Whereas most distributions have a single mode, or peak, this one has two. Note that a distribution like this which has two peaks, or two maxima, is called bimodal even if the two peaks are not identical in height.

Whenever you see a bimodal frequency distribution, it usually means that the population you have been studying can better be thought of as two distinct populations, mixed up together. In the light bulb example, those two populations are defective bulbs and nondefective bulbs. Each has its own maximum.

THE SIGMA NOTATION FOR SUMMATION

At the beginning of the section on Measures of Central Tendency, a data list was presented. Let's call it the X list and give names to each entry as follows:

7	X_1
29	X_2
3	X_3
5	X_4
6	X_5
10	X_6
21	X_7
14	X_8
3	X_9
4	X_{10}
9	X_{11}
5	X_{12}
3	X_{13}

These names are read "X sub 1," "X sub 2," and so on. Note carefully that the subscript refers to the *position* of an entry in the list, and not at all to its value. For example, the value of X_1 is 7, and the value of X_{13} is 3. When we wish to talk about one particular member of the list, without specifying which one, we use the notation X_i, which is read "X sub i" or the ith member of the list. Here's how we write in symbols the rule for computing the mean of this list of 13 numbers:

$$\text{Mean} = M = \frac{\Sigma X_i}{13}$$

$$= \frac{7 + 29 + 3 + 5 + 6 + 10 + 21 + 14 + 3 + 4 + 9 + 5 + 3}{13} = \frac{119}{13} = 9\frac{2}{13}$$

In this strange package, the X_i tells you that we're concerned with the data list called the X list, as indicated above. The capital sigma, Σ, tells you that you should add up the members of that list. Sigma is the Greek equivalent of the English letter S. It helps to remember: *sigma* (Σ) means *sum*.

Using this same notation, we can write a formula for finding the mean of the numbers in any data list, where the number of data values in the list is denoted by the letter N:

$$M = \frac{\Sigma X_i}{N}$$

It is very important that you learn to feel comfortable with this somewhat strange sigma notation.

MEASURING THE VARIABILITY OF A SET OF NUMBERS

We have looked at various measures of central tendency in a group of numbers, including the mean, the median, and the mode. Now we shall look at another important characteristic of any group of numbers, the *variability*. A set of numbers which are all very close to each other obviously has less variability than one in which there is a big difference between the small values and the large ones. This can be important. Consider the frequency distributions in Figure 9.2 for the observed miles-to-failure for two different kinds of tires in a standardized test.

One hundred samples of each kind of tire were driven until each tire blew out. Now if we consider the *mean* tire life, brand 2 tires are clearly superior. On the average, they last almost 30,000 miles. But variability is important too. If you buy brand 1 tires, you are confident that you can drive about 12,000 miles with almost no chance of a blowout. After that you should

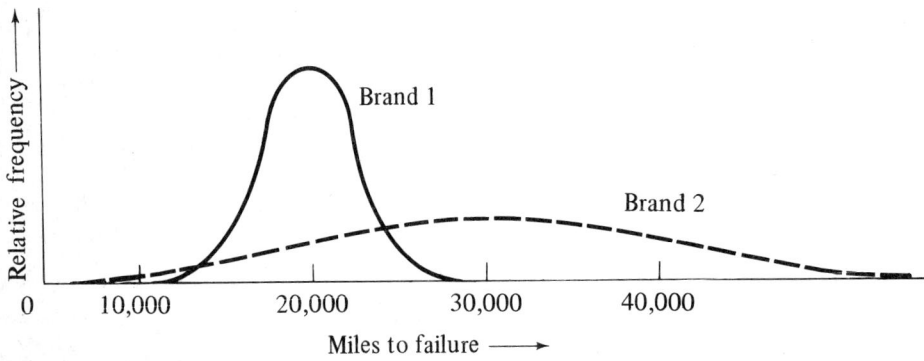

Figure 9.2 Two frequency distributions of different variability (variance).

change tires, since the chance of failure increases rapidly. For brand 2 tires, on the other hand, there is a noticeable chance of tire failure after about 6000 miles. Although the chances are that you could drive much farther than that, the danger of a blowout is so great that it would be foolish to choose brand 2. We would guess that the quality control for brand 2 tires is inferior. In any case, brand 2 tires are much more variable, and this variability is important.

As a second case in which dispersion on spread in numbers is important, consider the frequency distributions in Figure 9.3 for the maximum daily temperature, measured over a full year, in Detroit (where I was born) and in Santa Barbara (where I now live).

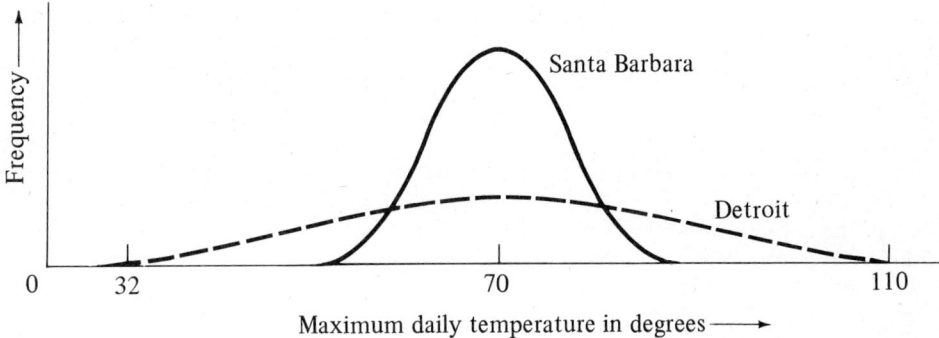

Figure 9.3 The weather in Detroit is much more variable than that in Santa Barbara.

As may be seen, the middle of each of these distributions is about the same. The average maximum daily temperature is indeed about the same in these two cities. But the variability is very different: In Detroit you freeze in the winter and roast in the summer; in Santa Barbara temperatures are much more even and less variable.

As a final example, recall that in Egypt, in the time of the Pharaohs, there was a famous 14-year period when on the average there was just enough grain produced for the population. But in this case the period contained seven years of plenty followed by seven years of famine. Good management of surplusses is required to avoid catastrophe in the face of such great variability.

These three examples should, I hope, convince you that variability is important. How about measuring it? How can we assign a *number* to a list of numbers, to indicate how variable they are? Consider the following two specific sets:

9, 10, 6, 14, 11   10, 10, 10, 10, 10

In each of these lists, the mean is 10: $M = 10$. But plainly, the first is more variable than the second. How can we *measure* variability?

One possibility, often used, is the *range*—the distance between the largest and the smallest numbers. For the first sample, the range is from 6 to 14, a spread of 8. For the second set, the spread or range is zero. Range is indeed a useful number, and is simple to compute. But it misses important features of data. For example, the following two samples have the same range, but clearly the second is more variable than the first:

1, 13, 13, 13, 13, 13, 13, 13, 13   1, 2, 2, 4, 6, 7, 9, 10, 10, 13

When we talk about the variability in a sample, we're really asking whether the numbers are closely grouped around the mean or are very spread out. One way of measuring the variability is to look at the deviation of each number from the mean. For each number, X_i in an X list, we could look at the deviation, X_i minus M. Consider a population with five numbers, shown above, 9, 10, 6, 14, 11.

To make it clear that it makes sense to talk about a *population* with just five numbers in it, imagine the following situation: Suppose that five cards are placed in a hat, with one number (9, 10, 6, 14, 11) on each card. Now a *sample* of size 1 may be taken from this population by drawing a card from the hat, recording the number, *and replacing it*. Similarly, a sample of size 100 could be taken from the population by repeating this entire operation 100 times. One could then find the mean of that sample of 100 numbers, if he wished.

For the moment, let's return to the population consisting of the five numbers 9, 10, 6, 4, 11. The mean of these numbers is 10.

$$M = \frac{\Sigma X_i}{5} = \frac{9 + 10 + 6 + 14 + 11}{5} = 10$$

For each number it is easy to find the deviation from the mean, $X_i - M$, as in Table 9.1.

TABLE 9.1

Number, X_i	Deviation, $X_i - M = X_i - 10$
9	−1
10	0
6	−4
14	+4
11	+1

The numbers below the mean have negative deviations. Those above the mean have positive deviations. The deviations range from −4 to +4. One way to indicate variability might be to look at deviations. If we wish to find the average deviation, we just add them up and divide by 5. This gives a result of 0, since the deviations cancel out. And the same would happen for any set of numbers.

We'd like to measure how big the deviations are. One way of doing this is to take the absolute value[1] of each deviation and then average these numbers, to produce the average absolute deviation from the mean. Using the traditional vertical lines to denote absolute value, this could be written as follows:

$$\text{Average absolute deviation} = \frac{\Sigma |X_i - M|}{5} = \frac{|-1| + |0| + |-4| + |4| + |1|}{5}$$

$$= \frac{1 + 0 + 4 + 4 + 1}{5} = \frac{10}{5} = 2$$

Now we have arrived at a single number, the average absolute deviation, to represent the variability in the population 9, 10, 6, 14, 11. This seems a very reasonable measure. When the

---

[1] The *absolute value* of a number is just the number written without any minus sign. Thus the absolute value of −461 is 461. The absolute value of −28 is 28. The absolute value of 46 is 46. The notation usually used for absolute value is a pair of vertical lines around the number. Thus we write |−461| to indicate the absolute value of −461. The other results above could be denoted as follows: |−28| = 28, |46| = 46.

numbers are close to the mean, the average absolute deviation will be small. And when there is great variability, the average absolute deviation will be large. This is a very appealing measure. However, it leads to great difficulties when you try to study it using the tools of mathematical analysis. This means that although it is very appealing, very little is known about this attractive measure of variability.

How else could one get rid of the minus signs in the deviations? One possibility is to square[2] each of the deviations from the mean and then find the *averaged squared deviation* from the mean. Since the square of any negative number is a positive number, every value which is different from the mean will contribute to this measure. In the case we have been working on, the squared deviations and their average are given in Table 9.2.

TABLE 9.2 STEPS IN COMPUTING THE AVERAGE SQUARED DEVIATION FROM THE MEAN, ALSO CALLED THE VARIANCE

Number	9	10	6	14	11
Deviation from the mean	9 − 10 = −1	10 − 10 = 0	6 − 10 = −4	14 − 10 = 4	11 − 10 = 1
Squared deviation	1	0	16	16	1

Average squared deviation from the mean = $\dfrac{1 + 0 + 16 + 16 + 1}{5} = \dfrac{34}{5} = 6.8$

This measure of a population, the average squared deviation from the mean, is called the *variance* of the population. You should memorize this definition:

The variance of a population is the average squared deviation from the mean, averaged over all the numbers in the population.

Be sure you understand that this definition agrees exactly with the computation carried out just above for the small population 9, 10, 6, 14, 11. There were four steps involved in calculating the variance:

1. Find the mean of the population, M.
2. Find the deviation from the mean, X_i minus M, for each member of the population.
3. Square each of these deviations.
4. Find the mean of the squared deviations. This gives you the average squared deviation from the mean, which is the variance of the population. Variance is often denoted SD^2.

Variance is probably the most important concept in this chapter. We shall be referring to it in almost every chapter to come. In preparation, you should be sure to know (*a*) how it is defined, (*b*) what it means, and (*c*) how to calculate it. The uncomfortable strangeness should not last long.

THE STANDARD DEVIATION

After computing the variance of a population, it is common to use it in order to find a single number which is representative of the deviations from the mean found in that distribution. This

---

[2] Remember that the square of a number is the result you get when you multiply that number by itself. Thus 6 squared is 36. A superscript 2 is used to denote the square of a number. We can write $6^2 = 6 \times 6 = 36$; similarly, $5^2 = 25$ and $8^2 = 64$. The square of any negative number is a positive number, since the product of any two negative numbers is always a positive number. Thus $(-7)^2 = (-7) \times (-7) = 49$. The square of −10 is 100: $(-10)^2 = 100$. And so on.

number is called the *standard deviation*, abbreviated *SD*. The standard deviation is just the square root[3] of the variance:

Standard deviation = $SD = \sqrt{\text{Variance}}$

Note that the standard deviation is, indeed, a *standard* deviation. It is a single number which tells you what kind of a deviation from the mean is typical of a given population. *For most populations you are likely to encounter, about two-thirds of all the numbers in the population are within 1 standard deviation of the mean.* In the example of miles-to-failure for two kinds of

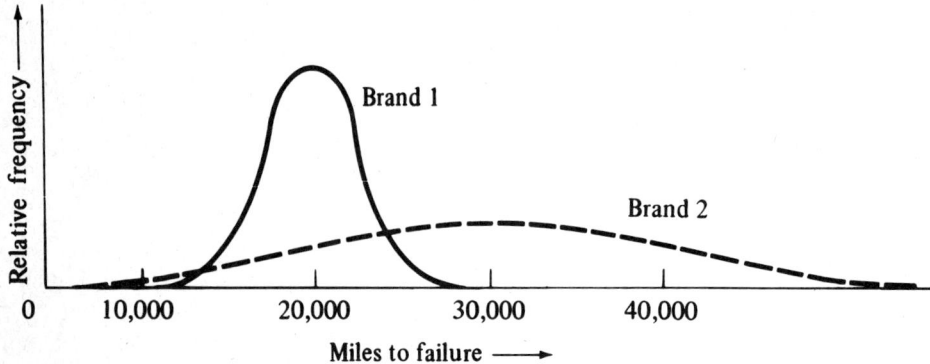

Figure 9.4 An example of a bimodal distribution.

tires, presented earlier in this chapter, brand 1 tire lives have a mean of 20,000 and a standard deviation of 2000. Since we have said that two-thirds of all numbers lie within 1 standard deviation of the mean, we can deduce that about two-thirds of all the brand 1 tires have a tire life between (20,000 − 2,000) and (20,000 + 2,000), that is, between 18,000 and 22,000 miles. For brand 2 tires, however, the mean is about 30,000 miles and the standard deviation (*SD*) is 9,000. For this population, two-thirds of the tires last between (30,000 − 9,000) and (30,000 + 9,000) or between 21,000 and 39,000. Plainly, the variation is much greater here.

We shall return to the very important standard deviation in the next chapter.

QUESTION AND ANSWER

Question: Since it's rarely used in statistical analyses, why did you introduce the average absolute deviation in this chapter?

Answer: Because if you understand the idea of an average distance from data values to the mean, or an average absolute deviation, then you should have little trouble understanding the *variance*, which is a crucial concept. Many students find it strange that in finding the variance you first square all the deviations and then take the square root of the result, to return to the standard deviation. But if you have understood the reasoning behind the average absolute deviation, you should be able to understand that the variance procedure serves a similar purpose: An observation 5 units below the mean contributes just as much to the measure of variability as does an observation 5 units above the mean. The sign (+ or −) of the deviation doesn't matter.

[3] In case you've forgotten just what a square root is, it is the opposite of the square of a number. Just as the square of 5 is 25, the square root of 25 is 5. Just as the square of 7 is 49, the square root of 49 is 7. In general, the square root of a number is that number which, multiplied by itself, will produce the original number. The symbol $\sqrt{}$ is used to indicate "the square root of." Thus you should be able to see that the following relations are true: $\sqrt{25} = 5$; $\sqrt{49} = 7$; $\sqrt{100} = 10$; $\sqrt{1} = 1$ (this last one is surprising). Not all square roots come out so simply. Thus $\sqrt{2} = 1.414$; $\sqrt{3} = 1.732$; $\sqrt{4} = 2$. (There is a table of square roots at the back of this book.)

Summary

Statistics can be divided into two parts. *Inferential statistics* is a set of tools to help make good decisions in the face of variable data. It is the main topic of this book. *Descriptive statistics* is a set of tools for dealing with large masses of data. This particular chapter deals with some simple descriptive statistics.

The number which occurs more often than any other number in a set of data is called the *mode*.

The *median* of a group of data is the middle number, when all the numbers are arranged in increasing order by size, from the smallest to the largest. (If there is an even number of entries in the data group, the median is the average of the two middle numbers.)

The *mean*, or arithmetic average, is the number obtained by adding up all the numbers in a data set and dividing by the number of entries in the set.

A *bimodal distribution* is one which has two distinct maxima, or peaks.

The *sigma notation* is used to describe the addition of the numbers in a data list. Suppose that the following list of numbers is known as the X list: 4, 2, 0, 5, 7, 9, 3. Then, for example,

$\Sigma X_i = 4 + 2 + 0 + 5 + 7 + 9 + 3 = 30$

This symbol complex is read "The sum of the members of the X list."

We often wish to measure the *variability* of a set of numbers. The variance of a set of numbers is the *average squared deviation from the mean*. To compute it you first find the mean of the set of numbers. Then, for each number, X_i, you find the deviation of that number from the mean: X_i minus M. You then square all these deviations, and finally you find the average squared deviation. This is called the *variance*.

The *standard deviation* of a population of numbers is the square root of the variance obtained by the method just described. The square root of a number is that number which, multiplied by itself, yields the original number. For example, the square root of 144 is 12.

For most populations, about two-thirds of the numbers in the population lie within 1 standard deviation of the mean.

Problems

1. Describe in your own words the difference between *inferential* and *descriptive* statistics. Then check the discussion at the start of this chapter to see how well you've done.
2. Consider the population 3, 2, 8, 5, 3, 4, 10. Compute the following:
   a. The mean.
   b. The median.
   c. The mode.
   d. The variance, SD^2, of the population.
3. Find the median of the following set of numbers: 142, 6, 9, 78.
4. For the data list 6, 4, 4, 8, 12, 7, 9, 9, 2, 0 calculate ΣX_i.
5. A skin diver with seven children has seven daughters and no sons. Run a sign test to determine whether this is a statistically significant departure from what we would expect if, for each child he had, there was a 50 : 50 chance that it would be a girl. Use a 0.05 significance level. Discuss your decision.
   a. Run the sign test, showing all steps.
   b. What would a Type One error be in this situation?
   c. What would a Type Two error be here?
6. In a bizarre gambling situation, the croupier first tosses a silver dollar and then rolls a single, six-sided die Draw a tree diagram to show all the possible *outcomes* of this experiment.

7. Consider the population 1, 5, 9, 4, 2, 3. Find the mean, M. Find the deviation of each number from the mean, and then find the average deviation. Then find the average squared deviation. This is called the *variance* of the population.

8. Consider the following two small populations:

   $A = 48, 49, 50, 51, 52$ $B = 30, 40, 50, 60, 70$

   Which one do you think has a larger *variance*? Show that your choice is correct by calculating the variance of each population.

9. Think of another example of a *bimodal* distribution, from your own experience. Sketch it, and explain why it has two peaks.

†10. Consider the bimodal distribution on page 63 as being made up of two separate populations, defective bulbs and nondefective bulbs. Now suppose that it was your job to look at each bulb tested and to determine, on the basis of its life, whether it was defective or regular. (This might be useful for analyzing the manufacturing process.) How would you decide which bulbs belong in which category? Could you be very confident for any single bulb? Could you be very confident for every bulb?

Variance; Populations and Samples; Z Scores

VARIANCE AND ESTIMATED VARIANCE

In the previous chapter we defined the variance of a population of numbers as the average squared deviation from the mean. We can write the following formula for finding the variance (denoted SD^2_{pop}):

Formula for finding the variance of a population containing N numbers, all of which are known, after the mean of the population, M_{pop}, has been calculated:

$$SD^2_{\text{pop}} = \frac{\Sigma\, (X_i - M_{\text{pop}})^2}{N} \qquad \text{where the summation is over all N numbers in the population}$$

Remember that in order to find the variance, you will find the deviation from the mean (X_i minus M_{pop}) for each number, X_i, in the population. Then you will square each of these deviations. Finally, you will add them all up and divide by N. This sequence of steps is outlined in the formula above.

It oftens happens that we are interested in the variance of a population, but we do not know all the numbers in the population. For example, in making decisions about financing education, it would be useful to know not only how much money the average student was able to earn in a summer but also how variable this amount was. For the population of summer earnings we would like to know the mean and the variance. For a large group of students this would be very difficult to determine. Rather than trying to study the entire population, we might decide to take a small random sample, and use the numbers in that sample to *estimate* the population variance. For reasons that are difficult to make clear in an introductory book, the proper rule for estimating population variance on the basis of the numbers in a sample is the following:

Formula for estimating the variance of a population on the basis of a sample of n numbers, after the mean of the sample, M_{samp}, has been calculated:

$$\text{Est. } SD^2_{pop} = \text{Estimated variance of the population} = \frac{\Sigma (X_i - M_{samp})^2}{n - 1}$$

This formula tells you to find the deviation of *each number in the sample* from the mean of the sample, M_{samp}. Then you must square each of the deviations, add up the squares, and divide the sum by n minus 1. As you can see, there is a very strong similarity between the two formulas we have just presented. But you should be aware of the differences, too. We shall be returning to these two formulas from time to time in the coming chapters, and eventually you should know both of them very well.

STATISTICS AND PARAMETERS

We have discussed two different calculations above: the first was a calculation of the variance of a population, based on all the numbers in the population. This is an example of a *parameter* of the population. A parameter is a measurement of an entire population. The second calculation above was a number based on the *numbers in a sample*. This number, Est. SD^2_{pop}, is an example of a *statistic*. A statistic is a measurement of a sample.

We have seen other examples of statistics in this book, including the mean of a sample, M_{samp}, and the number of green marbles in a sample of 10 marbles from the marble box. Any rule for assigning a number to a sample produces a statistic. The *range* of a sample, or the *mode* of a sample, are also statistics. Remember that sample and statistic both start with s: A statistic is a measure of a sample.

We have seen examples of parameters, too. For example, in the marble box there were 300 red marbles and 300 green marbles. The proportion of green marbles in the box, 300/600 or 0.5, is a *parameter* of the population of marbles. The mean of a population, M_{pop}, is another parameter. Remember that both parameter and population start with p. A parameter is a measurement of a population.

It often happens that we draw a sample from a population about which we need to make a decision. We compute a statistic on the basis of the sample, and make inferences about the true state of the population, or about its parameters. In the ESP experiment we observed a sample of 10 guesses from the population of all the guesses that a person might make in that situation. What we really would like to know is the overall proportion of correct guesses in the population of all the guesses the person might make. That proportion is a *parameter*. We shall return to these words later.

FURTHER COMMENTS ON THE STANDARD DEVIATION

In Chapter 9 we introduced the standard deviation, which is defined as the square root of the variance. This is the most commonly used single number when you want to indicate "how variable" the members of a population are. What good is it? The following rule of thumb is often helpful:

Rule of thumb: *For most populations which you will ever encounter, about 95 percent of the numbers in the population are within 2 standard deviations of the mean.*

We shall illustrate this principle repeatedly, and you will have occasion to use it. You should learn it by heart.

Example: Most college students, like it or not, are familiar with the College Board tests. The scores on these tests are arranged so that the population of all scores has a mean, M_{pop}, of 500 points, and a standard deviation of 100 points.

What does the rule of thumb tell us then about College Board scores? It tells us that 95 percent of all scores are within 200 points (2 standard deviations) of 500, the mean. That is, 95 percent of all scores are between 300 and 700. Assuming (correctly) that as many are above this range as are below it, then one-half of the remaining 5 percent (2½ percent) should be above 700, and the other 2½ percent should be below 300. This is in fact the case. Most of you have probably found that very few people in your high school had scores higher than 700. And probably just as few had scores lower than 300.

Example: If your roommate comes home and announces that he scored 82 on a test for which the class average was 76, is he justified in being unbearably pleased with himself? It depends on how variable the class data were. If almost all the scores were tightly clustered around the mean, then a score of 82 might be fantastically good. On the other hand, if the scores are quite variable, 82 may be only a little better than average. If the standard deviation is 2 for the class, then your roommate was 6/2 = 3 standard deviations above the mean (since he scored 6 points higher than the mean). That's impressive. But if the standard deviation was 10, then he was only about 6/10 = 0.6 standard deviations above the mean, which is good but not superb. It is often useful to consider how far a single score is removed from the mean, expressed in terms of standard deviations. We have seen in the rule of thumb that most scores are within 2 standard deviations of the mean. Indeed, scores as much as 3 standard deviations from the mean of almost any population are quite rare.

The procedure of expressing the position of a single score relative to the mean, and in terms of the standard deviation of a population, is called the assignment of a *standard score*, or Z score, defined as follows:

If one particular score from a population is denoted as X_i, the standard score, or Z score, corresponding to that observation is given by the following relationship:

The Z score corresponding to X_i is

$$Z_i = \frac{X_i - M}{SD}$$ where M is the mean of the population, and SD is the standard deviation

This formula just says to count how many standard deviations a given data value, X_i, is from the mean. Data values higher than the mean have positive Z scores, data values lower than the mean have negative Z scores. For example, consider the College Board score population already mentioned. The mean is 500 ($M = 500$). The standard deviation is 100 ($SD = 100$). What is the Z score corresponding to a test result of 600? It is +1, since

$$\frac{X_i - M}{SD} = \frac{600 - 500}{100} = \frac{100}{100} = +1$$

Similarly, the standard score (Z score) corresponding to a test result of 350 is $(350 - 500)/100 = -1.5$. The Z score corresponding to a test result of 500 is 0. And so on.

Z scores are very handy for getting a quick idea of where a given value lies in respect to an entire population. If your roommate tells you that his grade on a test was 10 points below the mean, that may not mean much. But if he tells you that he scored 5 standard deviations below the mean (a Z score of −5), that means a great deal. Alas.

QUESTION AND ANSWER

Question: In order to find the variance of a population, you find the squared deviation from the mean, $(X_i - M_{pop})^2$, for each number, X_i, in the population. Then you add them up and divide by N,[1] the number of elements in the population. But when you're estimating the variance of a population on the basis of a sample, you add up the squared deviations from the sample mean, $(X_i - M_{samp})^2$, and then divide by $(n - 1)$, where n is the number of elements in the sample. Why do you divide by N in one case and $(n - 1)$ in the other?

Answer: If there's one question that makes me want to abandon teaching introductory statistics, this is it. There are various elegant answers to the question, but they require too much mathematical sophistication for most of the people who ask the question. Here are some of the answers I give, trying to make this a bit clearer:

1. That's just the way things are. If students would spend more time memorizing these important formulas and less time pestering their teachers, the teachers would have a pleasanter life.
2. If you divide by n, instead of by $(n - 1)$ for the estimated variance, you end up *underestimating* the variance, on the average. But it works out so that your estimates are neither too large nor too small, on the average, if you use the formula with the $(n - 1)$ in the denominator.
3. Sometimes it helps to consider a very simple population, and look at all the possible samples of a given size one could derive from that population. Then one can compare two different estimators for population variance: the plausible estimator, $\Sigma (X_i - M)^2 /n$, and the somewhat less plausible estimator, $\Sigma (X_i - M)^2 /(n - 1)$. An example of this sort is examined in detail in the next question and answer.
4. Whenever you *estimate* the variance of a population, on the basis of a sample, you have to work with an estimate of the mean of the population, M_{samp}. You calculate the squared deviations from the *sample mean*. But what you really want to know is the squared deviation from the actual population mean. It works out that the sample mean is right in the middle of the sample, so individual data points have a small deviation from the sample mean compared with their deviation from the population mean.

   You can summarize the whole thing like this: Because you don't know the population mean, you have to work with the sample mean when estimating variance on the basis of a sample. But this makes all the deviations a little too small, on the average. To correct for this, when you're all through you divide by $(n - 1)$ rather than by n.

   If this entire answer served only to make the problem more confusing, I apologize. One option remains: Memorize the two formulas presented in this chapter, and take it on faith. If one day you take an advanced statistics course, you will probably be able to prove that the above rules are optimal and unbiased.

Question: You implied above that there was one more hope for convincing me that I should divide by $(n - 1)$ instead of by n in estimating the variance of a population on the basis of a sample. What is it?

Answer: Imagine a very simple population with two elements: 0 and 100. For example, you might write "0" on one piece of cardboard and "100" on another. Now to calculate the variance of that population, you find the average squared deviation from the mean. The mean is 50: $(0 + 100)/2$. The average squared deviation from the mean is $(50)^2$ or 2500.

$$SD^2_{pop} = \frac{\Sigma (X_i - M_{pop})^2}{N} = \frac{(0 - 50)^2 + (100 - 50)^2}{2} = (50)^2$$

---
[1] We are using capital N to indicate the total number of elements in a *population*. We use lowercase n to indicate the size of a *sample* used for an estimate.

TABLE 10.1 CALCULATIONS SHOWING WHAT HAPPENS IF YOU USE TWO DIFFERENT FORMULAS TO ESTIMATE POPULATION VARIANCE ON THE BASIS OF A SAMPLE

Sample	Sample mean M_{samp}	Estimated SD^2_{pop} based on sample using correct relationship $\text{Est.} SD^2_{pop} = \dfrac{\Sigma(X_i - M_{samp})^2}{n-1}$	Number you would get if you used the plausible but incorrect formula, $\dfrac{\Sigma(X_i - M_{samp})^2}{n}$
0, 0	0	$\dfrac{0+0}{1} = 0$	$\dfrac{0+0}{2} = 0$
0, 100	50	$\dfrac{2500 + 2500}{1} = 5000$	$\dfrac{2500 + 2500}{2} = 2500$
100, 0	50	$\dfrac{2500 + 2500}{1} = 5000$	$\dfrac{2500 + 2500}{2} = 2500$
100, 100	100	$\dfrac{0+0}{1} = 0$	$\dfrac{0+0}{2} = 0$

Average estimate:

Average $\text{Est.} SD^2_{pop} = \dfrac{0 + 5000 + 5000 + 0}{4}$

$= 2500$

The average estimate is exactly equal to the true SD^2_{pop}

Average result:

$\dfrac{0 + 2500 + 2500 + 0}{4} = 1250$

The average result produced by the above formula is 1250, which is far smaller than the actual population variance

so

$SD^2_{pop} = 2500$

Imagine, for the moment, that I have prepared a population like this by writing 0 on one piece of cardboard, writing 100 on a second, and putting the two in a hat. I then ask a student to take a *sample* from this population and to estimate the population variance on the basis of a sample. I don't let him look in the hat. I can take a sample of any size by withdrawing one card from the hat, recording the number, replacing the card, and then shaking the hat before taking another card. In this manner, I might take a sample of size 5: 0, 0, 100, 0, 100. The student has no way of knowing how many cards are in the hat. (Indeed, you should be able to see that the arguments in this section would be unchanged if the hat contained 500 cards marked "100," and another 500 cards marked "0.")

Now suppose the student decides to take a sample of size 2: I proceed as above, and write down the results: 100, 100. He then estimates the variance.

$\text{Est.} SD^2_{pop} = \dfrac{\Sigma(X_i - M_{samp})^2}{n-1} = \dfrac{0^2 + 0^2}{2-1} = 0$

He might, on the other hand, have observed the following sample: 0, 0. In this case, too, he would estimate the population variance to be zero. Or he might have observed 100, 0: in this case, he would estimate the population variance to be

$\text{Est.} SD^2_{pop} = \dfrac{\Sigma(X_i - M_{samp})^2}{n-1} = \dfrac{(100-50)^2 + (0-50)^2}{2-1} = 5000$

There is only one other sample he might have observed: 0, 100. In this case, he would obtain the same estimate of the population variance: 5000. Now *on the average*, if we consider the four possible samples of size 2 from this population, we see that the estimates average out to the correct value: The average estimate of SD^2_{pop}, averaged over all four possible samples, is

Average Est. $SD^2_{pop} = \dfrac{0 + 0 + 5000 + 5000}{4} = 2500$

If we had used n instead of $(n-1)$ in estimating the variance, we would have had two estimates of 0 and two estimates of 2500, for an average estimate of 1250. This would have been much smaller than the true population variance of 2500.

This argument is summarized in Table 10.1.

Summary

In coming chapters we will be talking about entire populations and about samples taken from populations. Since there will be places where this could get confusing, we have two special words to help remember whether we're talking about entire populations or just samples. A *parameter* is a measurement of an entire population. The mean of an entire population, which will sometimes be denoted M_{pop}, is a parameter. A *statistic*, on the other hand, is a measurement of a sample. The mean of a sample, which will sometimes be denoted M_{samp}, is an example of a statistic.

The variance of a population is defined as the average squared deviation from the mean of that population. If we denote the number of elements in the population by N, the mean of the population by M_{pop}, and the variance of the population by SD^2, then we can express the computation procedure as follows:

$$SD^2_{pop} = \frac{\Sigma(X_i - M_{pop})^2}{N}$$

Often we would like to know the variance of a population, but it is impractical or impossible to find out all the numbers in the population. For example, we might be interested in the number of days of sick leave used by each college teacher in the state of California. Rather than trying to get data on all of them, we might take a sample, say, of 50 of them and get information from these 50. Then we could use the following computation rule for estimating the variance of a population on the basis of a sample:

$$\text{Est. } SD^2_{pop} = \frac{\Sigma(X_i - M_{samp})^2}{n - 1}$$

where n is the number of cases in the sample (50 in the example), and M_{samp} is the mean of the numbers in the sample.

Rule of thumb: For most populations which you will ever encounter, about 95 percent of the numbers in the population are within 2 standard deviations of the mean.

A standard score, or Z score, is sometimes used to express the value of a single score from a population, relative to the mean of that population. If we denote the mean of the population by M and the standard deviation by SD, then the standard score corresponding to any particular data value, X_i, is given by

$$Z_i = \frac{X_i - M}{SD}$$

Problems

1. IQ scores are arranged to have a mean of 100 ($M = 100$) and a standard deviation of 15 ($SD = 15$).

    a. What interval will probably include about 95 percent of all IQ scores?
    b. Find the Z score corresponding to each of the following IQs: 70; 110; 83; 122; 100; 200.

2. The meat from 20 swordfish is inspected to check for dangerous levels of mercury. Is the resulting concentration level a statistic or a parameter? If you said statistic, what is the corresponding parameter? If you said parameter, what is the corresponding statistic?

3. Suppose that a *population* consists of just two numbers, 1 and 3. (For example, each number might be written on one side of a coin.) What is the mean of that population? the variance? the standard deviation?

4. For each of the following, say whether you think a sign test could appropriately be used. If you say Yes, what null hypothesis would you specify.

   a. A gardener finds an unmarked package of 7 seeds in a drawer. He plants them and notes that all 7 come up petunias; there are no zinnias.

   b. Twenty people in an ergonomics experiment hold their breath for as long as possible 5 minutes before and 5 minutes after vigorous exercises. It is found that 14 of the 20 held their breath longer on the second try, after the exercise.

   c. A car dealer notes that of the last 15 cars he sold, only 2 were convertibles. In past years about 25 percent of his sales have been convertibles.

   d. In a survey, 16 out of 20 people say they have "complete confidence" in their physician.

   e. Members of three foursomes keep track of their golf scores before and after reading *Power Golf* by Ben Hogan. Of the 12, 11 find that their scores have deteriorated as a result.

   f. Out of 13 people picked up for hard-drug violations in the last two months, 8 were high school dropouts.

   g. The teaching evaluations of 200 students were examined. Of these students, 146 had classes both in the morning and in the afternoon. Out of these, 91 gave higher ratings to the morning classes, while the rest gave higher ratings to the afternoon classes.

5. In order to test a student's knowledge of statistics, the professor gives him 10 problems to do on a midterm. The resulting 10 scores provide a sample of the student's performance on a much larger set of "appropriate" questions for such a test. Suppose the student scores 78 percent on the test. Is this a statistic or a parameter? If you said parameter, what is the corresponding statistic? If you said statistic, what is the corresponding parameter?

6†. In Chapter 9 we looked at the standard deviation and at the average absolute deviation from the mean. Try to prove that if we considered the signed (or algebraic) deviation $(X_i - M)$, which is sometimes a positive number and sometimes a negative number, the following is true:

$$\frac{\Sigma (X_i - M)}{n} = 0$$

for *any* group of numbers, where M is the mean of the group of numbers, n is the size of the group, and the summation is over the entire group.

The Sampling Distribution of the Mean; Probability Density Functions

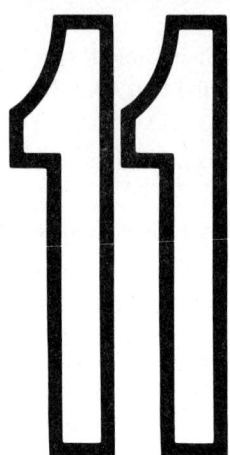

In this chapter we continue the process started in Chapter 9, which will go on for several more chapters. We are acquiring a set of conceptual and analytic tools that will make it possible to understand and use other statistical tests in addition to the sign test. The sign test is often useful. And as has been stressed, the logic underlying much of the subsequent development is exactly the same as the logic you have already seen in the sign test. Having understood that logic as it has been presented and with the help of the tools introduced in Chapters 9 through 14, you should have no trouble with the subsequent material.

THE SAMPLING DISTRIBUTION OF THE MEAN

Example: Suppose that at a leading West Coast university there are 13,000 students. If you average the grade point averages (GPA) of these students, the population mean is given by $M = 2.498$. The population variance (the average squared deviation from the mean) is given by $SD^2 = 0.36$. The administration announced that it had chosen a random sample of 40 students to form a student advisory board to represent student opinion in the administration. The student newspaper, noting that sample seemed to be overrepresented with good students, checked the GPAs of the 40 students and found a mean, $M_{40} = 3.21$, where the symbol M_{40} is used to indicate the mean of the GPAs of the 40 students. Is this value sufficiently higher than the overall mean of 2.498 to arouse suspicion? Are there grounds here for questioning the assertion that the sample was randomly chosen from the entire student body?

THE SAMPLING DISTRIBUTION OF THE MEAN; PROBABILITY DENSITY FUNCTIONS

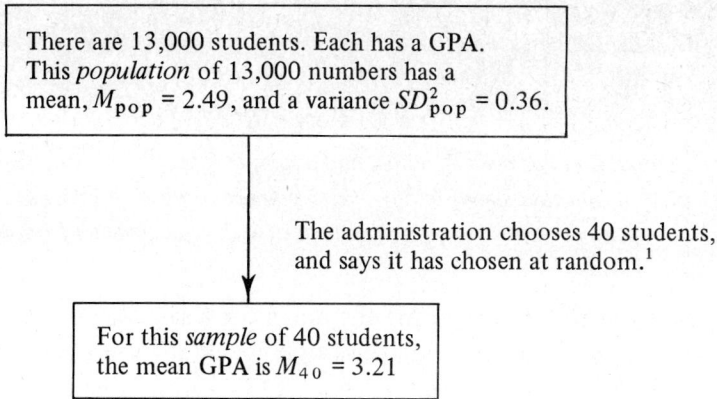

Here is our decision problem:

The 40 students in the sample, on the average, have higher GPAs than the student body as a whole. Is the discrepancy great enough to allow one to confidently accuse the administration of choosing a *biased sample* containing an unusually large number of people with good grades?

In order to answer this question, we must first find out some related information. If you took a random sample of 40 from this population, what is the chance that you would see a sample mean as extreme as 3.21? Would it happen 1 time in 10? 1 in 100? 1 time in 5, or what?

A question to consider: Confronted with this problem and knowing no more statistics than you do, what decision would you make? How could you go about finding out whether a sample mean of 3.21 is very improbable or quite plausible? Try to answer this question before reading further.

The fundamental way of finding out what kind of results you would obtain with samples of size 40 is to take a random sample of 40 students, find the GPA for each, and compute the average for 40. This would give you one possible value for M_{40}. Then repeat the entire process, over and over and over. Keep on doing this, until you have, say, 1000 different random samples of 40 students, and 1000 different sample means. Then plot a *frequency distribution* showing how likely various values are. It might look something like Figure 11.1.

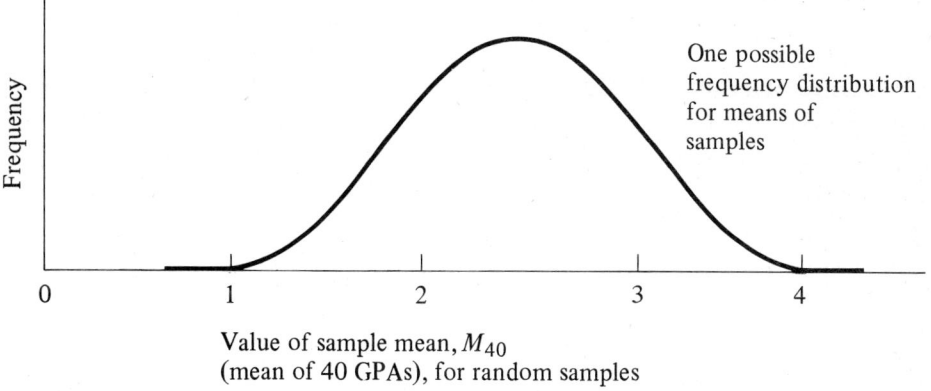

Figure 11.1 How often various values of the sample mean, M_{40}, were observed in 1000 samples of 40 students each.

---

[1] A *random* sampling process is one which gives every single member of a population an equal chance to be included in every sample. The sampling process cannot make it especially likely that one kind of person will be included or especially unlikely that another kind of person will be included.

This experimental method would give you a very good approximation to something called the *sampling distribution of the mean*, for samples of size 40, from the entire population of 13,000 students. Once you had discovered how likely are various possible results under these conditions, you would be in a good position to decide whether a value of 3.21 was suspicious.

Sampling Distribution of the Mean: *When a population has been specified and a sample size chosen, the sampling distribution of the mean is the probability distribution that specifies how likely you are to observe various possible values of the sample mean when you repeatedly draw samples of that size (with replacement) from the given parent population.*

The qualifier "with replacement" means that after computing the mean for a sample you return the sampled elements to the population, so that each successive sample is taken from the entire population.

How many different possible values for the sample mean could you possibly see in the situation described above? Recall that in the marble box, there were eleven possible events. There were eleven possible proportions of green marbles: 0/10, 1/10, 2/10, etc. In this situation there are vastly more possible values for the sample mean. You can get a feeling for the possible variety by considering the number of possible different samples of size 40 you could obtain from a population of size 13,000. Following the same sort of logic as was used in Chapter 6, the number of different samples can be seen to be

$$\underbrace{\frac{13{,}000}{1} \times \frac{12{,}999}{2} \times \frac{12{,}998}{3} \times \frac{12{,}997}{4} \text{ etc.} \times \frac{12{,}961}{40}}_{40 \text{ terms}}$$

This number is larger than

1,000,000,000,000,000,000,000,000,000,000,000,000,000,000,000,000,000,000,000,000,000,000,000

Since there is an almost infinite variety of possible samples, it is out of the question to look at each one and find a theoretical probability distribution in that way. And since there is almost an infinite number of possible values for the sample mean, we cannot organize the frequency distribution by considering each possible value as a different event. There would be an unmanageable number of events. Instead, we must look at the probability of seeing sample means in various *intervals* or *ranges*. For example, we might look at the chance of a sample mean between 2.5 and 2.6, the chance of a sample mean between 2.6 and 2.7, etc. In this fashion, we could define our *events* in terms of intervals and discuss only the chance of seeing sample means within various intervals. This will require a slightly new way of looking at probability distributions, which will be presented next. We shall return to this example in the next chapter.

To recapitulate, two important ideas have been presented above. The first is the concept of the sampling distribution of the mean. The second is the notion that because there is such a great variety of possible values of the sample mean it wouldn't make sense to try to know the probability of seeing each specific value; rather, we should look at the probabilities of various intervals, bands, or regions of sample means.

CONTINUOUS COMPARED TO DISCRETE PROBABILITY DISTRIBUTIONS

In the marble box example we computed the probability of each of eleven events. This told us all we needed to know about what would happen when the null hypothesis was true. In the example just discussed, however, there is an infinite variety of different possible GPAs, so a new procedure is required. This general procedure is the use of *probability density functions* to replace discrete probability distributions.

Example: Suppose that you have a wheel-of-fortune-like pointer, which spins over a dial-like clock face (see Figure 11.2). Suppose further that the pointer is very sharp, and the clock face

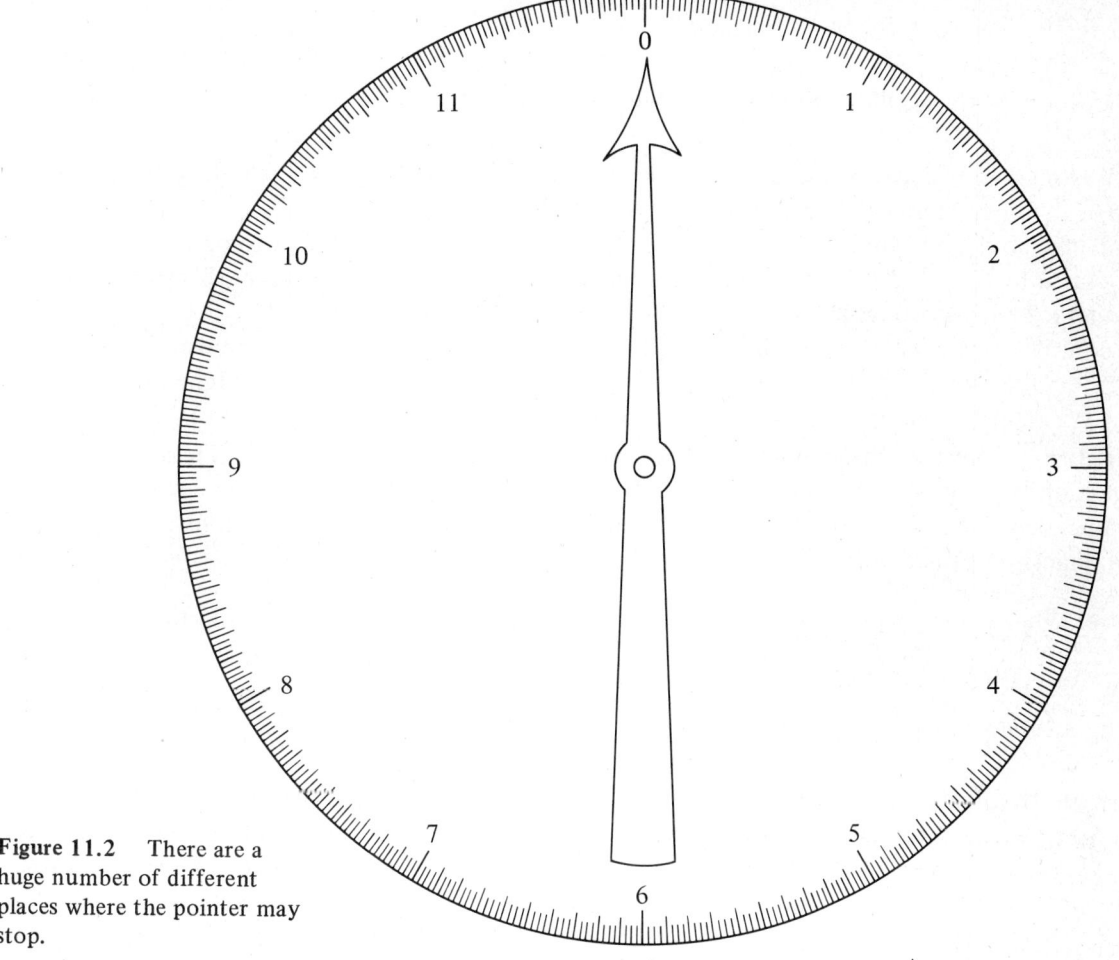

Figure 11.2 There are a huge number of different places where the pointer may stop.

has extremely find graduations, so that there are millions of possible "readings" which could result by spinning the pointer. These might range from 0.00000 to 11.99999. In a situation like this, it doesn't seem useful to talk about the probability of seeing the result "exactly 3." It all depends on how fine the graduations are; and in any case, the probability is minuscule. But it is easy and useful to find the probability of an interval: the probability of seeing a value between 1 and 2 is obviously about 1/12, assuming that the pointer is fair. Similarly, the chance of seeing a value between 3 and 9 must be just 6/12, or 1/2. The chance of seeing a value between 4.5 and 6.5 would be 2/12, or 1/6. And similarly we could find the probability of a reading in any other interval of interest. Now it turns out that there is a very simple way to represent the situation just described. We draw what is called a *probability density function* as in Figure 11.3.

For any probability density function the total area under the function is equal to exactly 1 square unit. The probability of seeing a result in any given interval is just the same as the area which lies over that interval. For example, the shaded section in Figure 11.3 is the area above the

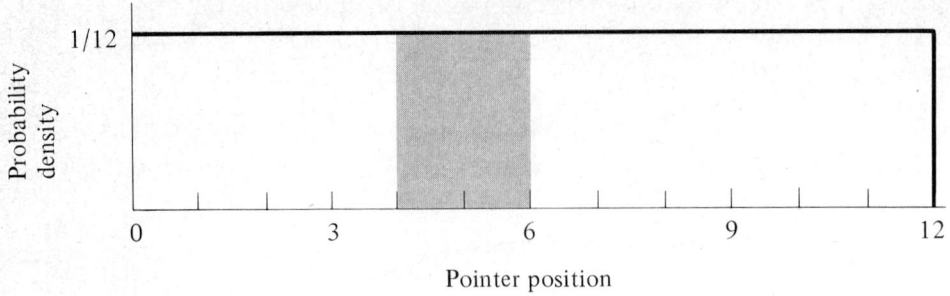

Figure 11.3 A probability distribution for the rotating pointer of Figure 11.2.

interval from 4 to 6. This section has an area[2] of 1/12 (the height) times 2 (the length) or 1/6. So the probability of an observation between 4 and 6 is 1/6.

The probability of observing a pointer position between 0 and 12 is just the area which lies above *that* interval: 1/12 x 12 = 1. There is a probability of 1 that the pointer will stop somewhere. That's reasonable.

A completely analogous procedure can be used to find the probability of a position in any other interval in which you are interested. The probability density function sketched in Figure 11.3 is, reasonably, known as a *rectangular* probability density function.

Figure 11.4 presents three more examples of probability density functions. These are drawn so that each one has an area of exactly 1 square inch (100 squares) bounded by the curve. Regardless of what scale is used for drawing them, all probability density functions contain an area of exactly 1 square unit.

Exercise: What is the probability of seeing an observation between 11 and 11.5 in the situation described by the probability density function in Figure 11.4b.

Answer: The probability of an observation between 11 and 11.5 is equal to the area bounded by the curve and lying over the interval in question. It is shaded in Figure 11.5.

The area of the shaded section is just the length times the height. Since the length is 0.5 (11.5 − 11.0), and the height is 0.5, the area is 0.5 x 0.5 = 0.25. This tells us that the probability of an observation between 11.0 and 11.5 is 0.25, in the above situation. There is 1 chance in 4 of an observation in that interval.

[2] Remember, from high school geometry, that the area of a rectangle is the height (H) multiplied by the length (L): $A = H \times L$. For example, if the height of a rectangle is 3 units and the length is 4 units, you can see that the area is 4 x 3 = 12 square units.

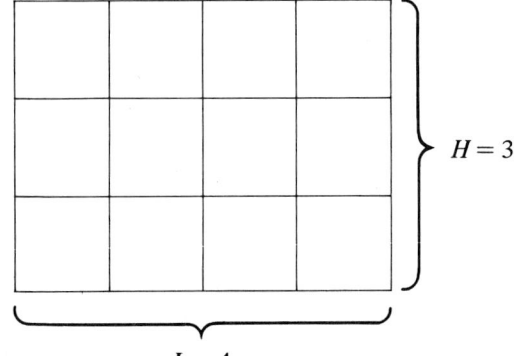

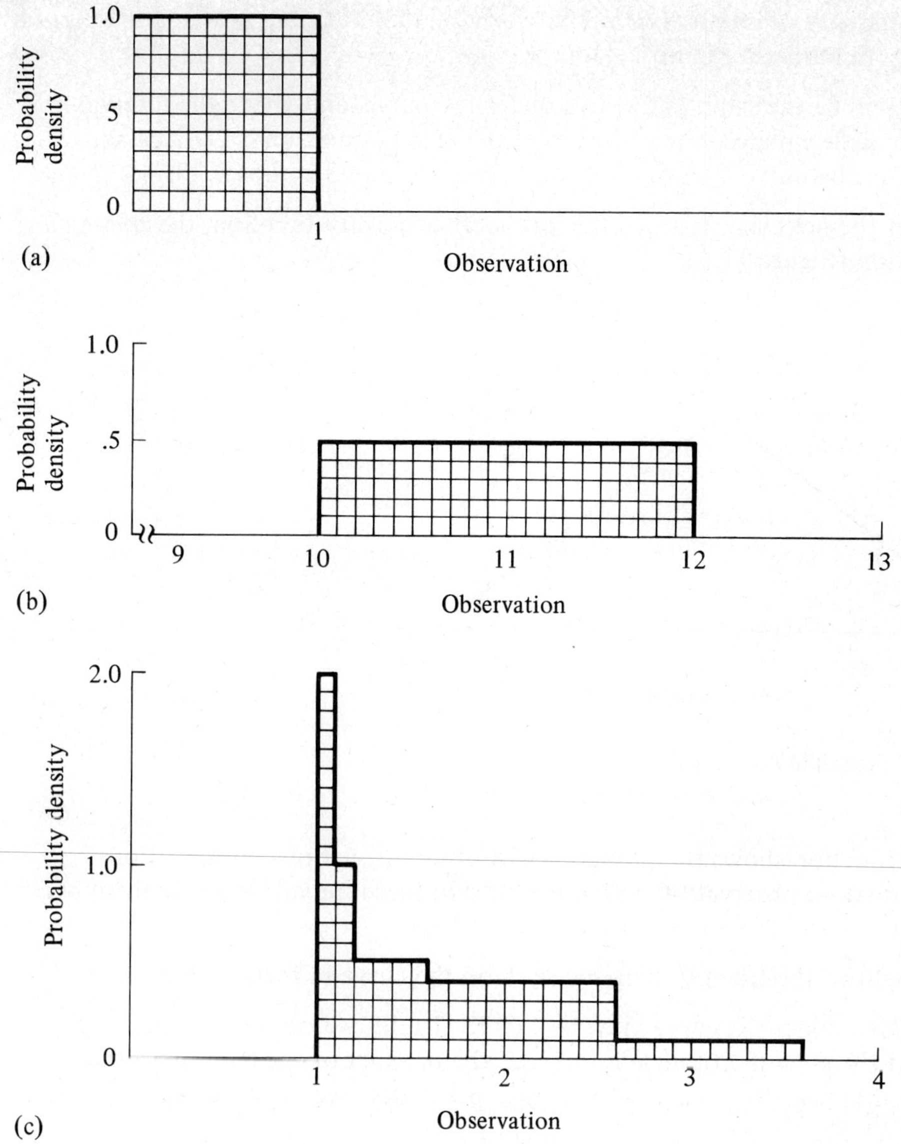

Figure 11.4 Here are three more examples of probability density functions. These are drawn so that each one has an area of exactly one large square, or 100 small squares, bounded by the curve. Regardless of what scale is used for drawing them, all probability density functions contain an area of exactly *one square unit*.

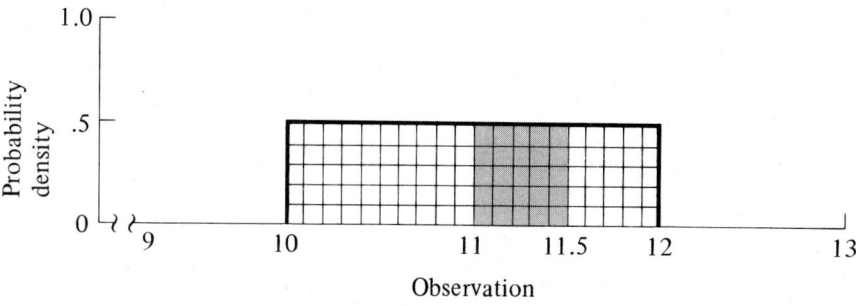

Figure 11.5 The probability of an observation between 11.0 and 11.5 is equal to the area shaded.

Exercise: What is the probability of an observation between 2 and 3 in the situation characterized by the probability density function in Figure 11.4c?

Answer: The probability will be the same as the area under the curve, and above the interval in question. By counting the little squares, one can see that the area in question is 28 little squares, or 0.28 square units. The probability of an observation between 2 and 3 is thus 0.28.

We shall now consider the next simplest possible probability density function, the *triangular* probability density function (Figure 11.6).

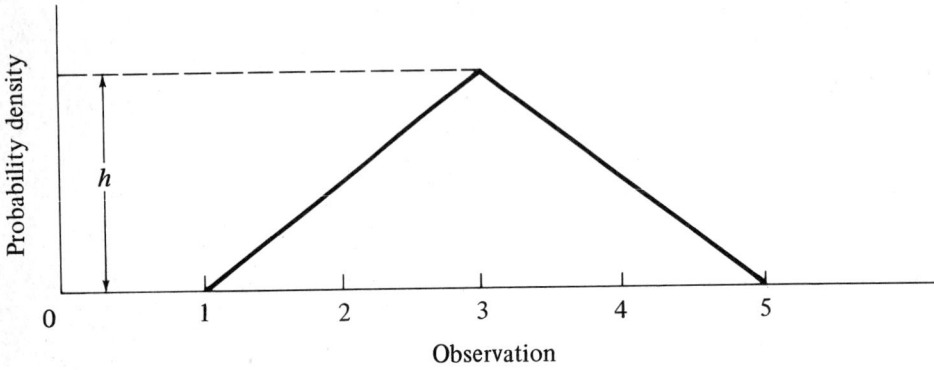

Figure 11.6 A "triangular" probability density function.

This probability density function shows the probability of observing various "observations" between 1 and 5. It says that no observation will occur outside that interval, and that the most likely observation is 3.

Exercise: What is the height of the triangle, denoted by h on the curve in Figure 11.6?

Answer: Since the total area under the curve must be exactly 1 unit, we can deduce the height using the simple rule that the area of a triangle is one-half the product of the base times the height: $A = \frac{1}{2}(b \times h)$. In this case, the base is 4 units long. Since the area must be exactly 1 unit, we know that $1 = \frac{1}{2}(4 \times h)$. So the height must be equal to $1/2$.

Exercise: Suppose that observations are distributed according to the probability density function shown in Figure 11.6: what is the probability of an observation between 3 and 5?

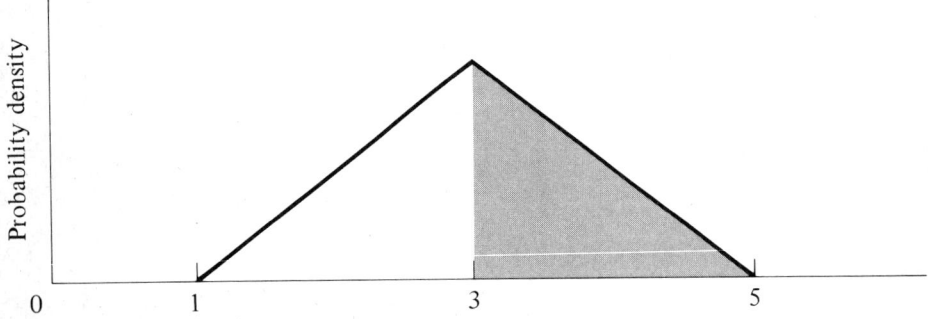

Figure 11.7

Answer: The area over this interval is just one-half of the total area. Since the total area is 1 unit, the area over this interval is 1/2. So the probability of such an observation is 1/2 (Figure 11.7).

Exercise: Suppose that observations are distributed according to the probability density function shown in Figure 11.7. What is the probability of an observation between 1 and 2?

Answer: Again, we must find the area over the interval involved.

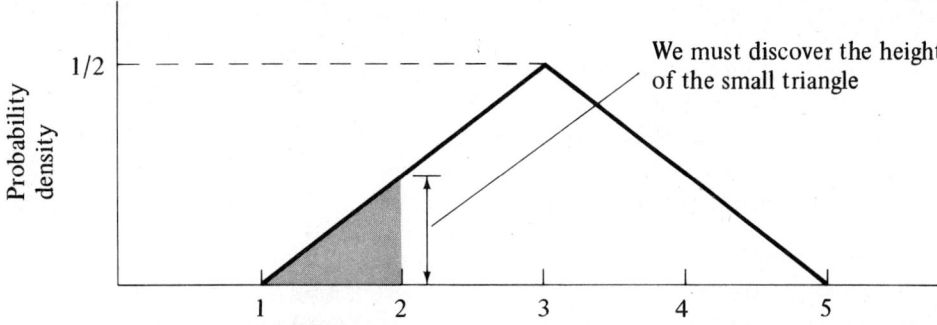

Figure 11.8

In order to find the area of the small shaded triangle over the interval between 1 and 2 in Figure 11.8, we must first figure out how high this triangle is. Since the entire distribution is 1/2 unit high, we can deduce by a similar triangles argument that the small triangle must be 1/4 unit high. If you have forgotten about similar triangles, check with a ruler. You will see that the small triangle is 1/4 unit high. Therefore the area of the small triangle is $A = \frac{1}{2}(b \times h)$, $A = \frac{1}{2}(1 \times 1/4) = 1/8$. So the probability of an observation between 1 and 2 must be 1/8.

Exercise: What is the probability of an observation between 2 and 3 (Figure 11.9)?

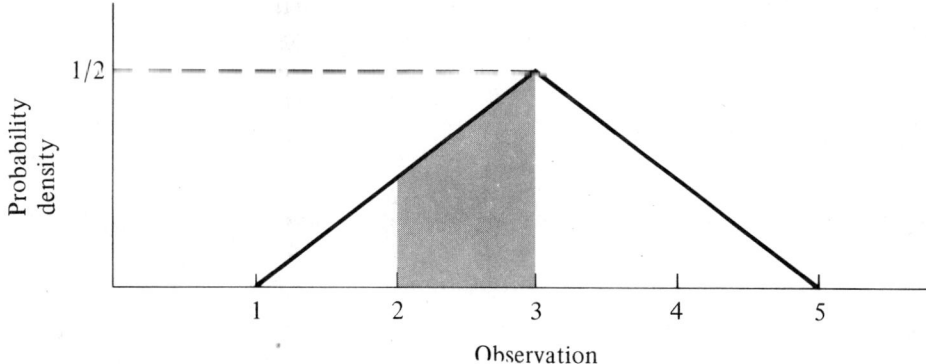

Figure 11.9

Answer: Again, we must find the probability over the interval in question. The simplest way of finding this area is to note that the area in question is the difference between the areas of the triangles (see Figure 11.10).

Procedures very similar to this, in concept, are used whenever you wish to find the probability of an observation in a specified interval and you know the probability density function which applies. Triangles are especially convenient, since it is easy to calculate the areas

86   THE NORMAL DISTRIBUTION AND THE Z TEST

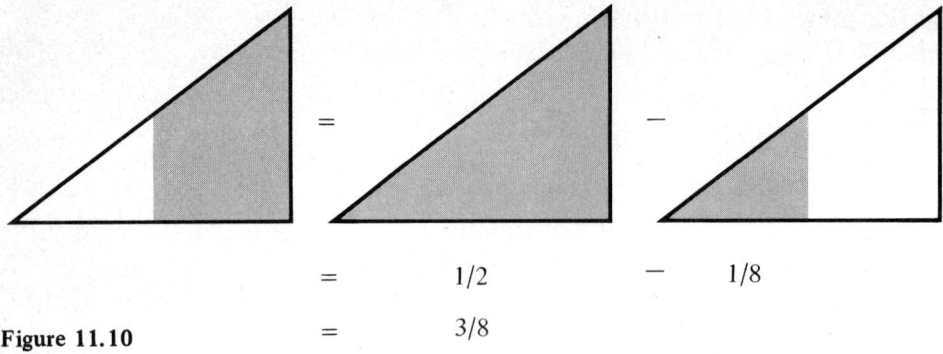

Figure 11.10

involved. For other probability density functions, this is not so simple. But one can always find an area by one means or another. If you have understood these steps, you will be well prepared for the study of the normal distribution which we shall get to soon.

The mean and the variance completely specify the triangular distribution. If you know the mean and the spread of a triangular distribution, you know all there is to know about it. Figure 11.11 presents some examples of triangular distributions with different *means* and the same variance.

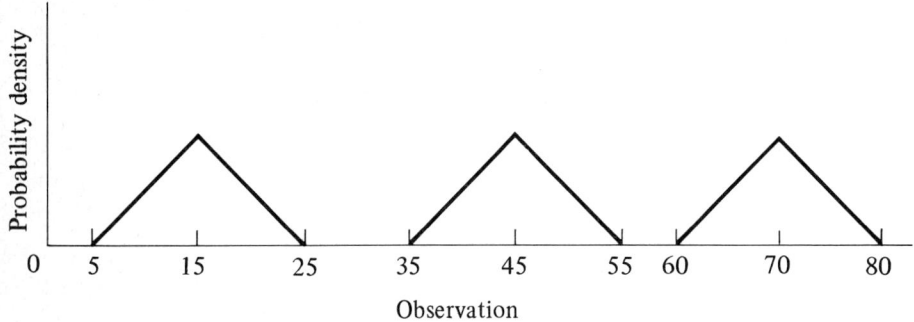

Figure 11.11 Three probability density functions with different means but the same variance.

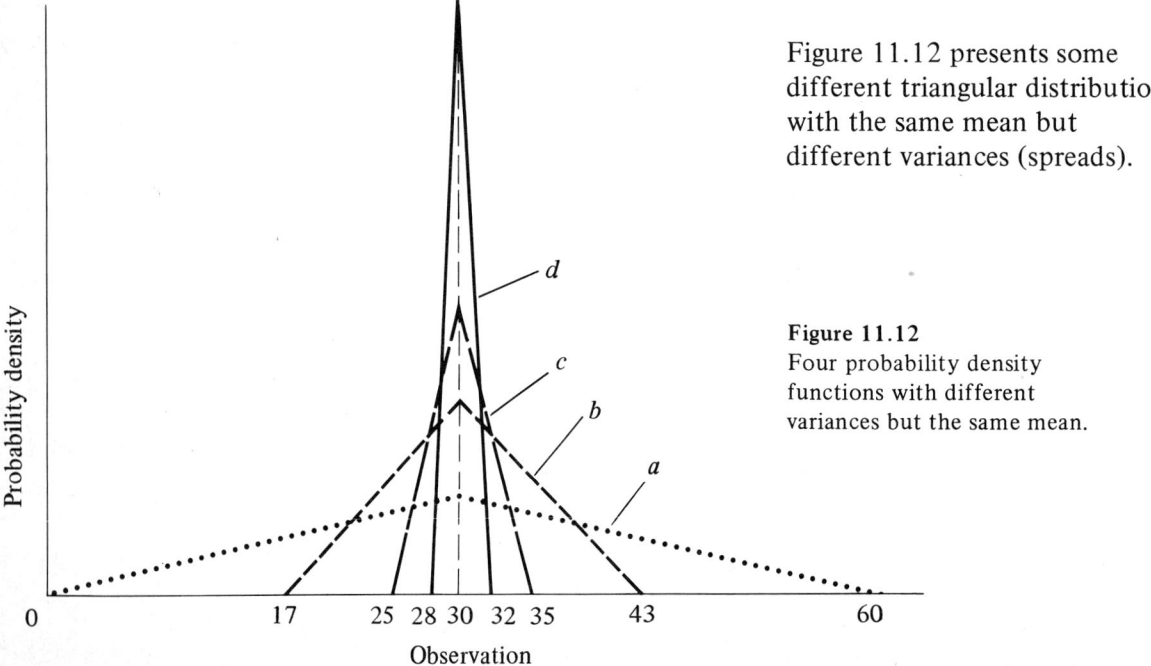

Figure 11.12 presents some different triangular distributions with the same mean but different variances (spreads).

Figure 11.12
Four probability density functions with different variances but the same mean.

TRIANGULAR AND NORMAL DISTRIBUTIONS

Now we get to the point of this discussion of triangular probability density functions. Almost everything we have said about these distributions can be said about the most important probability distribution in statistics: the normal distribution. Many kinds of bizarre and common observations are normally distributed. Such diverse data as the chest measurements of the Scottish regiments; the length of the third segment of a lobster carapace; the force with which a carpenter strikes a nail, under standard conditions; the weight in micrograms of chocolate candies; and lots of others are normally distributed. Also, as will be seen below, sample means for large samples from any population whatsoever are normally distributed.

Like the triangular distribution, a normal distribution is completely specified when the mean and the variance have been given. And in a completely parallel manner, we can draw a series of normal distributions with the same variance and different means, as in Figure 11.13, or with the same mean and different variances, as in Figure 11.14.

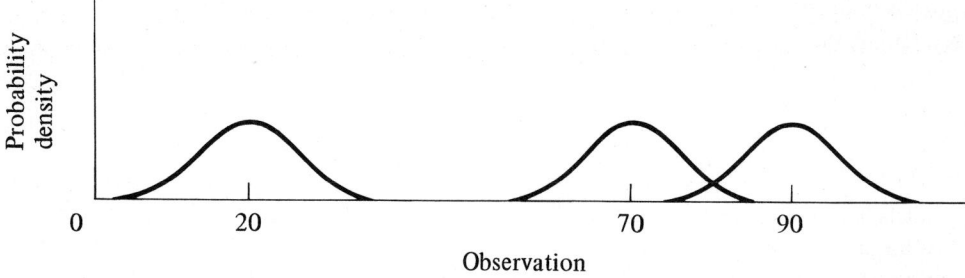

Figure 11.13 Three normal distributions with the same variance and different means.

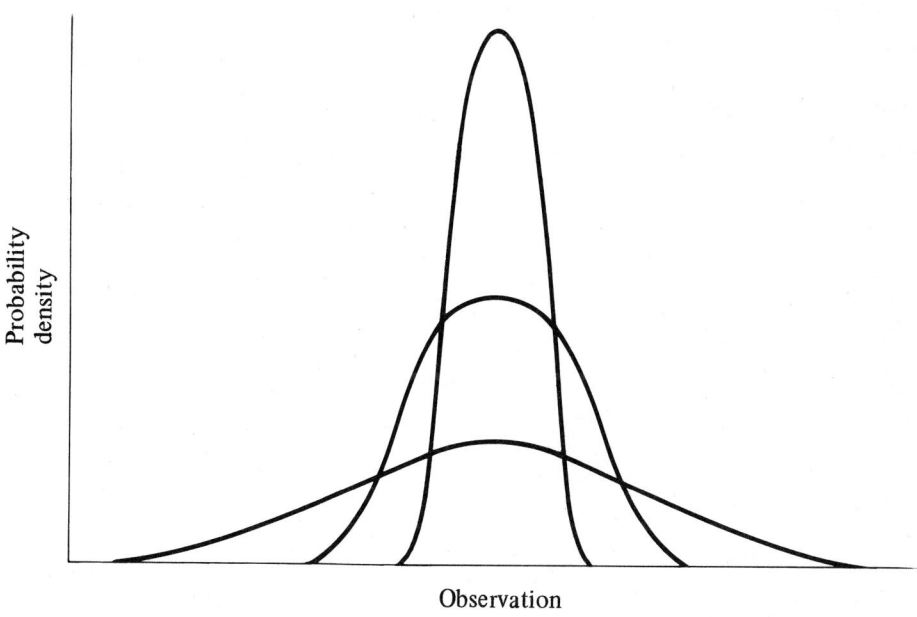

Figure 11.14 Three normal distributions with the same mean but different variances.

QUESTION AND ANSWER

Question: How could you take 1,000 different samples of 40 students each, from a population with only 13,000 students? 40 × 1,000 = 40,000.

Answer: After you have taken a sample of 40 students, you calculate the mean of the 40 GPAs, and then "replace" the students in the population, so the same student could end up in several different samples. This sort of a sampling system is called "sampling with replacement." The name makes sense if you think of the whole population as being made up of cards in a hat. Then, after drawing a sample of cards, you *replace* them in the hat before drawing additional cards.

Question: You talk about "random sampling", but since any sample is possible, it's conceivable that a person could take a random sample of 40 students from a student body of 13,000, and by pure chance end up with the 40 best students in the school. How can we say the sampling process was *not* random, since you know that any result is possible even when it *is* random?

Answer: The crux of the problem is that you can never be *certain* that a sampling process is not random. It is *possible* that you might just hit the 40 best students, as suggested. But it is very, very unlikely. The assertion that the sampling process in question is nonrandom might conceivably be wrong: but it is about as certain as anything in this uncertain world.

Summary

In this chapter, the crucial concept of the sampling distribution of the mean is introduced. When both the population and the size of the sample are specified, and we then take sample after sample and compute the mean of each sample, the frequency distribution that we draw using these means is called the *sampling distribution of the mean*. It is a probability distribution which tells us how likely the various values of the sample mean are. It is the *distribution* we get when we show all of the *sample means* we computed; hence the name, sampling distribution of the mean. In the marble box we experimentally approximated another sampling distribution: the *sampling distribution of the proportion* of green marbles. And in Chapters 7 and 8 we calculated a theoretical *sampling distribution of the proportion* of heads in a ten-coin-toss experiment.

The notion of a continuous probability distribution was introduced. In the ESP example there were only eleven different possible events, or scores: 0/10, 1/10, 2/10, etc., 10/10. In looking at all the possible values for the sample mean, there are often millions of possible results. To avoid the impossible task of finding the probability of every different value, we look at entire intervals. Continuous probability distributions always have an area equal to 1 square unit. Then the probability of an observation in any specific interval is just equal to the area bounded by the distribution, on top of that interval (see Figure 11.15).

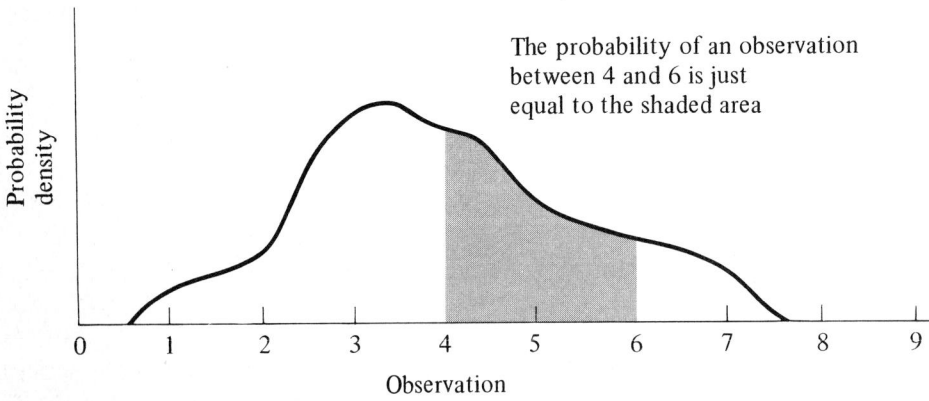

Figure 11.15 The probability of an observation between 4 and 6 is just equal to the shaded area.

Problems

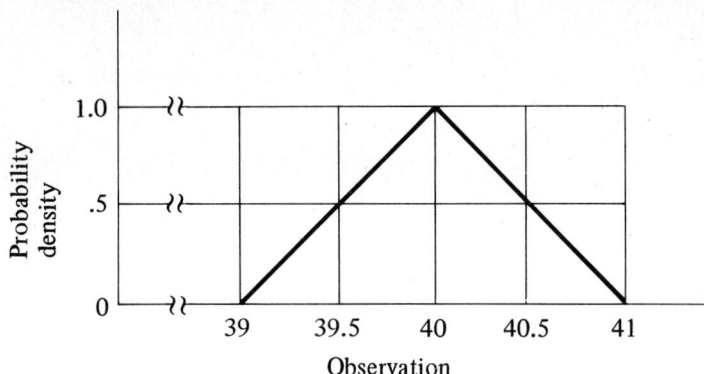

1. For the probability density function sketched above, find the following probabilities:
   a. What is the probability of an observation between 39 and 41?
   b. What is the probability of an observation between 39 and 40?
   c. What is the probability of an observation which is either smaller than 39.5 or larger than 40.5?
   d. What is the probability of an observation between 40.0 and 40.5?
   e. What is the probability of an observation larger than 41?
2. Which of the two probability distributions below has the greater variance (SD^2)? How do you know?

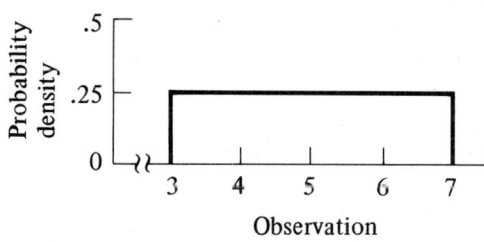 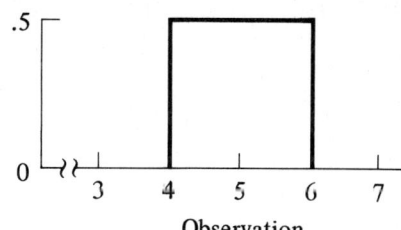

3. The idea of the sampling distribution of the mean is a very important one in this chapter. *Describe a procedure* you could use to find the sampling distribution of the mean, for samples of size 10, from the following population: The IQs of each of the 240 students in the tenth grade at Melville High School in Cleveland. A solution appears below.
4. It was asserted that there is a vast number of different possible values for the sample mean for samples of size 40 taken from a population of 13,000 GPAs. Illustrate this by listing 12 *different* possible sample mean values between 2.50 and 2.51.
5. Look at the probability distribution in Figure 11.5c.
   a. What is the chance of an observation between 2.4 and 3.0?
   b. What is the chance of an observation between 3.0 and 4.0?
   c. What is the chance of an observation between 1 and 3?
6. Look at each of the six statistical decision situations on page 17. For each, say exactly what would have to happen in order for you to make a Type One error.
   a. What would your decision be?
   b. What would the actual state of the world be?

Then look at the same situations, and say for each in turn exactly what would have to happen in order for you to make a Type Two error:

a. What would your decision be?

b. What would the actual state of the world be?

†7. Suppose that the sampling distribution of the mean of samples of size 40 from a large population looks like the following. What does this tell you about the population?

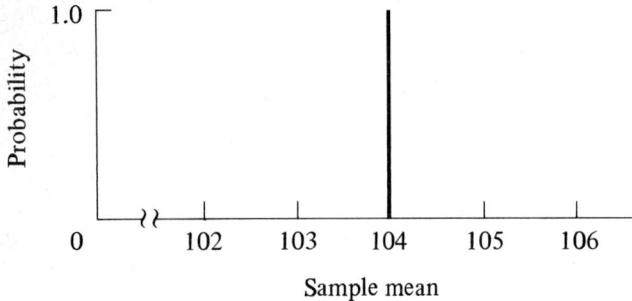

Solution to Problem 3: We assume that each of the 240 IQ scores is known, and we can represent the situation by writing each score on a piece of paper and putting the 240 pieces of paper in a large box. Then we take a sample of 10 IQs, find the mean of those 10 numbers, and write it down. This gives one possible value of the sample mean. Now replace these ten pieces of paper in the box, and take a new sample of 10 scores at random.[3] Find the mean. Write it down. Now repeat this whole process (take a sample of 10 scores; find the mean; write it down; replace the sample) many times, over and over. After, say, 500 repetitions, you could make a frequency distribution indicating in visual form how likely various possible values of the sample mean are. This would be a good approximation to the sampling distribution of the mean. The more times you repeated the entire process, the better your approximation would be.

---

[3] Strictly speaking, as an alert student point out, this isn't exactly right. Instead of taking a sample of 10 pieces of paper at once, one should take a single piece of paper, write down the score, and replace the single piece before drawing the second; likewise, each piece should be replaced before the next is drawn. This would be a pure case of *sampling with replacement*. In practice, when the population is large relative to the sample size, there will almost never be an important difference.

The Addition Rule for Probabilities; A Self-Test for Review

THE ADDITION RULE FOR PROBABILITIES OF NONOVERLAPPING EVENTS

In Chapter 6 the following table of probabilities was generated for the ESP experiment:

Event	0/10	1/10	2/10	3/10	4/10	5/10	6/10	7/10	8/10	9/10	10/10
Probability	0.001	0.010	0.044	0.118	0.205	0.246	0.205	0.118	0.044	0.010	0.001

Then assertions like the following ones were made:

1. The probability of observing 9/10 *or* 10/10 correct is 0.010 + 0.001 = 0.011.
2. The probability of seeing one of the four best scores, 7/10 or 8/10 or 9/10 or 10/10, is 0.118 + 0.044 + 0.010 + 0.001 = 0.173.

These statements were perfectly valid: In order to find the probability that one of several events would occur, we added up the probabilities of the separate events. Can we always use this procedure? Please try to answer this question before reading on.

Suppose that the student body of a college is 50 percent male and 50 percent female. Suppose, further, that 65 percent of the students receive financial aid of some sort, and 35 percent do not. We can summarise these facts in two tables:

Event	Male	Female
Probability	0.50	0.50

Event	Receive financial aid	Receive none
Probability	0.65	0.35

92 THE NORMAL DISTRIBUTION AND THE Z TEST

Now what do you think is the probability that a person chosen at random will be either male *or* a person who receives financial aid? By analogy with the above statements, we might reason that the probability of (Male or Receive financial aid) would be 0.50 + 0.65 = 1.15. This clearly won't do, since a probability larger than 1 is impossible. And it is surely not the case that everyone either is Male or Receives financial aid. Why did we get into trouble by adding probabilities? Because the two categories used here, Male and Receives-financial-aid, are overlapping. Many males receive financial aid. To make the difficulty even more clear, suppose that 39 percent of the students are males under 21 years of age. Then the probability that student is either Male or Male under 21 years of age is certainly not 0.50 plus 0.39. Here the categories overlap totally, and the probability of seeing one of these two events is of course just 0.50.

For nonoverlapping events, however, we can add probabilities. For example, the probability that a person chosen at random will be either Female or Male-under-21-years-of-age is indeed 0.50 plus 0.39. There is no possibility of a person belonging to both of the two categories.

At this point we can state the general rule which we have already used and illustrated: Any two events in a sample space are said to be *mutually exclusive* (or nonoverlapping) if it is impossible for both of them to occur at the same time; that is, if there is no single outcome corresponding to both of these events.

If any two events, event A and event B, are mutually exclusive, then the probability that we will observe one or the other of these two events is the sum of the probabilities of the two events.

The probability of seeing event A or event B is the probability of event A added to the probability of event B. Further examples of this rule are provided in the problems.

A SELF-TEST FOR REVIEW

Almost everyone learns best by *doing* things. In the past eleven chapters you have seen the most important ideas underlying the use of statistics to aid decision making. These ideas form the foundation on which the rest of the book is built. The rest of this chapter is a review test for you to work. Solutions are given after the test. You will be cheating yourself, however, if you look at the solutions before working the problems. If you do just about all the problems correctly, you have good reason to be pleased. You are well along the road to a solid understanding of statistics.

1. Give a short, accurate definition of each of the following:

    a. Type One error.

    b. Type Two error.

    c. Event.

d. Sample space.

e. Rejection region.

f. Significance level.

2. a. If the probability of Type One error is 0.05, what do we know about the probability of a Type Two error?

   b. If the probability of a Type One error is 1, what do we know about the probability of a Type Two error?

3. Consider the probability density function in Figure 12.1.

   a. What is the probability of a result, X_i, between 5 and 6? _____

   b. What is the probability of an observation between 8 and 11? _____

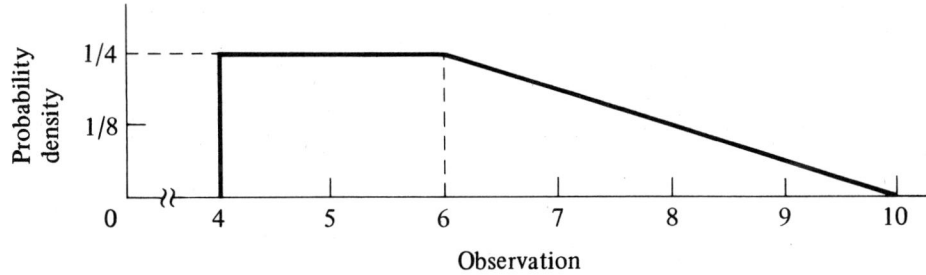

Figure 12.1

4. Nine subjects were run in an experiment to test the effect of baroque music on reading comprehension. Each subject took a series of reading tests, half with baroque music being

played while he read the material, and half in silence. The overall comprehension scores are given below:

Subject	Score for reading with music	Score for reading without music	Difference
1	87	84	+ 3
2	51	41	+10
3	92	90	+ 2
4	77	79	− 2
5	64	60	+ 4
6	80	72	+ 8
7	75	74	+ 1
8	75	68	+ 7
9	78	75	+ 3

a. Analyze these results using a sign test. Show all your steps clearly. Use a significance level of 0.05.

b. What would a Type One error be in this situation?

c. A Type Two error?

d. Given the decision you made, which kind of error might have happened?

5. How many different outcomes are there in the following experiment: A red die is rolled, and the result is recorded; then a blue die is rolled, and the result is recorded; then a green die is rolled, and the result is recorded.

6. The grades on an introductory psychology final exam constituted a population of 500 numbers. The mean of all the grades was 78. The standard deviation was about 5. For each of the following grades, give the corresponding Z score:

    a.   61 _____

          72 _____

         100 _____

          78 _____

b. About how many of the 500 grades would you expect to have Z scores less than -2? (*Note*: -3 is less than -2.)

7. a. For the following sample of numbers, compute the mean, the median, the mode, and the best estimate of the population variance: 7, 15, 13, 6, 8, 7, 14. Is the mode a parameter or a statistic?

   b. For this list, find

   $\Sigma X_i =$ _____

8. There are four aces in a hat, spades (S), hearts (H), diamonds (D), and clubs (C). You first draw a card at random, and write down whether it is S, H, D, or C. Then replace the card in the hat and make another draw and write S, H, D, or C to indicate which ace you drew.
   a. Draw a probability tree for this experiment.

   b. What is the probability of both cards being red?

Answers to self-test

1. a. A Type One error is like a *false alarm*: saying something interesting is happening when in fact you're observing only the results of chance variation. A Type One error is rejecting the null hypothesis when you shouldn't.

   b. A Type Two error occurs when you *miss* something: you fail to detect a real effect, and conclude that there is no convincing departure from chance. A Type Two error is failing to reject the null hypothesis when you *should* reject it.

   c. An event is a collection of outcomes in a sample space.

   d. A sample space is the set of all the possible outcomes of an experiment.

   e. A rejection region is a set of possible experimental results such that you will reject the null hypothesis if one of those results occurs.

   f. The significance level is the maximum risk of a Type One error that you are willing to accept.

2. a. If we are told that the probability of a Type One error is 0.05, we don't know anything precise about the probability of a Type Two error.

   b. If the probability of rejecting the null hypothesis when it is true is 1, then the probability of rejecting it must be just as high when the null hypothesis is false. If you always reject the null hypothesis when you should, you never make a Type Two error. So the chance of a Type Two error is zero.

3. a. The probability of an observation between 5 and 6 is equal to the shaded area in Figure 12.2. This rectangle has height 1/4 and length 1 (since 6 − 5 = 1), so its area is 1/4 × 1 = 1/4 = 0.25. The probability of an observation between 5 and 6 is thus 0.25.

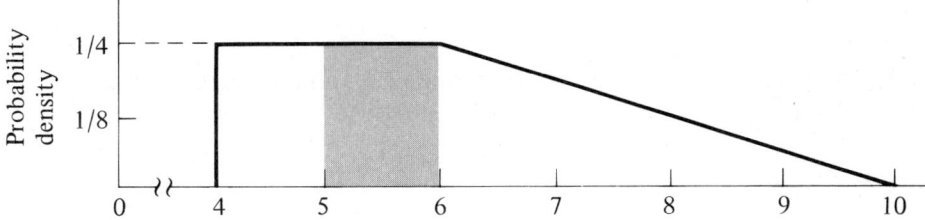

Figure 12.2

   b. The probability of an observation between 8 and 11 is equal to the shaded area in Figure 12.3. This triangle has a height of 1/8 and a base of 2 units. So its area is given by area = 1/2 base × height or area = 1/2 × 2 × 1/8 = 1/8. So the chance of an observation between 8 and 11 is just 1/8.

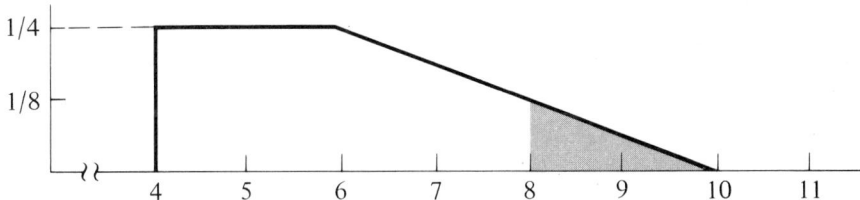

Figure 12.3

4. a. In order to run a sign test, we must be interested in testing a null hypothesis which states that each of two alternatives is equally likely. Here we are interested in knowing if positive

difference scores are just as likely as negative difference scores (in which case the music has no effect). When nine subjects are tested, there are ten possible scores: 0/9 may do better with music; 1/9 may do better with music; 2/9 may do better with music; etc. These are the events of interest.

Step 1: List all possible events: 0/9; 1/9; 2/9; 3/9; 4/9; 5/9; 6/9; 7/9; 8/9; 9/9 plusses.

Step 2: Make a null hypothesis: For any given student, there is a 50 : 50 chance he will do better with music (and have a +), and an equal chance that he will do better without music (and have a −). This is what would obtain if the music had no effect.

Step 3: Study the null hypothesis and find the probabilities of the ten possible events if the null hypothesis is true.

With nine subjects, there are 2^9 or 512 possible outcomes. That is, there are 512 different patterns of plusses and minuses like $+ - + + - - - + -$. Use of the counting rules, as in the accompanying table, tells us how many of these outcomes correspond to each possible event.

Event	No. of corresponding outcomes	Probability of event
0/9 plusses	1	1/512 = 0.002
1/9 plusses	9	9/512 = 0.018
2/9 plusses	$\frac{9}{1} \times \frac{8}{2} = 36$	36/512 = 0.070
3/9 plusses	$\frac{9}{1} \times \frac{8}{2} \times \frac{7}{3} = 84$	84/512
4/9 plusses	$\frac{9}{1} \times \frac{8}{2} \times \frac{7}{3} \times \frac{6}{4} = 126$	126/512
5/9 plusses	$\frac{9}{1} \times \frac{8}{2} \times \frac{7}{3} \times \frac{6}{4} \times \frac{5}{5} = 126$	126/512
6/9 plusses	$\frac{9}{1} \times \frac{8}{2} \times \frac{7}{3} \times \frac{6}{4} \times \frac{5}{5} \times \frac{4}{6} = 84$	84/512
7/9 plusses	$\frac{9}{1} \times \frac{8}{2} \times \frac{7}{3} \times \frac{6}{4} \times \frac{5}{5} \times \frac{4}{6} \times \frac{3}{7} = 36$	36/512 = 0.070
8/9 plusses	$\frac{9}{1} \times \frac{8}{2} \times \frac{7}{3} \times \frac{6}{4} \times \frac{5}{5} \times \frac{4}{6} \times \frac{3}{7} \times \frac{2}{8} = 9$	9/512 = 0.018
9/9 plusses	1	1/512 = 0.002

Step 4: Set a significance level. In the problem we are told to use a significance level of 0.05. Thus our rejection region must be composed of events with total probability less than 0.05 when the null hypothesis is true.

Step 5: Pick a rejection region. We can include the two most extreme scores at each end: 0/9, 1/9, 8/9, and 9/9. Together these have a probability of $0.002 + 0.018 + 0.018 + 0.002 = 0.040$. This is less than our cut off value of 0.05. Including any other events would result in too great a risk of a Type One error.

Step 6: Run the experiment and make a decision. In the data eight of the nine scores were plusses: we observed the event 8/9 plusses. This is in the rejection region, so we reject the null hypothesis and conclude that we have significant evidence that music helps people do this kind of reading.

b. In this situation, a Type One error occurs if we say music has an effect when in fact it has none and we are seeing only the results of random variation.

c. A Type Two error occurs if we say music has no effect and, in fact, it *does* influence reading comprehension.

d. Given the decision above, we may have made a Type One error.

5. $6 \times 6 \times 6 = 216$ possible outcomes. An example: red 4, blue 2, green 5.

6. a. Since $Z = \dfrac{X_i - M}{SD}$, we can compute:

When $X_i = 61$, $Z = \dfrac{61 - 78}{5} = \dfrac{-17}{5} = -3.4$

When $X_i = 72$, $Z = \dfrac{72 - 78}{5} = \dfrac{-6}{5} = -1.2$

When $X_i = 100$, $Z = \dfrac{100 - 78}{5} = \dfrac{22}{5} = 4.4$

When $X_i = 78$, $Z = \dfrac{78 - 78}{5} = \dfrac{0}{5} = 0$

b. Since about 5 percent of all observations are more than 2 standard deviations from the mean, about 5 percent of all Z scores will be greater than $+2$ or less than -2. About 2½ percent will be less than -2. Out of 500 people, then, about 2½ percent or 12 to 13 people would be expected to have Z scores less than -2.

7. a. The mean is $\dfrac{7 + 15 + 13 + 6 + 8 + 7 + 14}{7} = 10$.

The mode is 7.
The median can be found by arranging the numbers in increasing order: 6, 7, 7, 8, 13, 14, 15. The middle number, 8, is the median.
The best estimate of the population variance is given by

$$\text{Est.}SD^2_{pop} = \dfrac{\sum(X_i - M_{samp})^2}{n - 1}$$

so

$$\text{Est. }SD^2_{pop} = \dfrac{(7-10)^2 + (15-10)^2 + (13-10)^2 + (6-10)^2 + (8-10)^2 + (7-10)^2 + (14-10)^2}{7 - 1}$$

$$= \dfrac{9 + 25 + 9 + 16 + 4 + 9 + 16}{6}$$

$$= \dfrac{88}{6} = 14\tfrac{2}{3}$$

Since we are dealing with a *sample,* the mode, like all the numbers you found in this problem, is a statistic.

b. For this list, $\Sigma X_i = 70$.

8. a. See Figure 12.4.
   b. The probability of the event "both cards red" is 4/16 (see Figure 12.4).

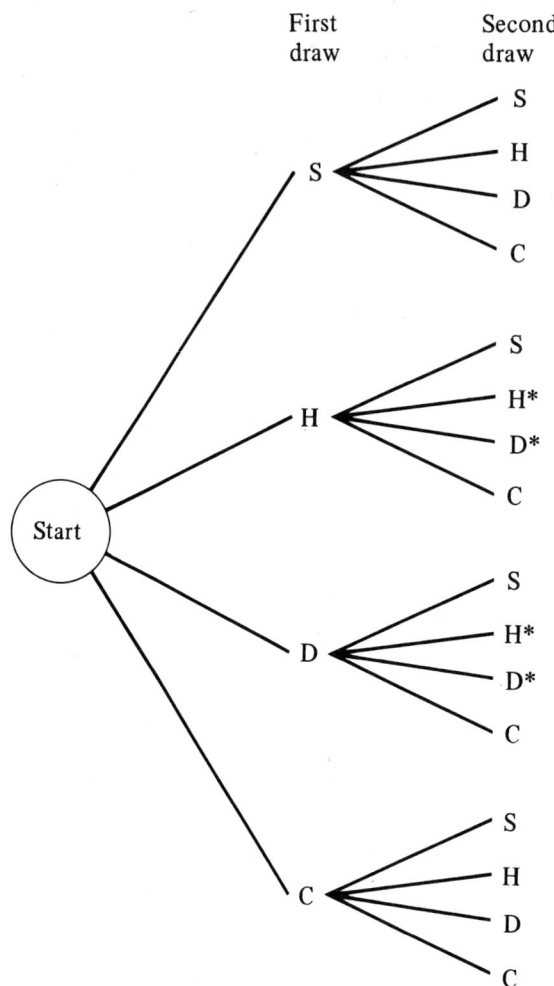

Figure 12.4 The outcomes marked with an asterisk (*) are in the event "both cards red."

Summary

The only new material presented in this chapter is the addition rule for probabilities of nonoverlapping events. Any two events in a sample space are said to be mutually exclusive (or nonoverlapping) if it is impossible for both of them to occur at the same time.

If any two events, event **A** *and event* **B**, *are mutually exclusive, then the probability that we will observe one* or *the other of these two events is the sum of the probabilities of the two events.*

This may be written

Probability (A *or* B) = Probability(A) + Probability(B)

Problems

1. Consider the experiment in which one card is chosen at random from a standard deck. The following events are defined:

   A. A heart is drawn.
   B. An ace is drawn.
   C. A face card (jack, queen, or king) is drawn.
   D. A card with eight spots or fewer is drawn (A, 2, 3, 4, 5, 6, 7, or 8).
   E. A red card is drawn.
   F. The jack of diamonds is drawn.

   For each pair of events listed below, state whether they are mutually exclusive or overlapping:

   A and B _____

   A and C _____

   A and D _____

   A and F _____

   B and C _____

   B and F _____

   C and D _____

   C and E _____

   C and F _____

†2. Suppose that 35 people participate in a week long reading improvement clinic. Their reading ability is measured before and after the program on a 15-point achievement scale. A posttest–pretest difference score is computed for each person, and the results are as follows: 10 people had positive (+) difference scores; 1 person had a negative (−) difference score; and 24 people showed no change (0 difference score).

   a. Analyze these results with a sign test. (You may wish to reread the question and answer in Chapter 8.) Use a 0.05 significance level.

   b. Discuss whether this is a meaningful result: Do you think we have strong evidence about the usefulness of the clinic?

The Normal Distribution; The Central Limit Theorem

In Chapter 11, at the end of the discussion of continuous probability density functions, we briefly mentioned the normal distribution. This curve can be described by the mathematical formula on page 104. It is important because a great many empirical frequency distributions have a form very close to the form of the normal distribution. When we learn the properties of the pure normal distribution, it turns out that we can use those properties to help understand a variety of processes in the real world.

As an example of the ubiquity of this sort of frequency distribution, consider the curve in Figure 13.1, which shows how often various sizes of chest measurement were observed in a sample of 5732 soldiers in the Scottish regiments.

This empirical frequency distribution is very similar in form to the theoretical normal distribution. We shall see further examples, as time goes on, of measurements which are normally distributed. Because such distributions are very common, it is helpful to know a few of the basic properties common to all normal distributions.

SOME USEFUL FACTS ABOUT NORMAL DISTRIBUTIONS

Recall that according to the rule of thumb introduced in Chapter 10, for most situations, 95 percent of the observations in a population will be within 2 standard deviations of the mean. For the normal distribution, the rule of thumb is almost exact: 95.4 percent of all the observations in a normal distribution will be within 2 standard deviations of the mean. The shaded part of the normal distribution in Figure 13.2 contains all the observations within 2 standard deviations of the mean, M. Thus 95.4 percent of all observations lie in the shaded area.

Exercise: What percentage of the observations in a normal distribution will be greater than 2 standard deviations *above* the mean? Try to answer this before reading on.

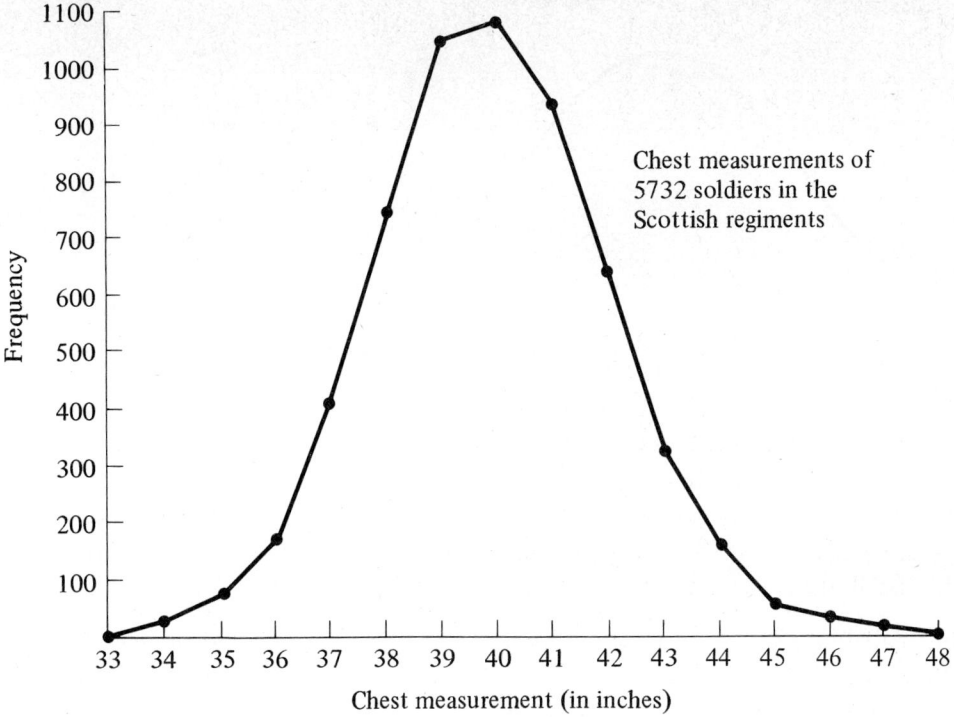

Figure 13.1 An example of a measurement, the chest measurements of several thousand soldiers, which is approximately normally distributed.

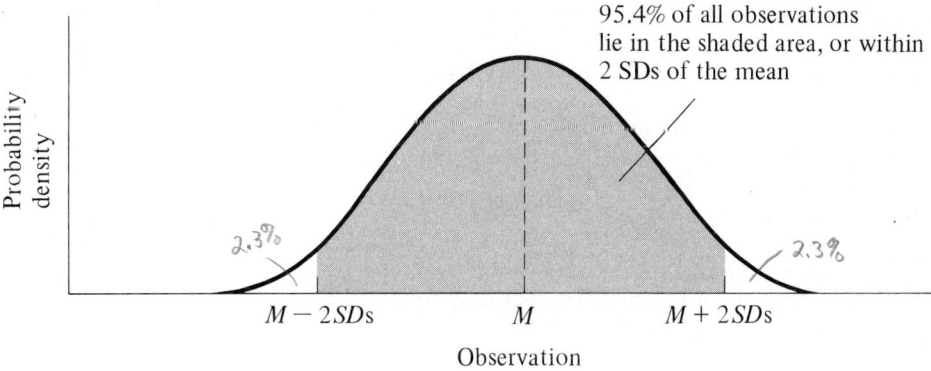

Figure 13.2 The shaded area contains 95.4 percent of the area under the curve. That is, 95.4 percent of all observations lie within 2 standard deviations of the mean.

Answer: Since 95.4 percent of the observations are in the shaded area, the rest, 4.6 percent, must be outside of it. Since the normal distribution is perfectly symmetrical, half of that 4.6 percent is on each end, and the little white area at the right in Figure 13.2 is 2.3 percent of the whole distribution. Thus 2.3 percent of all the observations are greater than 2 standard deviations above the mean.

For any normal distribution, we know how many of the observations will lie within 1 standard deviation of the mean (see Figure 13.3).

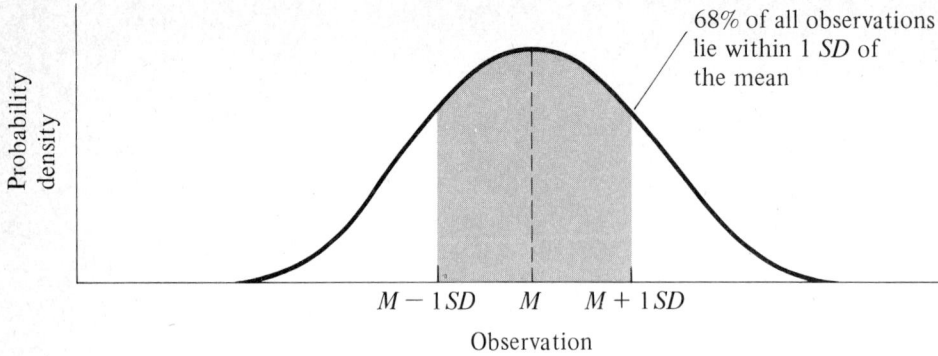

Figure 13.3 The shaded are contains 68 percent of the area under the curve. That is, 68 percent of all observations are within 1 standard deviation of the mean.

The probability of an observation between $M - 1\ SD$ and $M + 1\ SD$, which is equal to the area shaded, is 0.68. That is, 68 percent of all observations are within 1 standard deviation of the mean for any normal distribution.

Exercise: What is the probability of an observation greater than 1 standard deviation above the mean in a normal distribution?

Answer: Since 68 percent of Figure 13.3 is shaded, 32 percent is not. Half of that, or 16 percent, is on each tail of the distribution. Therefore the probability of an observation more than 1 standard deviation above the mean is 0.16.

We have said that 68 percent of all observations in a normal distribution are within 1 standard deviation of the mean, and 95.4 percent of all observations are within 2 standard deviations of the mean. It is sometimes also useful to know that 99.7 *percent of all observations are within 3 standard deviations of the mean, and more than* 99.99 *percent of all observations are within 4 standard deviations of the mean.*

WHAT IS A NORMAL DISTRIBUTION?

Some people erroneously assume that any bell shaped probability distribution is a normal distribution. This is wrong. The mean and the variance remain to be specified, but the height of the probability density curve is always related to the value of the observation, X_i, as shown in Figure 13.4. At the observation X_i the probability density (the height of the solid line) is given by

$$\text{Probability density} = \frac{1}{\sqrt{2\pi} \times SD}\, e^{-\frac{1}{2}\left(\frac{X_i - M}{SD}\right)^2}$$

There is no reason whatsoever for you to memorize this equation. But it is important that you appreciate that it specifies a very precise mathematical form for the normal distribution. The symbol π, greek pi, which you may remember from the days when you studied geometry, stands for the number 3.14159. The symbol e stands for the number 2.71828. SD is the standard deviation of the normal distribution in question, and M is the mean of that distribution. You should realize that once you have specified the mean, M, and the standard deviation, SD, you have specified the exact height for that normal distribution at every point. It is intriguing, and most useful, that for any normal distribution whatsoever the rules about "shaded areas" are true.

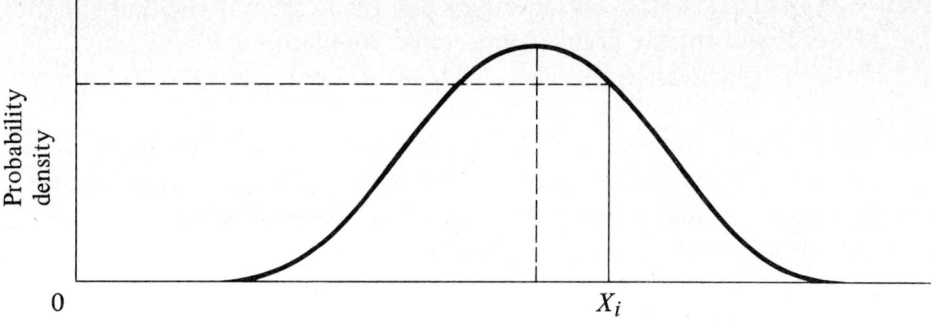

Figure 13.4

THE CENTRAL LIMIT THEOREM

There is a most remarkable and almost magical theorem in statistics which gives some very useful information about sample means from a specified population. It is a theorem about the sampling distribution of the mean for a specified population. (Remember what *that* is?)

Think back to the example introduced at the start of Chapter 11. We were concerned with a sample of size 40 taken from a population of 13,000 students. What we really wanted to know was the chance of various possible results for the *sample mean* when we take sample after sample of size 40. Remember that the sampling distribution of the mean is the distribution that tells you how likely you are to observe various possible numbers as the *mean* GPA of the 40 students in the sample.

Let's call the mean of a sample of 40 GPAs M_{40}. How can we determine the chance of various possible values for this number? The most direct way is to take a sample of 40 students, calculate the mean GPA, M_{40}, return the students to the population and take another sample, calculate another value for M_{40}, and continue the process for a long, long time — say for 1000 samples. Then we could draw a frequency distribution, showing the likelihood of various values for M_{40} on the basis of our sampling exercise. This distribution would be the sampling distribution of the mean for samples of size 40 for this population.

Here's the hooker: That sampling distribution would be a *normal distribution* no matter what kind of distribution of GPAs was found in the population of 13,000 students.

A student asked the following astute question: Suppose the distribution of grades was not smooth but was bimodal?

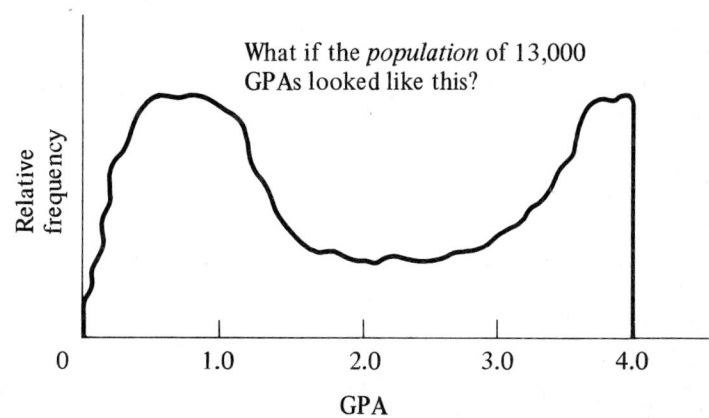

Figure 13.5 A very improbable distribution of GPAs in a large population.

What if the *population* of 13,000 GPAs looked something like Figure 13.5. Now that would be a pretty wild population, but that's not the point. Suppose the population *did* look like that. Would the sampling distribution of the mean of samples of size 40 *still* be a normal distribution? *Yes it would!*

To take an even more extreme case, suppose that you put 200 cards in a box and wrote the number 6 on half of them and the number 80 on the other half. Now suppose you take samples of size 40 from this population, and compute the sample mean, M_{40}, for each sample. Will the resulting sampling distribution of the mean be a normal distribution?

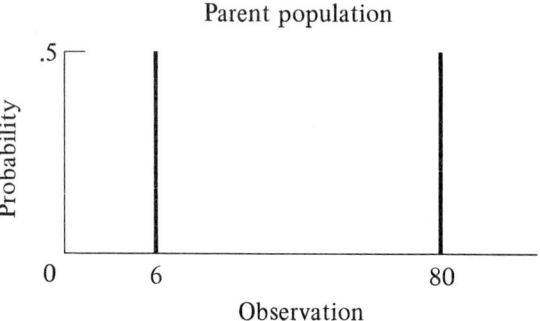

Figure 13.6 Suppose you took samples of size 40 from this extreme population: How do you think sample means would be distributed when the population looks like this?

What do you think? How will the means of samples of size 40 be distributed when the population looks like Figure 13.6? *They will be normally distributed*!

Now if this fails to astound and amaze you, you are missing one of the most exciting parts of this course. Why should the sample means for this population, or for any other one, be distributed exactly according to the normal distribution; that is, according to the equation for probability density given earlier?

The normal distribution pops up all over the place. Errors in astronomical observation are normally distributed. The width of the second shell segment in the tail of a shrimp is, too. So are the chest measurements of 5742 members of the Scottish Regiments. If the weights of 1000 brand-new pennies were determined with a very precise analytical balance, chances are that the weights, too, would be normally distributed. So would the weights of a thousand 1934 dimes. And on, and on. Finally, not least marvelously, the sample means for *large samples* from any population whatsoever will be normally distributed.

Sir Francis Galton, a cousin of Charles Darwin, wrote the following concerning the Law of Frequency of Error, as the normal distribution was then known:

> I know of scarcely anything so apt to impress the imagination as the wonderful form of cosmic order expressed by the "Law of Frequency of Error." The law would have been personified by the Greeks and deified, if they had known it. It reigns with serenity and in complete self-effacement amidst the wildest confusion. The huger the mob and the greater the apparent anarchy, the more perfect its sway. It is the supreme law of Unreason. Whenever a large sample of chaotic elements are taken in hand and marshalled in the order of their magnitude, an unsuspected and most beautiful form of regularity proves to have been latent all along.

It was stressed above that the means of *large samples* will be normally distributed. What about small ones?

The smallest possible sample is a sample of size 1: a sample with only a single number in it. The mean of such a sample is a sort of degenerate case, but it is of course equal to the number. Now consider samples of size 1 from the population sketched in Figure 13.6. How will sample means be distributed? We know the sample mean will be 6 half of the time, and 80 the other half of the time. So for samples of size 1, the sampling distribution of the mean will be identical to the parent population. (See Figure 13.7.) As you should be able to see, this is true for any parent population whatsoever.

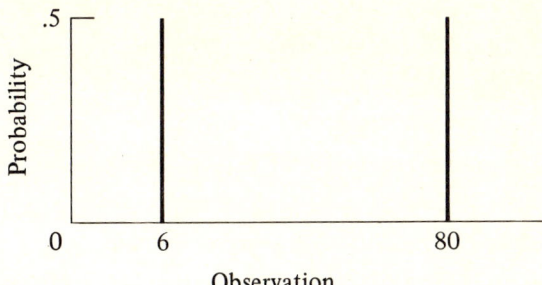

Figure 13.7 The sampling distribution of the mean *for samples of size* 1 is identical to the parent population.

We shall return later to intermediate sample sizes. For most purposes a sample of size 30 or greater is a *large* sample, and the sample means for such samples will be normally distributed.

QUESTION AND ANSWER

Question: What can we say about the *mean* of the sampling distribution of the mean? When we take sample after sample after sample indefinitely from a population, what will be the average value of the sample means?

Answer: Not unreasonably, it turns out that the average value will be equal to the population mean. For example, in the case mentioned in Chapter 11, where the average GPA for 13,000 students was 2.498, the mean of the sampling distribution of the mean is 2.498, too. On the average, then, the sample mean will come out exactly equal to the population mean.

Question: For the population sketched in Figures 13.6 and 13.7, what will the mean of the sampling distribution of the mean be?

Answer: It will be equal to the mean of the parent population. Since in that population you observe 6 half of the time and 80 the other half of the time, the population mean will be 6 plus 80 divided by 2, or 43. Thus 43 will be the mean of the sampling distribution of the mean, too.

Summary

The normal distribution crops up all over the place.

For any normal distribution, 95.4 percent of all the observations lie within 2 standard deviations (*SD*s) of the mean.

For any normal distribution, 68 percent of all observations lie within 1 standard deviation of the mean.

For any normal distribution, 99.7 percent of all observations lie within 3 standard deviations of the mean.

In order to be a normal distribution, a continuous probability density function must be described by one specific mathematical formula.

When you look at large samples (30 or more numbers in each) from any population whatsoever, the sample means are normally distributed.

The Central Limit Theorem, Continued

In the previous chapter considerable time was spent describing the sampling distribution of the mean. Since this is such a crucial concept, we will give another example here. Consider a population consisting of the heights of each of 100 students in a statistics class. What would the sampling distribution of the mean be for samples of size 10?

To see what it would be, imagine taking a sample of size 10 and finding the average of the 10 numbers in the sample. Write it down. Now take a new sample of size 10 from the full population, and again find the mean. Continue this process for a long, long time, say for 1000 samples. You will have observed a great variety of different values for M_{10}, the mean of a sample of size 10. The overall frequency distribution for these 1000 numbers would be the sampling distribution of the mean.

Remember, as mentioned in Chapter 13, that the mean of the sampling distribution of the mean is exactly equal to the population mean. We write $M_M = M_{pop}$. In this case, the average of all the different values of M_{10} would be equal to the average of the 100 heights of the students in the class. (When you see M_M, you can think, "the mean of the means.")

HOW VARIABLE ARE SAMPLE MEANS?

What can we say about the variability of the sample means? What is the *variance* of the sampling distribution of the mean? It turns out that we can make a very precise statement about this. Before doing so, we can make two partial statements:

1. *The variability of the sample mean should be related to the variability of the population.* For example, consider a hat containing 500 poker chips, each with the number 87 on it, and another 500 poker chips, each marked with the number 88. The variance of such a population would be very small. And if you were to take samples of size 10, the sample means wouldn't be very variable either. On the other hand, if the hat

contained 1000 chips, marked with the numbers from 1, 2, 3, 4, 5, etc., up to 1000, then we would expect the means of samples of size 10 to be far more variable.

2. *The variability of the sample mean should be inversely related to the size of the sample:* We would expect sample means for *large* samples to be less variable than sample means for *small* samples. Do you agree?

It turns out that the true relationship between the population variance and the variance of the sampling distribution of the mean shows clearly the two effects mentioned above:

For samples of size n, from a population with variance SD^2_{pop}, the variance of the sampling distribution of all the means of samples of size n is given by

Variance of the sampling distribution of the mean $= \dfrac{SD^2_{pop}}{n}$

Example: Think back to the example introduced in Chapter 11 (page 78). The situation described involved a population of 13,000 students, with the average GPA, the mean of the population, being $M_{pop} = 2.498$, and the variance of the population of GPAs being $SD^2_{pop} = 0.36$. A sample of size 40, supposedly taken at random from that population, was found to have a mean, $M_{40} = 3.21$. Is this plausible? Or is it much higher than one would expect?

A question to consider: What do we know about the sampling distribution of the mean for samples of size 40 for this population? We know that the mean of the sampling distribution, M_M, will be equal to the mean of the parent population, or 2.498. And we know that the variance of the sampling distribution, SD^2_M, will be 1/40 of the variance of the parent population, according to the formula above. The variance of the sampling distribution of the mean, SD^2_M, is thus given by

$$SD^2_M = \dfrac{SD^2_{pop}}{n} = \dfrac{SD^2_{pop}}{40} = \dfrac{0.36}{40} = 0.009$$

Since this sample is a *large* one (greater than 30), we know that the means will be normally distributed. So we can say with confidence that the means of samples of size 40 will be normally distributed, with the mean of the normal distribution equal to 2.498 and the variance equal to 0.009. We are now ready to answer a question: In this normal distribution would an observed value (mean) of 3.21 be very surprising?

To help you understand the question (and to anticipate the answer) look at two pictures. First, consider Figure 14.1, a sketch of a GPA distribution with a mean of 2.498 and a variance

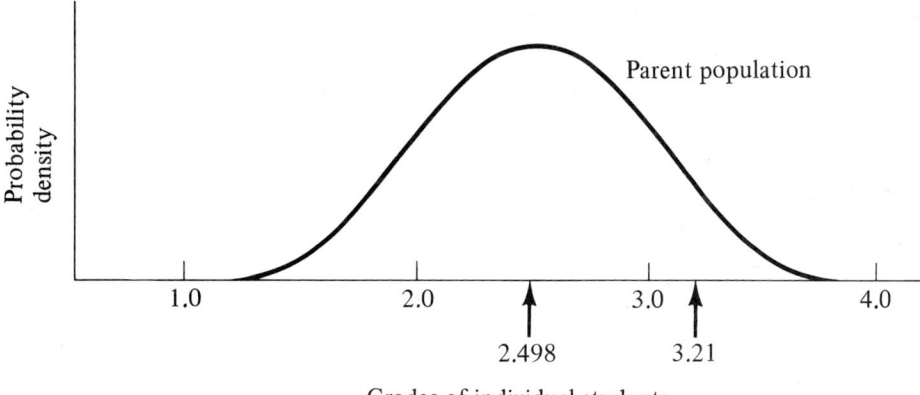

Figure 14.1 This probability distribution shows the likelihood of seeing various GPAs in the entire student population of 13,000 students.

of 0.36. These parameters, you will recall, describe the parent population of 13,000 GPAs. Note that seeing a GPA of 3.21 for a single student wouldn't be very surprising. Quite a few people have GPAs higher than that.

The sketch of the sampling distribution of the mean for samples of size 40 from the above population (Figure 14.2) looks very different: this distribution has the same mean, 2.498; but it has a much smaller variance, 0.36/40 = 0.009. Figure 14.2 shows that the sample means are very

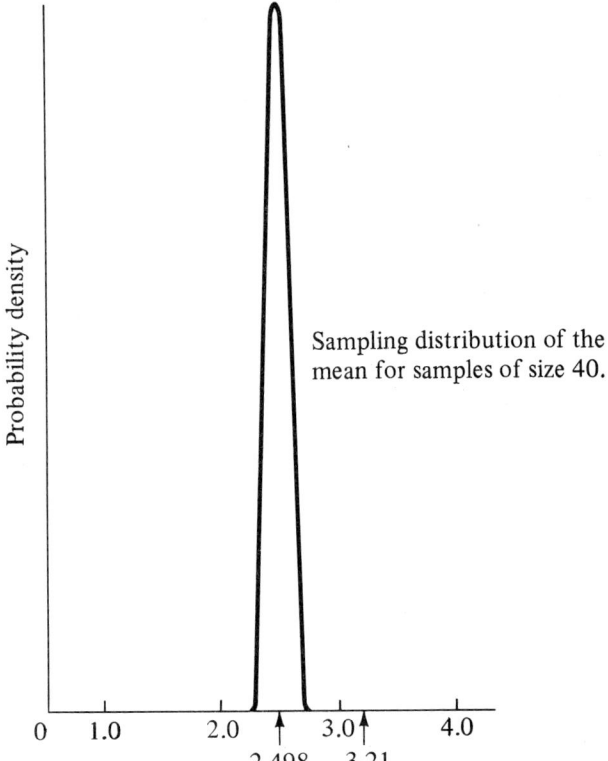

Figure 14.2 This probability distribution shows the likelihood of seeing various values for M_{40}, the mean GPA for a sample of 40 students from the parent population.

closely grouped around 2.498, the mean of the sampling distribution of the mean, M_M. The fact that the variance is so small (0.009) shows that observations of M_{40} that are far from the mean are very unlikely. Note that in this distribution, an observation of 3.21 (a *sample mean* of 3.21) would be most unusual. Almost all the values are much closer to 2.498. We'll now confirm this deduction by computing a standard score.

Remember that we started off by asking whether a sample mean of 3.21 was suspiciously high for a random sample taken from the parent population. Recall that in a distribution with mean M and variance SD^2, the Z score for an observation X_i is $Z = (X_i - M)/SD$. We are concerned with an observation, 3.21, in respect to the entire distribution of the sample means. Here, in the sampling distribution of the mean, $M = 2.498$ and $SD^2 = 0.009$. Thus

$$SD = \sqrt{SD^2} = \sqrt{0.009} \cong 0.1 \qquad \text{since } 0.1 \times 0.1 = 0.010 \cong 0.009 \text{ (the symbol } \cong \text{ means "is approximately equal to")}$$

The Z score corresponding to an observation of 3.21 is

$$Z = \frac{X_i - M}{SD} = \frac{3.21 - 2.498}{0.1} \cong \frac{0.7}{0.1} = 7$$

Now, a Z score of 2 would be somewhat surprising. A Z score of 3 is remarkable: recall that one sees such extreme Z scores only about 3 times in 1000.[1] But a Z score of 7 is fantastic. The chance of a random sample of size 40 producing a sample mean of 3.21, given the above population, is so small as to be impossible. We can say with almost total confidence that this was *not* a random sample; the students were chosen in a way that made it more likely for a student with a high GPA to be chosen than for a student with a low GPA.

You should note that very few people without some knowledge of statistics would consider an observed mean of 3.21 to be terribly high. The central limit theorem is a useful friend.

THE CENTRAL LIMIT THEOREM

We can now combine the several facts just mentioned and state the central limit theorem:

For any population you encounter, with mean, M_{pop}, and variance, SD^2_{pop}, the sampling distribution of the mean for samples of size n (where n is at least 30) will be a normal distribution with a mean the same as the population mean, M_{pop}, and with a variance equal to SD^2_{pop}/n.

It is furthermore the case that when the parent population is itself normally distributed, then the above relationships hold regardless of the size of the sample. In this case, we do not need to require that the sample size must be at least 30.

Means of samples from a uniform probability distribution.

Example: Suppose you have a ten-sided cylinder, with the digits 0, 1, 2, 3, etc., to 9 marked on the sides. When you roll it on a smooth surface, each of the ten outcomes is equally likely. The probability distribution corresponding to this experiment looks like Figure 14.3.

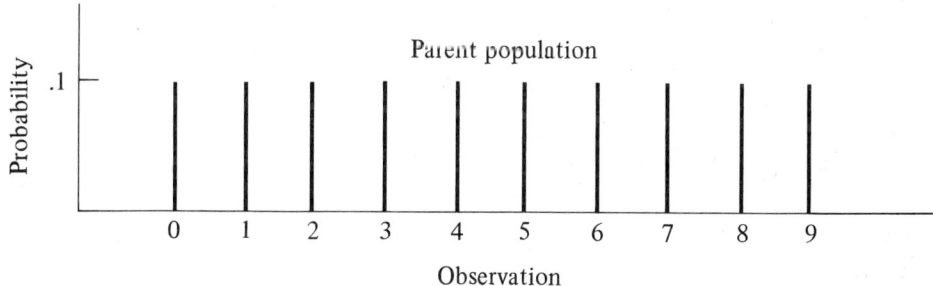

Figure 14.3 Here is the probability distribution for the parent population of all possible rolls of the 10-sided cylinder.

Samples of Size One:
Now, suppose that you take samples of size 1, and look at the means of these samples. What will the sampling distribution of the mean look like? It will look like Figure 14.4. It will look exactly like the parent distribution, since the mean of a sample of size 1 is the same as the number sampled. And each number is equally likely.

---

[1] This follows from the fact that 99.7 percent of all observations in a normal distribution are within 3 standard deviations of the mean.

114   THE NORMAL DISTRIBUTION AND THE Z TEST

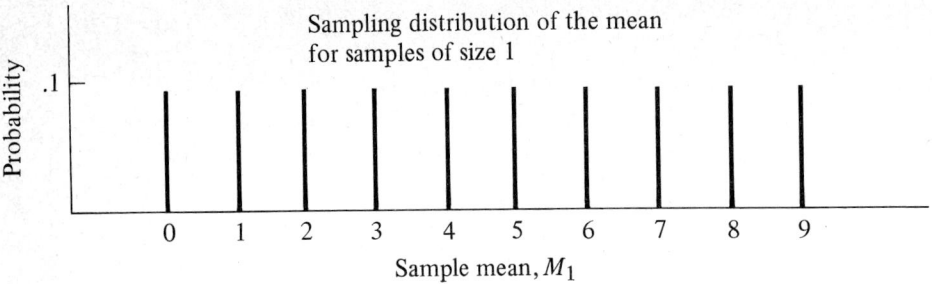

Figure 14.4 For samples of size 1 from the above mentioned population, the sample means will be distributed like this.

Samples of Size 2.
Suppose you rolled the cylinder twice, and wrote down the results. How many possible outcomes are there? You should be able, by now, to draw a 100-branch probability tree in your mind's eye. (If you can't, ask someone to help you.) If you take a sample of size 2 from the cylinder population, there are 10 × 10 = 100 possible different samples. Let's compute the sampling distribution of the mean for samples of size 2. To do this, we list all the possible values for the mean, M, and find the probability of observing each value, as in Table 14.1.

TABLE 14.1

M	List of outcomes which will produce this value of M	Number of different outcomes which will give this value of M	Probability of M
0	(0, 0)	1	1/100 = 0.01
1/2	(0, 1), (1, 0)	2	2/100 = 0.02
1	(2, 0), (0, 2), (1, 1)	3	3/100 = 0.03
1 1/2	(3, 0), (0, 3), (2, 1), (1, 2)	4	4/100 = 0.04
	Etc.		

Explanation of one line of the table: We are taking a sample of two numbers from the population 0, 1, 2, 3, 4, 5, 6, 7, 8, 9. How can we get a *sample mean* of exactly 1? We must observe a total of 2 in the two numbers we take from the above population. How can we get this total? The first number could be 2, and the second 0 (2, 0), or vice versa (0, 2). Or we could observe 1 on each of the two throws (1, 1). No other outcome will yield a sample mean of 1. So the probability of this event (M = 1) is just 3/100 or 0.03.

After finishing this table, one can draw the Figure 14.5. Now this doesn't look exactly like a normal distribution. But it is impressive how much closer it is to a normal distribution than to the original (parent) probability distribution from which the samples were drawn. (See above.)

Samples of Size Three.
If you take samples of size 3 and proceed in a similar manner, you see the results presented in Figure 14.6. Notice that simply by taking samples of size 3, the sampling distribution of the mean looks quite a lot like a normal distribution for this particular parent population. For samples of size 4, the similarity in shape is even more striking.

We have looked at the sampling distribution of the mean for samples of size 1, 2, and 3 taken from the population 0, 1, 2, 3, 4, 5, 6, 7, 8, 9. The central limit theorem states that as sample size

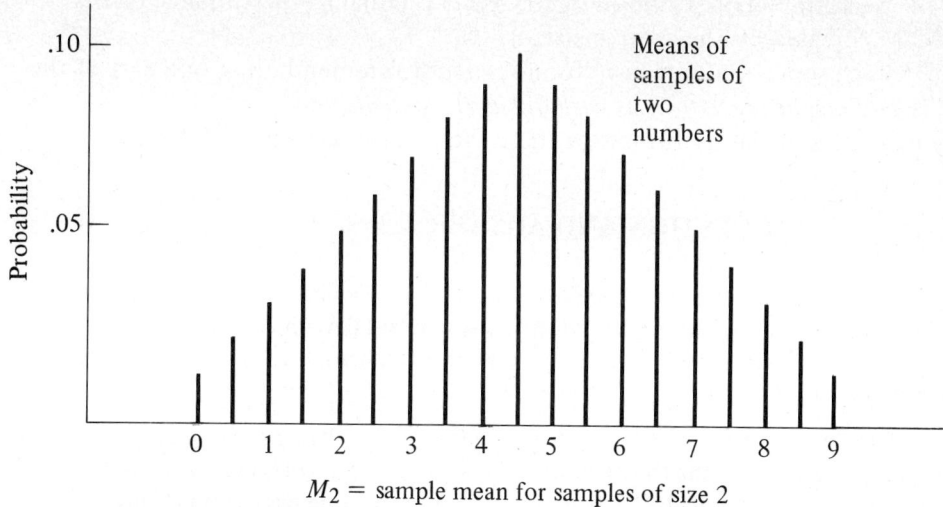

Figure 14.5 Sampling distribution of the mean for samples of size 2 from the population (0, 1, 2, 3, 4, 5, 6, 7, 8, 9).

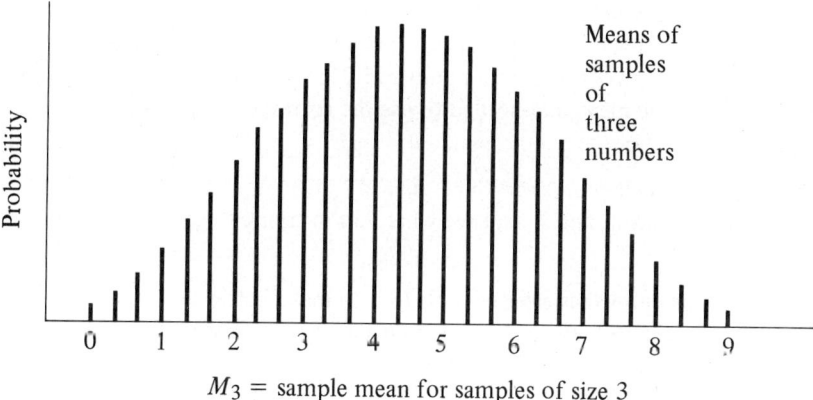

Figure 14.6 Sampling distribution of the mean for samples of size 3 from the population (0, 1, 2, 3, 4, 5, 6, 7, 8, 9).

increases, the sampling distribution of the mean from any population whatsoever gets closer and closer to a normal distribution. We have just seen a simple but dramatic example in which the shape of the sampling distribution of the mean changes in the direction of normality with increasing sample size. If you can remember these three pictures, you should have no trouble remembering an important part of the central limit theorem: For large samples (30 or more numbers in each sample) the sampling distribution of the mean is a normal distribution, regardless of the population from which the samples were taken.

Sampling from a normal distribution: A special situation exists when the parent population is itself a normal distribution: then the sampling distribution of the mean is a normal distribution for small samples as well as large ones.

In addition to these two parts of the central limit theorem, you should also remember that the mean of the sampling distribution of the mean is always the same as the population mean, and that the *variance* of the sampling distribution of the mean is equal to the variance of the parent population (SD^2_{pop}) divided by the sample size, n.

   Note: A mnemonic which some students have found useful for remembering one part of the central limit theorem is *The means are $1/n$ th as variable as the population.*

   We shall see many examples of the usefulness of these facts in subsequent chapters.

QUESTION AND ANSWER

Question: In working through the GPA example, you calculated a Z score for the observation 3.21, using the formula $Z = (X_i - M)/SD$. Then you used $SD = 0.01$, which is the standard deviation of the sampling distribution of the mean for samples of size 40. Why did you use this number rather than the standard deviation of the parent population?

Answer: Unfortunately it is fairly easy to get confused about this. In the coming chapters, this may get you into trouble. So please try hard to understand the following discussion. A standard score, or Z score, summarizes in a single number how one particular number, X_i, stands in relation to an entire distribution. This distribution, *from which the number X_i was presumably taken*, has a mean, M, and standard deviation, SD. Now in the example considered in this chapter, the number 3.21 is a *sample mean*. It is based on a sample of 40 GPAs. We are interested in comparing it to the entire distribution of possible sample means, which is known as the *sampling distribution of the mean*. And the sampling distribution of the mean, as argued in the chapter, has a variance of 0.009, and a standard deviation of about 0.01.

   The important thing to remember, here, is which population you're working with at any point: the population of 13,000 GPAs (the parent population), or the population of all possible sample means (the sampling distribution of the mean). Since we were considering a number (3.21) which was a *sample mean*, we were working in the sampling distribution of the mean and needed to use the corresponding value for SD in computing a Z score.

Question: Where is this all going? For the first eight chapters everything seemed to hang together. Since then, we've been memorizing lots of new terms and ideas, and I can't see what you're trying to get at. Can you help?

Answer: Way back in the beginning, statistics was defined as a set of tools to help make better decisions on the basis of variable data or information. In this chapter, we made a decision about an observation (3.21) which was supposedly the mean of a random sample of 40 GPAs from a population of 13,000. We decided that it ain't necessarily so. What we've been doing since Chapter 9 is to build the conceptual and terminology skills necessary to move ahead to decisions of this sort — which are quite common in the real world. You'll see more examples in the next chapter.

Question: Do we know that the university administration, in choosing 40 students, expressly picked people on the basis of their grades?

Answer: No. But we do know that whatever system they used was especially likely to select good students. They might, for instance, have asked the faculty to recommend "typical students." The faculty, unwittingly, might have picked the students it knew best: the faculty is often biased toward high scholastic achievement.

Summary

   The sampling distribution of the mean was discussed some more. This is the probability distribution which tells you how likely you are to observe various values for the sample mean after the sample size and the parent population have been specified.

   The *central limit theorem* includes the following parts:

1. The mean of the sampling distribution of the mean is equal to the mean of the parent population from which the samples are taken: $M_M = M_{pop}$.

2. The variance of the sampling distribution of the mean is equal to the variance of the parent population, SD_{pop}^2, divided by the sample size, n: $SD_M^2 = SD_{pop}^2/n$.
3. The sampling distribution of the mean is always a normal distribution when the samples are large (30 or more). This is true regardless of the parent population from which the samples are taken.
4. When the parent population is itself a normal distribution, then the sampling distribution of the mean is always a normal distribution, even for small samples.

As we increase our understanding of the behavior of sample means from a given population, we will become more skillful at *deciding,* on the basis of a sample mean, whether it is plausible to assume that the sample was taken from a specified population. In this chapter, for example, we decided that it was very very unlikely that a sample of 40 GPAs with a mean, M_{40}, of 3.21 was taken at random from a population which had a mean of 2.498 and a variance of 0.36. We were thus able to conclude that the sampling process was almost certainly biased in such a way that students with good grades were more likely to be included in the sample.

Problems

1. a. Each of 50 cards in a hat is marked with the number 36. What is the variance of this population? (Think back to the definition of variance before you start a long calculation.)

   b. Describe the sampling distribution of the mean for samples of size 5.

   c. How does the variance of the sampling distribution of the mean relate to the population variance? Is this an exception to point 2 in the Summary above?

2. A language student who is a statistician notes that the number of words he can learn in an hour is normally distributed, with a mean of 14 and a standard deviation of $\sqrt{12}$. One Sunday he switches to a health food diet and for the next *three hours* averages 19.3 words per hour. Interpret this number, 19.3, as the mean of a sample of size 3 from a population. Should he conclude that his learning rate has changed?

   Hints: What would be the probability distribution for means of samples of size 3, taken from the population, "the number of words learned in 1 hour," for this student before his dietary change? Is it likely that you would see a number as extreme as 19.3 in this derived probability distribution? Compute a standard score to support your argument.

3. In Table 14.1 are computed the probabilities of four different values for M, the sample mean, for samples of size 2 from the population 0, 1, 2, 3, 4, 5, 6, 7, 8, 9. Write down the next three lines of that table, and find the probabilities of sample means of 2, 2 1/2, and 3. Do your results agree with the graph in Figure 14.5?

4. I have been looking at a particular sampling distribution of the mean for a population I'm studying and for samples of size 10. The variance of the parent population is 150. How big a sample size do I need if I want the variance of my new sampling distribution of the mean to be half as large as for the sample size of 10?

5. Describe an experimental procedure you could use to check the sampling distributions shown in Figures 14.5 and 14.6.

The Z Test, or Normal Test

In the preceding two chapters the following parts of the central limit theorem were presented:

1. For samples of any size, the mean of the sampling distribution of the mean, M_M, is the same as the mean of the parent population, $M_M = M_{pop}$.
2. For large samples (30 or more numbers in each sample) the sampling distribution of the mean is essentially a normal distribution, no matter how the parent population is distributed.
3. When the parent population from which the samples are taken is itself a normal distribution, then the sampling distribution of the mean is also a normal distribution, no matter what the sample size is, even for small samples.
4. If the variance of the parent population from which the samples are taken is SD_{pop}^2, and if there are exactly n numbers in each sample, then the *variance* of the sampling distribution of the mean is given by

Variance of the sampling distribution of the mean = $SD_M^2 = \dfrac{SD_{pop}^2}{n}$

For example, if the samples are of size 10, then the variance of the sampling distribution of the mean will be 1/10 of the population variance. If the samples each contain 94 numbers, then the variance of the sampling distribution of the mean will be 1/94 of the population variance, and so on.

These four points are extremely important. If you have not already done so, you should commit them to memory.

AN EXAMPLE: GAS MILEAGE

I have been using Shell gasoline in my car for over a year, and whenever I buy gas, I compute the mileage (mpg) for the tank of gas just finished. After recording data for a year, I analyze the results and note that I can very accurately summarize the data as follows: the mileage observed is normally distributed, with a mean of 23.5 mpg and a variance of 5. This is an exact statement about how likely various values of mpg are. Now, suppose that a Texaco station opens, which is more convenient than the Shell station, and I buy gas there four times and compute the mpg each time. For these four observations I find a mean mileage, M_4, of 25. (The actual numbers were 24.4, 26.2, 25.1, and 24.3 mpg, but that doesn't matter.)

The decision to be made: This mean value is higher than the mileage I had been getting with Shell. Is it enough higher so that I can be quite confident that there's a real difference and that this discrepancy isn't just due to random variation?

In order to answer this question we have to know how likely such data would be if the gasolines were the same — if my four Texaco mpg values were drawn from a normal distribution with a mean of 23.5 and a variance of 5. To answer this we use the central limit theorem, applied to this particular case with a sample size of 4 (since the observed mean, 25, is based on four numbers).

Once we know what to do, what can we say about the way that the means of samples of size 4 will be distributed? How likely will various possible values of M_4 be under these conditions?

1. From point 1 above we know that the mean of the sampling distribution of the mean will be 23.5, the same as the population mean. On the average, then, the means of samples of size 4 will be 23.5.
2. From point 3 above we know that the sampling distribution of the mean will be a normal distribution, since the parent population (which I studied for a year) is a normal distribution.
3. From point 4 above we can deduce that the variance of the sampling distribution of the mean for samples of size 4 is 5/4, since

Variance of the sampling distribution of the mean = $SD_M^2 = \dfrac{SD_{pop}^2}{n} = \dfrac{5}{4}$ (since $SD_{pop}^2 = 5$ and n, the sample size, is 4)

Putting this all together, and drawing a picture (Figure 15.1), we have deduced that if samples of size 4 are taken from a normal distribution with a mean of 23.5 and a variance SD_{pop}^2 of 5, the sampling distribution of the mean will be a normal distribution with a mean, M_M, of 23.5 and a variance of 5/4.

Now we can return to the original decision whether the observed sample mean, 25, is so high that we would be surprised to find it in this sampling distribution of the mean. To answer the question, let's compute a Z score, or standard score. *This is the standard score corresponding to the observation* of 25, on the basis of the probability distribution in Figure 15.1, the sampling

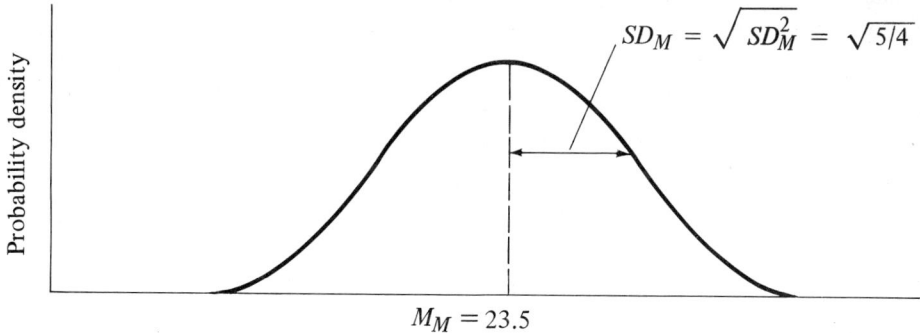

Figure 15.1 The sampling distribution of the mean, for samples of size 4, from a normal distribution with $M_{pop} = 23.5$ and with $SD_{pop}^2 = 5$.

distribution of the mean. You will (or *should*) remember that the Z score is defined as follows: The Z score corresponding to an observation, X_i, from a distribution with mean M and variance SD^2 is

$$Z = \frac{X_i - M}{SD}$$

In this case

$X_i = 25.0$
$M = 23.5$
$SD = \sqrt{5/4} = \sqrt{1.25} = 1.1$

so

$$Z = \frac{25.0 - 23.5}{1.1} = \frac{1.5}{1.1}$$

$$\cong 1.4$$

You might be tempted, here, to say $SD = \sqrt{SD^2_{pop}} = \sqrt{5}$. That is, you might want to use the *population* standard deviation in computing a Z score. This would be an important error. The Z score tells where one number, X_i, stands *with respect to a specified distribution of such numbers*. Here X_i is a sample mean ($M_4 = 25$), so we must compare it to other sample means. This means we compute the Z score on the basis of the mean and standard deviation of an entire distribution of *sample means*: the sampling distribution of the mean.

Now, to go back, is 1.4 so extreme a value of the Z score as to cause us to reject the idea that the observed sample mean (25) is based on numbers taken from the same population as the Shell numbers, that is, a normal distribution with a mean of 23.5 and a variance of 5? It is not. For remember the rule of thumb which stated that 5 percent of the time you observe numbers more than 2 standard deviations away from the mean, and 95 percent of the time they are within 2 standard deviations. This means that 95 percent of the time the absolute value of the Z score is less than 2 (the Z score is between −2 and +2) and that 5 percent of the time it is more extreme. We can deduce, then, that our observed value of the Z score, 1.4, is really not very extreme. We can expect more extreme values quite often, by pure chance. So on the basis of the four tanks of gas so far, we can't conclude that Texaco is superior to Shell. In this sort of situation, many people would decide to gather some more data and make a second analysis.

Important note: It is easy to get confused in this kind of example, because there are two different but closely related distributions being studied. One is the parent population: the distribution of mpg values using Shell; the second is the distribution of *sample means* for samples of size 4 taken from that distribution: the sampling distribution of the mean seen in Figure 15.1.

The question we are really asking is: Could the observed value, 25, have been taken from the sampling distribution of the mean for samples of size 4 from the Shell parent distribution? If it was, then there is no difference in the two gasolines. If we decide that it could not plausibly have come from that sampling distribution, then we conclude that there is a significant difference.

THE Z TEST COMPARED WITH THE SIGN TEST

What we have just done, in essence, is a Z test (or normal test). Now let's review the logic, comparing what we have done with the six steps we used in the sign test:

THE Z TEST

Sign test

Step 1:
List all the possible events (0/10, 1/10, etc.)

Z test

In this case we cannot list all the possible values we might obtain for the mean of a sample of size 4, but we can think of an infinite distribution of possible values.

Step 2:
State a null hypothesis.

The null hypothesis (H_0) in this case is that the observed sample mean is based on numbers taken from the same normal distribution as the Shell mpgs: a normal distribution with a mean of 23.5 and a variance of 5.

Note that just as in the sign test, the null hypothesis is a precise statement of no effect or no difference. In this case, the null hypothesis is a precise statement of no difference between the Shell distribution and the distribution from which the sample was taken, or no difference between gasolines.

Step 3:
Study the null hypothesis. Deduce how likely each possible event (result) would be if the null hypothesis were true.

In this case we deduce how likely various values for M_4, the mean of samples of size 4, would be if the null hypothesis were true: if the two gasolines were the same. We calculate what the *sampling distribution of the mean* would be like.

Step 4:
Pick a significance level (chance of Type One error) and study various possible decision rules.

By referring to the rule of thumb we have implicitly picked a significance level of 0.05. We have not done anything analogous to the sign-test procedure of listing a variety of possible decision rules. As will become clear, choosing a significance level amounts to choosing a decision rule.

Step 5:
Pick a rejection region which fits the significance level. Determine which observations (events) will lead you to reject the null hypothesis and which will not.

We didn't do this in so many words, but basically we decided that if the observation was more than 2 standard deviations away from the mean, we would conclude we had a difference. This amounts to a rejection region as shown in Figure 15.2.

Whereas the rejection region for the sign test is a specific list of events, in the case of the Z test it is a specific *pair of intervals*. Each interval specifies a set of different values that would lead to rejection of the idea that the two brands of gasoline are the same.

When the null hypothesis is true, we will observe Z scores between −2 and +2 95 percent of the time. So we decide to reject the null hypothesis (and conclude that Texaco gas gives mileage significantly different from Shell gas) only if we observe a Z score below

Sign test (*Continued*)

Step 6:
Gather data and make a decision.

Z test (*Continued*)

−2 or above +2. Those two intervals are shown by the shaded bars under the curve in Figure 15.2. Putting it formally, the rejection region has two parts:

All values of M_4 smaller than 21.3 mpg and
All values of M_4 larger than 25.7 mpg

Equivalently, we can express the rejection region in terms of the Z statistic:

All values of Z smaller than −2   and
All values of Z larger than +2

In this case our data are the result that $M_4 = 25$. That value is indicated in Figure 15.2 by an asterisk. As may be seen, the asterisk does not lie in the rejection region. The observed value of 25 is not so extreme as to lead us to conclude that it couldn't reasonably have come from the sampling distribution of the mean, mentioned in Steps 3 and 5.

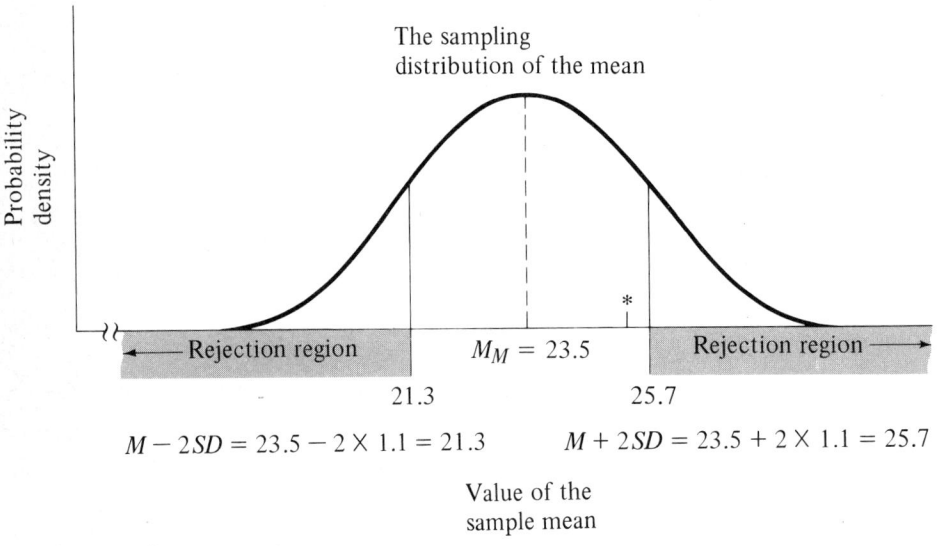

Figure 15.2 The *rejection region*, indicated by the two shaded bars, is really a *pair* of intervals. It contains all observations less than 21.3 and all observations greater than 25.7.

The Z test, or normal test, is really quite simple. The one dangerous point of possible confusion has already been mentioned. It is essential to be sure that you keep straight the two different probability distributions which are involved, and that you avoid confusing the two.

Five Steps for the Z Test

In place of the six steps used in the sign test, it will be helpful to reorganize the steps of the Z test into five steps, as follows:

A. *State a null hypothesis.* As before, this is a precise statement that there is no difference: that a sample *was* taken from a population with specified mean, M_{pop}, and variance, SD^2_{pop}.
B. *Choose a significance level.* As before, this involves deciding how great a risk of Type One error we are willing to accept. Significance level is often denoted by the Greek letter *alpha*, α.
C. *Determine the rejection region for our test statistic.* In this case, our test statistic is a Z score. And once we have chosen a significance level, we can immediately find the rejection region on the basis of our knowledge of the normal distribution. With a significance level of 0.05, we reject the null hypothesis whenever we observe a Z score less than -2 or greater than $+2$. This is appropriate, since the chance of so extreme a Z score is just 0.05 when the null hypothesis is true.
D. *Compute the particular value of the test statistic, Z, which results from the data we have gathered.* In order to interpret a Z score appropriately, we must be concerned with a particular observation (for example, a sample mean) from a normal distribution with a known mean and variance. Often that normal distribution is a sampling distribution of the mean. Remember that the sampling distribution of the mean is always a normal distribution for large samples (30 or more numbers in a sample), or for any size sample from a normally distributed parent population. An intermediate step often required is to determine the mean M_M, and variance, SD^2_M, of the sampling distribution of the mean for the given sample size. Once we have reduced our experiment to a single observation from a known normal distribution with mean M and variance SD^2, we can compute a Z statistic: $Z = (X_i - M)/SD$. In the gasoline mileage example, the single observation was the mean of four numbers, 25. And the SD was the standard deviation of the sampling distribution of the mean for samples of size 4. The mean, M, was the mean of the sampling distribution of the mean. But since it is equal to the mean of the parent population, a misinterpretation here will only muddy your mind; it will not, as it happens, hurt your computation.
E. *Determine whether your computed test statistic lies in the rejection region specified in step C, and make the appropriate decision.* This last step presents no problems.

Sample Problem Involving the Z Test

Sinistralics, a midwestern society of left-handed people, was interested in studying the influence of handedness on economic position. In one city of 10,000 people the average income per head of family was found to be $8500. The variance of this measure (income per head of family) was found to be 3,600,000. A random sample of 90 people was taken from the entire group of left-handed heads of family. For these 90 it was found that the average income was $9134. Is this different enough from the city-wide average of $8500 to constitute convincing proof that left-handed people are different from the population as a whole? *Suggestion:* Try to solve this before reading on.

Solution: First we must be sure that this is an appropriate place for using a Z test. Can we reformulate the question so that it is in the following form: Is it plausible to assume that the observation, X_i, came from a normal distribution with mean M and variance SD^2? If we can rework the given information into a question of this form, with numbers substituted for the symbols X_i, M, and SD^2, then this is an appropriate place for a normal test.

First we must find a normal distribution. None is mentioned in the problem. (We do not know how income is distributed in the city, but chances are that it is not normally distributed.) Knowing the central limit theorem, however, we can (right?) see a normal distribution just below the surface. For we are looking at a mean of a sample of size 90 and asking how it relates to the means of other samples of size 90 which might be taken from the city-wide income population. The sampling distribution of the mean for samples of size 90 is a normal distribution (see Chapter 14, Summary, point 3). Furthermore, using the central limit theorem we can determine the mean and variance of the sampling distribution of the mean. So this *is* an appropriate place for a normal test. Now we can go through the five steps in order.

A. *State a null hypothesis.* As usual, our null hypothesis, H_0, is a statement that there is *no difference*. H_0 : *The income of left-handed people is just the same as the income of right-handed people.* More specifically, the sample with mean, $M_{90} = \$9134$, observed among left-handed people, can plausibly be assumed to have come from a population like the general city-wide population with a mean of \$8500 and a variance of 3,600,000.
B. *Choose a significance level.* Let's assume that we are willing to risk a Type One error 1 time in 20: our significance level is 0.05.
C. *Determine a rejection region for our test statistic.* We know that when a number is taken at random from a normal distribution and a Z score computed, there is 1 chance in 20 of a Z score below -2 or above $+2$. So our rejection region is all values of Z smaller than -2 or larger than $+2$. (Note there are two intervals in the rejection region.)
D. *Compute the particular value of the test statistic, Z, which results from the data we have gathered.* This is the most complicated step; but if the null hypothesis was properly stated, this step should present no problem. We are going to compute a Z score: $Z = (X_i - M)/SD$. To do this we must determine the proper numbers for X_i, M, and SD. If the null hypothesis is true, then the observation X_i came from a *normal distribution* with mean M and standard deviation SD. Our observation, X_i, is 9134, the mean of a sample of 90 left-handed person's incomes. The normal distribution from which it hypothetically came is the sampling distribution of the mean, for samples of size 90 from the city-wide population specified in step A above. The mean of the sampling distribution, according to the central limit theorem, will be the same as the mean of the parent population, or 8500, so $M_M = 8500$. And the variance of the sampling distribution of the mean will be SD_{pop}^2/n, where SD_{pop}^2 is the variance of the parent population, or 3,600,000, and n is the sample size, or 90. Using these numbers we can find the variance of the sampling distribution of the mean:

$$\text{Variance of the sampling distribution of the mean} = SD_M^2 = \frac{SD_{pop}^2}{n} = \frac{3,600,000}{90} = 40,000$$

Taking the square root of the variance, we can get the standard deviation:

$$SD_M = \sqrt{SD_M^2} = \sqrt{40,000} = 200$$

This is the value for SD we will need to compute the Z statistic, since it is the standard deviation of the normal distribution from which our observation, 9134, is hypothesized to have come. $SD = 200$.

We can now compute the Z statistic:

$$Z = \frac{X_i - M_M}{SD_M} = \frac{9134 - 8500}{200} = \frac{634}{200} = 3.17$$

E. *Determine whether your computed test statistic lies in the rejection region specified in step C, and make the appropriate decision.* In step C we said we would reject the null hypothesis if we observed a Z score greater than $+2$ or less than -2. We observed 3.17, which is quite a bit greater than $+2$. Thus we reject the null hypothesis that the sample of left-handed people was taken from a population with a mean of 8500 and a variance of 3,600,000. We conclude that left-handed people have higher earning power than the general population, a conclusion which should be gratifying to the Sinistralics. In the jargon of statistical testing, we can say: We conclude that the observed sample mean is significantly different from the general population mean, at the 0.05 significance level; or in English: We can be very confident that left-handed people earn more than the general public in the town in question.

Figure 15.3 may help to review the logic of this test and clarify your thinking. You should make sketches like this whenever you do a Z test, to help avoid confusion. We started with a general population with a mean of 8500 and a variance of 3,600,000. Nothing was specified about the form of this distribution.

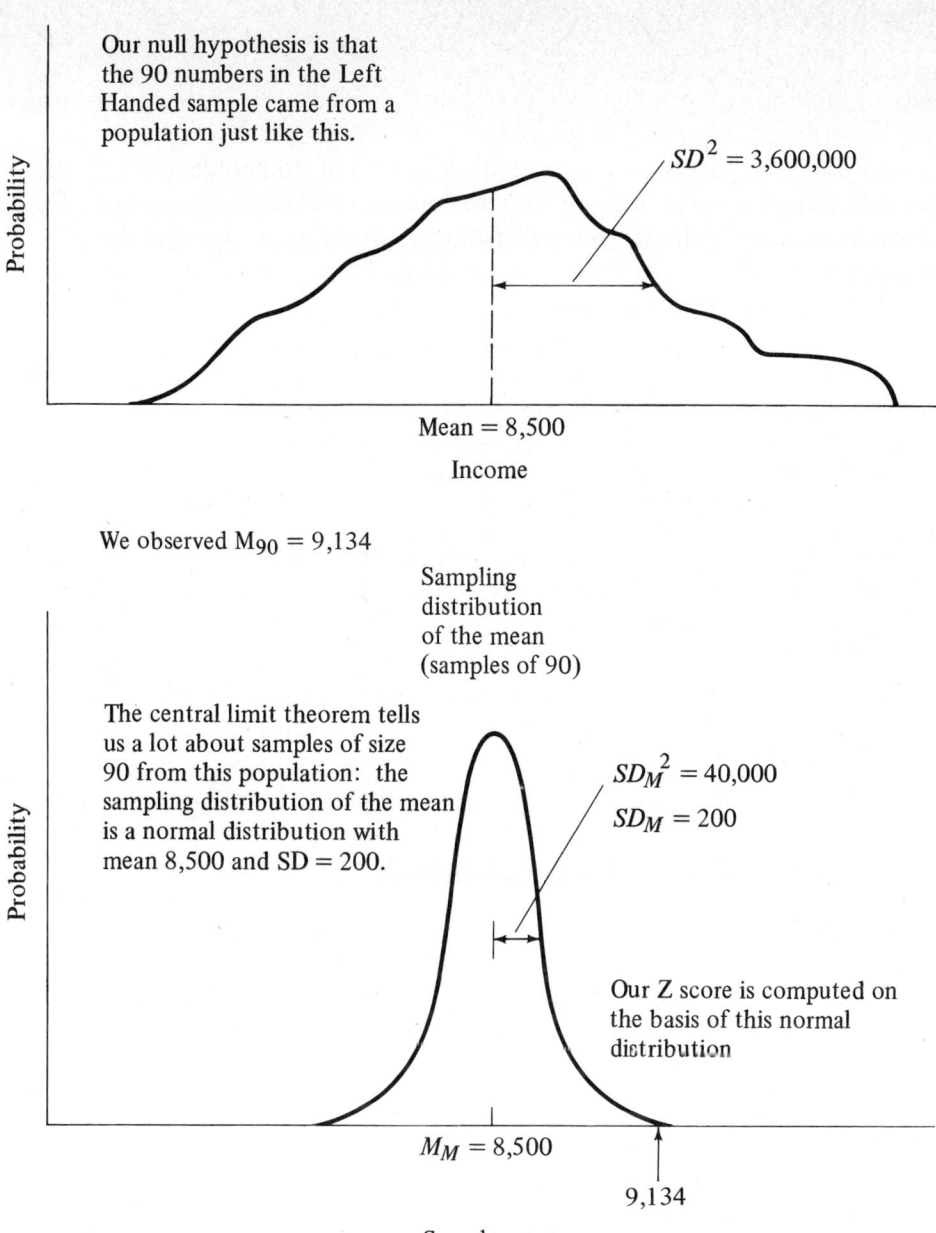

Figure 15.3 The parent population, a distribution of incomes, is sketched above. The derived sampling distribution of the mean is shown below. The Z score is computed on the basis of the mean and standard deviation of this derived distribution.

If the null hypothesis is true and left-handed people are no different, then we have a set of numbers just like any other sample of size 90 from the parent population in Figure 15.3. And our mean, $M_{90} = 9134$, is one observation taken at random from the sampling distribution at the bottom of Figure 15.3. We just happened to observe a very unlikely sample mean: 9134. This observation has a Z score within this distribution of 3.17, as computed in step D. It is not impossible that we should make such an observation; but it is very unlikely, and this leads us to conclude that left-handed people earn more than the average.

THE SAMPLING DISTRIBUTION OF THE Z STATISTIC

There is another way of looking at the Z test, which makes it more clearly similar to the sign test we used for the ESP experiment. Suppose, in the town of 10,000 people discussed in the previous sample problem, a statistician were to take a random sample of 90 people and find the average income for those 90. Suppose, next, that he were to compute a Z score, showing where that average income stood in relation to the sampling distribution of the mean for samples of size 90, just as we computed such a Z score in the sample problem. He would obtain a value, say 0.57, for that Z score. Now suppose that he took *another* such random sample of 90, found the mean income, and proceeded through all the same steps to get a new Z score. It might be -1.41. And now, assuming that he is a most patient and persistent statistician, suppose he continued to take sample after sample, finding a Z score for each, until he had 1000 such Z scores. How common would various values for the Z score be? In different words, what would he find to be the (experimentally derived) *sampling distribution for the Z statistic*, based on the mean of a large sample from a population with a known mean and variance?

It turns out that the collection of Z scores so generated would be normally distributed. The mean of their normal distribution is zero, and the variance and standard deviation are both equal to 1. This special normal distribution is discussed more fully in Chapter 17. It has a special name—the *unit normal distribution,* or the *Z distribution.* In this distribution the chance of a score greater than $+2$ or less than -2 is 0.05; 68 percent of all observations are between the values of -1 and $+1$. Knowing this normal distribution, we can interpret the resulting Z score. When we observe a very high score, such as the 3.17 in the previous example, we are justified in doubting that our observation really came from the specified sampling distribution. Therefore, we reject the null hypothesis.

The sampling experiment of our persistent statistician is exactly analogous to the sampling experiment with the marble box described in Part I of the book. It gives us information about how likely it is that we shall observe various values of the Z score *when the null hypothesis is true.* The statement that the statistician took random samples from the population before finding the sample mean and the Z statistic is the same as saying that the null hypothesis is true for a Z test based on that sample. The null hypothesis for the Z test is that the sample in question did come at random from the specified parent population. And that is the condition satisfied in the sampling experiment of the statistician.

In the ESP analysis we started with a null hypothesis: The subject was just guessing, and had a 50 : 50 chance of being correct on each guess. We then used the marble box to approximate the sampling distribution of the score (number correct: 0/10, 1/10, etc.). And once we know that sampling distribution, we know how likely it is that we shall see various scores when the null hypothesis is true. If we see a very extreme score, like 10/10, we realize that it is very improbable that we could see such a score in the sampling distribution. That is, it is very unlikely that we would see such a score when the null hypothesis is true. So we reject the null hypothesis. A very similar argument can be used here: In the hypothetical sampling experiment just described, which would produce an approximation to the unit normal distribution, we can obtain information about the probability of various values of the Z statistic when the null hypothesis is true. In that sampling distribution (the unit normal distribution) it is very unlikely that we will see a score of 3.17 or higher. So when we observe such a score, we decide that it very probably didn't come from the unit normal distribution. That is, it probably didn't come from the sampling distribution of the Z statistic, for H_0 true. On this basis, we conclude that the null hypothesis probably is not true, and we reject it. We shall see a very similar pattern in each of the other statistical tests we encounter in the coming pages.

Summary

The Z test, or normal test, is appropriate whenever we are interested in a question which can be reduced to the following form: Is it plausible to assume that the observation X_i came from a normal distribution with mean M and variance SD^2? Often the normal distribution in question is a sampling distribution of the mean for samples from a parent population with known mean and variance. (If the sample size is greater than 30, we do not need to know the shape of the parent population.) The observation X_i in such a case is the *mean* of a sample hypothetically taken from the parent population.

The Z test can be reduced to the following five steps, which are closely related to the six steps for the sign test:

A. State a null hypothesis: A sample is taken from a population with specified mean, M_{pop}, and variance, SD^2_{pop}.
B. Choose a significance level.
C. Determine the rejection region for the test statistic, Z.
D. Compute the particular value of the test statistic, Z, which results from the data gathered. This often involves finding the mean, M_M, and standard deviation, SD_M, of a sampling distribution of the mean for samples from the parent population.
E. Determine whether your computed test statistic lies in the rejection region specified in step C, and make the appropriate decision.

Problems

1. Suppose that after completing the analysis described in the early part of this chapter, I continue to buy Texaco, and after a while I've made 20 measurements of mpg (bought 20 tanks of gas). I have found an average, $M_{20} = 25.2$ mpg. Analyze these results with a Z test, going through the five steps we have outlined. Draw a sketch of the sampling distribution at step D. Use a significance level, α, of 0.05. Do you now have grounds for concluding that there is a significant difference between the two kinds of gasoline? Show all your steps.

2. Loaves of bread are supposed to weigh 454 grams (1 pound) each. I buy a case of 20 and find an average weight per loaf of 431 grams. The company claims that this is a random fluctuation and says that the actual weight of loaves is normally distributed with mean, M, 454 grams, and variance, SD^2, 80. Do you think they are telling the truth? Check this by carrying out a normal test (Z test), using a significance level of 0.01. (*Note*: There is 1 chance in 100 of seeing a Z score greater than +2.58 or smaller than −2.58.) Given the decision you have made, have you accepted a risk of a Type One error or a risk of a Type Two error? Do you think there is much risk?

3. a. In the mileage example at the start of this chapter, we decided not to reject the null hypothesis: we decided we had no conclusive evidence that Texaco gas was different from Shell gas. What kind of error (Type One or Type Two) did we risk by making that decision?

   b. In the left-handed people's income example, we decided that left-handed people earn more than the general population. What kind of error (Type One or Type Two) did we risk by making that decision?

4. Look at the sketch in Figure 15.2. What is the total area (in square units) under the curve? How much of this area is above the two shaded sections? How do you know?

5. In the left-handed people's income example it is mentioned that the population of the city is 10,000. What is the relevance of this fact to the problem?

6. What are the five steps of a Z test?

†7. Are Z scores always normally distributed? If not, when *are* they so distributed?

The Z Test for a Single Observation; The Binomial Distribution

EXAMPLE OF A Z TEST FOR A SINGLE OBSERVATION

Chlorine at the Waterworks

Suppose you are night watchman at the waterworks. You take measurements of the chlorine content of the water once each hour. You know that when all the machinery is functioning properly, the concentration of chlorine is normally distributed, with mean $M = 1400$ ppm (parts per million), and standard deviation of 90. At 4:00 AM, you take a measurement and observe 1597 ppm.

Question: Do you push the panic button? We know the probability distribution of observations when the machinery is working perfectly. It is a normal distribution which looks like Figure 16.1. The question at hand is whether the observation 1597 is so extreme that we are willing to conclude, No, I don't believe that observation came from that distribution.

A simple next step is to compute the Z score corresponding to the observation $X_i = 1597$. You should by now have memorized the definition of a Z score:

$$Z = \frac{X_i - M}{SD} = \frac{1597 - 1400}{90} = \frac{197}{90} \cong 2.2$$

The Z score corresponding to the observed number is 2.2. So what should we decide? Recall (the rule of thumb) that only 5 percent of all Z scores are more than 2 standard deviations away from the mean, in either direction. So we know that the chance of seeing a number like this is less than 5 in 100, or 1 in 20. Under these conditions we probably would decide that this observation was suspicious, and begin a complete check of the chlorination machinery to try to find the problem.

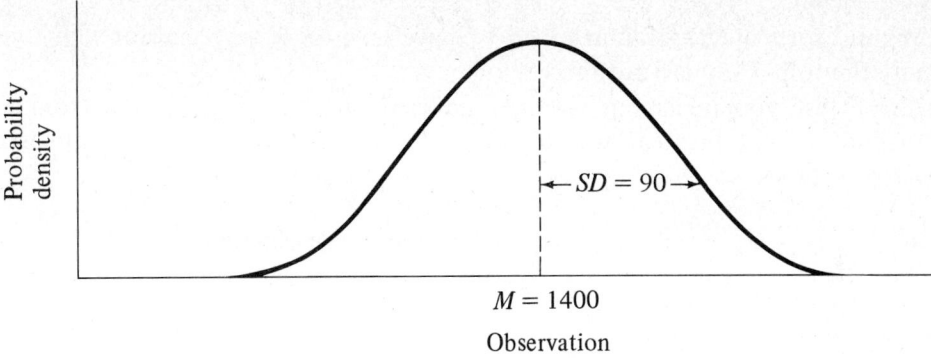

Figure 16.1 The distribution of chlorine-concentration observations when all is well.

A formal, five-step Z test: The logic in these steps is the same as the reasoning we just completed, but the individual pieces are more explicit.

Step A: State the null hypothesis. Stated verbally, the null hypothesis is: Nothing is wrong; the machinery is working perfectly. Stated mathematically, that is the same as saying: The observation 1597 was taken from the normal distribution with $M = 1400$ and $SD = 90$.

Step B: Choose a significance level. Decide what chance of a Type One error you are willing to accept. Let's say $\alpha = 0.05$; but we'll want to return to this question later. It may be that it's so important to detect problems before it's too late that we'll accept a *larger* risk of a Type One error, or false alarm.

Step C: Determine the rejection region for the test statistic, Z. This is fixed by the significance level. For $\alpha = 0.05$, the rejection region, a pair of intervals, contains all values of Z less than -2 or greater than $+2$.

Step D: Compute the particular value of the test statistic, Z, which results from the data we've gathered. Since we're working directly with a single observation ($X_i = 1597$) and asking whether it came from a specified normal distribution with $M = 1400$ and $SD = 90$, we can proceed immediately to the computation:

$$Z = \frac{X_i - M}{SD} = \frac{1597 - 1400}{90} = \frac{197}{90} \cong 2.2$$

Step E: Determine whether the computed test statistic lies in the rejection region, and make a decision. It does, so we reject the null hypothesis and decide that something is disturbed. We decide that something is wrong with the machinery. It's time for a thorough check of everything.

Discussion of the two types of error: In this situation two types of mistake are possible: one could push the panic button when in fact the waterworks was functioning normally (Type One error), or one could fail to push the button when in fact something was wrong (Type Two error). The costs of the two kinds of error are very different here. Pushing the panic button too soon may involve only some simple diagnostic procedures to check the chlorination apparatus. But pushing it too late could result in all kinds of mean, ugly, nasty, poisonous water being sent out to peoples homes. Under these conditions, one might prefer to be "trigger happy" and adopt a 0.10 significance level. This would result in more Type One errors, but it would reduce the chance of missing a reading which really meant something was wrong.

The point is that there is a trade-off between the two types of error. If you never reject the null hypothesis, you will never make a Type One error. But obviously you'll make lots of Type Two errors. On the other hand, if you reject the null hypothesis regardless of the data, you will never make a Type Two error. You will never miss a situation in which the machinery is broken, but you will spend a formidable amount of time checking things. The probability of Type One error (saying something is wrong when nothing is wrong) will be 1!

Two Situations for the Z Test

There are two different sorts of situation in which we have seen the Z test, and for which you should be able to apply the test. They are outlined below.

Whenever you run a Z test, you are asking whether one particular number, X_i, came from a particular normal distribution, with mean M and variance SD^2. We have seen two different sorts of situations, however, as follows:

Situation I, Sample Mean

Often the observation in question is a sample mean, and the normal distribution in question is the sampling distribution of the mean for samples of that size. In this case we have to figure out the standard deviation and the mean of that normal distribution on the basis of the parameters of the population from which the sample was drawn. (Examples include the gas mileage example in Chapter 15, the GPA example in Chapter 14, the left-handed income example in Chapter 15.)

Situation II, Single Observation

In the case just presented we started with a single observation and a well-specified normal distribution. This is the simplest kind of Z test.

The distinction between these two situations is important. Be sure you understand it well.

THE BINOMIAL DISTRIBUTION

Suppose that a friend with whom you had discussed ESP performed a 20-guess experiment, like the one we analyzed in Part I of this book. Having observed 15/20 correct guesses, he came to you, a budding statistician, to ask for advice: Was his evidence convincing enough for an effect?

Conceptually, this poses no problem. After all the sign tests you have carried out, it should not be difficult. You list the 21 possible events: 0/20, 1/20, etc., up to 20/20 correct. Having stated a null hypothesis, you start to compute the probability of each event when H_0 is true and the subject is guessing randomly. There are 2^{20} possible outcomes. That's a bit frightening, but works out to be 1,048,576 outcomes. (It's a good thing he didn't ask you to draw a probability tree.) Four nasty hours later you have done the necessary calculations and have the following table:

Event (number correct)	Probability of event (to 4 decimal places)
0	.0000
1	.0000
2	.0000
3	.0011
4	.0046
5	.0148
6	.0370
7	.0739
8	.1201
9	.1602
10	.1762

Event (number correct)	Probability of event (to 4 decimal places)
11	.1602
12	.1201
13	.0739
14	.0370
15	.0148
16	.0046
17	.0011
18	.0000
19	.0000
20	.0000

This is all hypothetical. But of course you *could* do it if you really wanted to. Once you know the numbers, it is easy to make a graph of the probability distribution (see Figure 16.2).

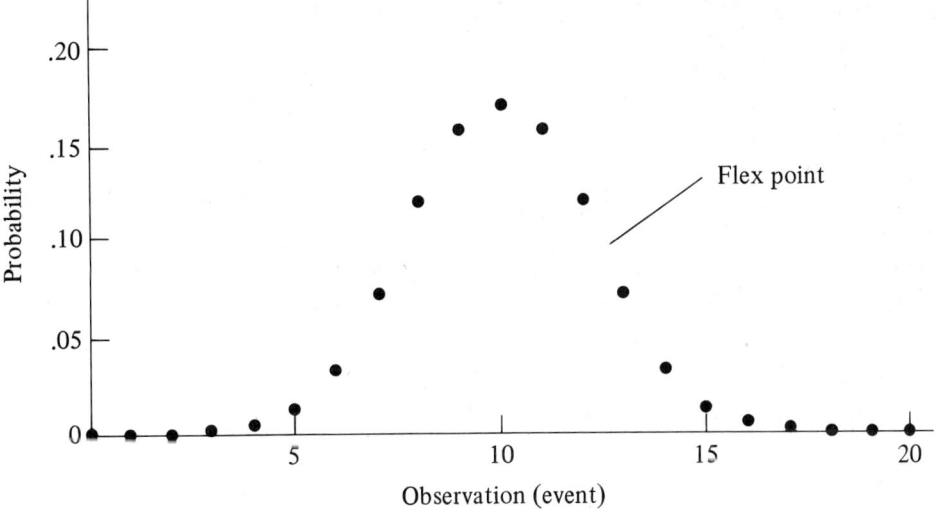

Figure 16.2 Does this probability distribution remind you of anything?

Does Figure 16.2 remind you of anything? Have you ever seen another curve with a similar shape? Plainly, it is reminiscent of the normal distribution. And, as will become very clear in the following chapter, that similarity can be extremely useful. If you had known about it earlier, it might have saved you the four hours of calculation.

Finding the Most Similar Normal Distribution

Suppose you want to find the normal distribution that most closely matches the above probability distribution. How would you proceed? Recall that if you know the mean and the variance of a normal distribution, you have completely specified that distribution. A reasonable guess would be to find the mean and variance of the above sign-test distribution and use those values for the normal distribution. But how do you find the mean and variance of the sign-test distribution? Try to answer this question before continuing.

The mean is easy. On the average, a person guessing blindly for 20 envelopes should get 10 correct. And by looking at Figure 16.2 you can confirm that the mean of this distribution looks like 10. The variance is more of a problem. Can you figure out how you could do a coin-toss or marble-box experiment to estimate the variance of this distribution? That is, can you estimate the variance in the population of all the possible scores (number correct in 20 guesses) that a person without ESP would get in an infinite series of repetitions of the experiment?

You might toss 20 coins, and record the number of heads; and do it again; and again; for, say, 500 times. You would then have a list of 500 numbers, which might start out something like this: 11, 14, 8, 10, 9, 9, etc. Then, using the familiar formula for estimating the variance of a population on the basis of a sample, you could find the mean of your sample, M_{samp}, and the estimated variance of the population, Est. SD^2_{pop}.

By looking at the probability distribution in Figure 16.2 again, there's a visual trick you can use to check your variance estimate. It just happens that the point on a normal distribution that is exactly 1 standard deviation from the mean can be identified with your eye: It is the *flex point* in the curve, that is, the point at which it stops curving down and starts curving out towards the tails. The flex point in the dotted curve of Figure 16.2 is indicated by a pointer. If this were a normal distribution, then the observation directly below the flex point (and below the end of the pointer) would be just 1 standard deviation greater than the mean. From Figure 16.2 you can see that the flex point is about halfway between 10 and 15, at roughly 12½. Since the mean is at 10, this suggests that the standard deviation of this distribution is about 12½ − 10 = 2½. (Some further examples of flex points are given in the Question and Answer section at the end of this chapter.)

Finally, if you want to determine the mean and variance of the above sign-test distribution, in order to identify the most similar normal distribution, you can use a very simple mathematical rule that gives the exact mean and variance for any sign-test distribution:

Where both alternatives are equally likely (for example, heads and tails or correct and wrong), the mean of the distribution will be N/2, where N is the number of trials. And for any such distribution the variance is exactly equal to N/4.

In this case, where $N = 20$, we guessed correctly that the mean is 20/2, or 10. In this case, with $N = 20$, the variance is equal to 20/4, which is 5. Since the square root of 5 is about 2.24, we were not far off when we guessed that the standard deviation of the distribution in Figure 16.2 was 2.5.

In the next chapter you will see how helpful it is to use these two simple rules to find the mean and variance of a sign-test distribution and to study the closely similar normal distribution with the same mean and variance. For the moment, suppose that you had known these two rules when your friend asked you to interpret his result of 15/20 correct guesses. You might have reasoned that when the null hypothesis is true, the distribution of scores (shown in Figure 16.2) has a mean of $N/2$, or 10, and a variance of $N/4$, or 5. And you might have gone on to compute a Z score based on your friend's observation of 15 correct. Since $X_i = 15$, $M = 10$, and $SD = \sqrt{5}$, or about 2.24, it follows that the Z score corresponding to your friend's result is:

$$Z = \frac{X_i - M}{SD} = \frac{15 - 10}{2.24} \cong 2.2$$

Knowing that such large Z scores happen less than 1 time in 20 in any normal distribution, you might have concluded that your friend had an interesting result. And, as you will see in the next chapter, you would have been correct. The normal distribution is sufficiently similar to the sign-test distribution to permit such reasoning.

Another Name for the Sign-Test Distribution

From the early pages of this book we have been looking at probability distributions that result (1) when coins are tossed repeatedly, (2) when persons with no special powers make guesses in a two alternative ESP experiment, or (3) when holes in a paddle are filled with marbles that are either red or green. In every case, we were interested in the score that results when a simple two-alternative process is repeated several times. The probability distribution that results formed the basis for our decision in the sign test, and we have called this the sign-test distribution. Beyond the confines of this book, however, this is usually known as an example of a *binomial distribution*. (The name binomial can be broken down into *bi-* meaning *two,* and *nomial,* meaning *name;* the two-name distribution: heads and tails, red and green, correct and wrong, plus and minus, etc.) Since you already know a great deal about one particular kind of binomial distribution, it is nice to know its name.

Recall that we have always been interested in studying null hypotheses of the following form: *For* each _____ *there's a 50 : 50 chance that one alternative will be observed, and an equal chance that the other alternative will be observed.* We have always started by assuming that heads and tails were equally likely or that there were an equal number of red and green marbles or that the subject was as likely to guess right as to guess wrong. We might have looked at some asymmetric situations. For example, if each card in the ESP experiment could have any one of three letters, X, O, or T, the subject would only have 1 chance in 3 of guessing correctly on each card. Nothing you have learned so far, except for a revised marble box, would help you find the probabilities of each event in an asymmetric situation such as this. But the resulting probability distribution is still called a binomial distribution. To keep things clear, the special case where each of two alternatives is equally likely is known as the *symmetric binomial distribution.* In this book it is the only one we will use.

Why the Binomial Distribution Looks Like a Normal Distribution

Suppose you write the number 1 on each of 500 red poker chips and the number 0 on another batch of 500 green chips, and you put them all in a big box. Then you draw a sample of 20 numbers, which might look like this: 1, 1, 0, 1, 0, 0, 0, 1, 0, 1, 0, 1, 1, 1, 1, 0, 1, 0, 0, 1. It takes only a few seconds to find the mean of this sample: 11/20. Indeed, if you had some time, you could repeat this process a few hundred times to approximate the sampling distribution of the mean for this population and for sample size 20. Your observed values might include 0/20, 1/20, 2/20, and so on, up to 20/20. And if you continued for a long enough time, your resulting relative frequency distribution would look very much like the probability distribution in Figure 16.2.

So it's not really surprising that binomial distributions often resemble normal distributions. If you had not numbered the chips, but had simply put 500 red and 500 green chips in the box, you could have recorded the score or event which happened for each sample of 20: 0/20 green, 1/20 green, etc., up to 20/20 green. And your frequency distribution for each of these 21 events would look exactly like the sampling distribution of the mean, which showed the chance of each value of the sample mean: 0/20, 1/20, etc. You learned several chapters ago that the sampling distribution of the mean is a normal distribution for large samples. The fact that binomial distributions resemble the normal distribution is just another example of the general result. And, as before, we need to have fairly large samples. The normal distribution is not very similar to the five-coin-toss binomial distribution.

The point is that there is an exact correspondence between any binomial experiment and a similar sampling experiment where you sample the numbers one and zero. Observing 13 heads in

20 tosses is precisely like observing the number one 13 times (and zero 7 times) in a sample of 20 numbers. And the score 13 heads out of 20 corresponds precisely to the sample mean, $\Sigma X_i/n = 13/20$, based on your sample of 20 numbers. Since the binomial distribution will then have exactly the same form as a sampling distribution of the mean for your specified sample size, it is inevitable that the binomial distribution for large samples, like the sampling distribution of the mean, should be very close to a normal distribution.

I hope you can feel a twinge of intellectual wonder to learn how the similarity of the binomial (sign-test) distribution to the normal distribution can be explained as a special application of the central limit theorem. For some fortunate people this sort of realization can be intensely satisfying.

QUESTION AND ANSWER

Question: You talk about the symmetric binomial, when each of the two alternatives is equally likely. What is the asymmetric binomial?

Answer: Suppose you have a loaded dime, which comes up heads 70 percent of the time and tails 30 percent of the time. If you flip it ten times, and write down the number of heads, you will observe one of eleven events: 0/10, 1/10, etc., up to 10/10. If you repeat the experiment over and over, you will generate a frequency distribution with a maximum at 7/10, and a long tail to the left, as in Figure 16.3. This is an example of the binomial distribution with ten trials and probability $p = 0.70$.

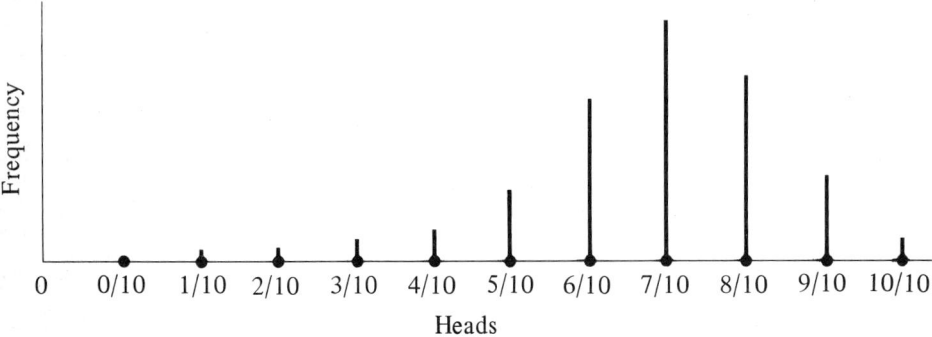

Figure 16.3 The binomial distribution for a ten-trial experiment where the probability of heads on each trial is .70 ($p = .70$).

(*Note:* For *any* distribution with n trials and probability p, the mean of the distribution is given by $M = Np$, and the variance is given by $SD^2 = Np(1 - p)$. In the case of the symmetric binomial, with $p = 1/2$, these reduce to the familiar $M = N/2$ and $SD^2 = N/4$.)

Question: On page 132 you mentioned the flex point in the normal distribution. That's a new term for me. Could you show some other examples?

Answer: Sure. A flex point on a curve is a point where it stops curving in one direction, and starts curving in another. In Figure 16.4 every flex point is indicated by an arrow. A normal distribution contains one flex point on each side of the mean. And it is sometimes useful to know that the flex point occurs exactly 1 standard deviation away from the mean. A clear illustration of this will be found in Figure 17.1.

Summary

A specially simple sort of Z test was introduced: the Z test for a single observation from a known normal distribution. Sometimes we are interested in deciding directly whether a single observation could plausibly have

Figure 16.4 Every *flex point* is marked with an arrow.

come from a specified normal distribution with known mean and variance. In this case, the Z test is very simple, and no preparatory calculations are needed in step D.

A *binomial distribution* is a probability distribution giving the probability of each of $n + 1$ possible events, when a simple dichotomous experiment is carried out n times. For example, in the ESP experiment we found a binomial distribution which told us the probabilities of eleven $(10 + 1)$ events when a person guessed correctly or incorrectly on each of ten successive envelopes. When each of the two outcomes (for example, correct or wrong) is equally likely, we call the distribution the *symmetric binomial distribution*.

Problems

1. Using the rules just presented for the binomial distribution, together with the rule of thumb, find out the following:

    a. Suppose in the families of several skin divers there are 64 children. Find two numbers which are appropriate to fill in the blanks in the following sentences:

    If for each child that a skin diver has, there is a 0.50 chance that the child will be a girl, then out of 64 children there is a 95 percent probability that the number of girls will be greater than _____ but less than _____.

    b. Work the same problem for a much larger group, containing 6400 children:

If boys and girls are equally likely, there is a 95 percent probability that the number of girls will be greater than _____ but less than _____.

2. Life expectancy for people who have reached the age of 30 in one particular country is normally distributed with a mean of 71 years and a standard deviation of 9 years. A group of 26 garbage collectors is studied, and on the average they live to the age of 75.3 years. Should you conclude that garbage collectors live significantly longer than the general populace? Do a Z test, showing all the steps. Use a significance level of 0.05.

3. Medical survey teams have determined that the fructose metabolism coefficient for healthy humans is normally distributed with mean 32.6 and standard deviation 0.8. For diabetics it is much less consistent. During a routine physical examination, Alex Williams is found to have a coefficient of 34.1. Run a Z test to see if this is significantly different from what one would expect if Williams were healthy. Use a 0.05 significance level. Show all five steps. Discuss your decision.

†4. During World War II a French statistician kept data on the weight of the bread he bought. He found that it averaged 950 grams, instead of 1 kilogram, as it should. He complained to the baker, who said he would fix his scale. In the subsequent months the statistician continued to keep track of the weight of his bread, and observed the distribution seen below.

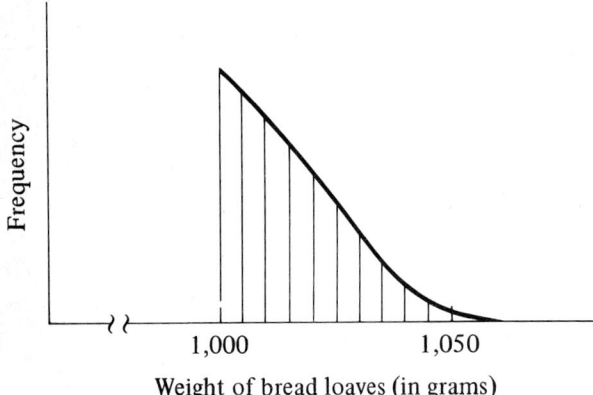

He then called the police. Why?[1]

---

[1] I am indebted to Professor Stephen Simons for suggesting this problem.

The Normal Approximation to the Binomial; The Normal Distribution Table

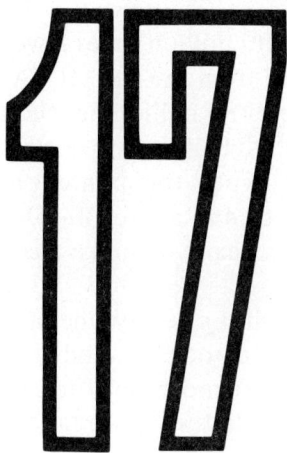

As was indicated in Chapter 16, there is a strong similarity between the shape of the sign-test probability distribution, which you now know how to compute, and the shape of the normal distribution. In order to choose the particular normal distribution which best fits a given binomial distribution, we choose one with the same mean and variance. Suppose you are interested in a sign-test situation with N trials. The marble box situation represented a case with ten trials: $N = 10$. In the self-test in Chapter 12 you saw a case with nine different scores, each of which had a plus or a minus; there, $N = 9$. Regardless of the value of N, it turns out that the *mean* of the symmetric binomial distribution is given by $M = N/2$, and the variance of the symmetric binomial distribution is given by $SD^2 = N/4$. In order to make these assertions more plausible, let's work through the simplest possible specific case.

THE MEAN AND VARIANCE FOR THE ONE-TRIAL SYMMETRIC BINOMIAL

Consider a one-trial coin tossing experiment. There are two possible outcomes, H and T. In terms of the *events* we have looked at in the past, we can describe these as *one head* and *zero heads*. Flipping a coin, then, is equivalent to putting two numbers (1 and 0) in a hat, and drawing one at random.

Now we can find the mean and the variance of this simple population containing the elements (1, 0).

First, $M = (1 + 0)/2 = 1/2$. This agrees with the formula $M = N/2$, since here $N = 1$.

SD^2 = Average squared deviation from the mean

$$= \frac{(1 - M)^2 + (0 - M)^2}{2} = \frac{(1 - 1/2)^2 + (0 - 1/2)^2}{2} = \frac{1/4 + 1/4}{2} = \frac{1}{4}$$

This is just what we'd get using the formula $SD^2 = N/4$. In this particular case with one trial, $N = 1$, so $SD^2 = N/4 = 1/4$.

In the problems for this chapter you will have a chance to see that the formulas work for the cases where $N = 2$, and $N = 3$, too.

EXAMPLE OF THE USEFULNESS OF THE NORMAL DISTRIBUTION AS AN APPROXIMATION TO THE SYMMETRIC BINOMIAL (SIGN-TEST) DISTRIBUTION

Suppose that in a national election 80 million votes have been cast. While they are being counted you take a true random sample from the full set. (A true random sample is one in which every single member of the population of 80 million has the same chance of being selected as any other.) The random sample contains 1000 ballots, of which 922 are either Republican (R) or Democratic (D). The other 78 are either votes for splinter parties, for no candidate, or improperly marked. Of these 922 ballots, 497 are for the Democrat and 425 are for the Republican. On the basis of these results, can you make a confident prediction about the outcome of the election?

This is a place for a sign test. The ballot of every one of the 922 voters may be categorized into one of two groups: Republican or Democratic. And we are interested in knowing if there is a statistically significant predominance of Democratic votes. Put differently, we are interested in whether it would be plausible to assume that when we take a voter at random there is exactly a 50 : 50 chance that he will be a Democrat and a 50 : 50 chance that he will be a Republican.[1] If it is implausible to assume that R and D are equally likely, then we can conclude that the chance of a person voting for the Democrats is greater than the chance of his voting for the Republicans. You are asking whether there is any reasonable chance of observing a result of 497 Democratic votes and 425 Republican votes, if the probability of each is 0.5. This is like asking: If you tossed a fair coin 922 times, is it at all likely that you would observe 497 heads and 425 tails? In principle, you could work through this problem using the standard sign-test procedure, testing the null hypothesis that the chance of a vote being Republican is equal to the probability of it being Democratic, or one half. (*Note:* We consider only those people who voted either R or D.) But it would take you a year and a day. We shall take advantage of the fact that there is a strong similarity between the binomial distribution for 922 trials and the normal distribution with a mean of 922/2 and a variance of 922/4 ($N/2$ and $N/4$, respectively). Now let's do the arithmetic.

For the binomial distribution with both alternatives equally likely and 922 observations:

$$M = \frac{N}{2} = \frac{922}{2} = 461$$

$$SD^2 = \frac{922}{4} = 230.5$$

$$SD = \sqrt{SD^2} \cong 15 \quad \text{(since } 15 \times 15 = 225\text{)}[2]$$

Now if the results of the sampling experiment, assuming that the chance of a Republican vote is the same as the chance of a Democratic vote, were normally distributed, with a mean of 461 and SD of 15, we know that 95 percent of the possible results would lie within 2 standard deviations

---

[1] Remember that we have limited our analysis to voters who vote for one of the two major parties. A somewhat more precise null hypothesis is: H_0: For every voter sampled, who is going to vote for one of the two major parties, there is an even chance that he will vote for the Democrats and an identical 50 : 50 chance he will vote for the Republicans.

[2] More accurately, as may be seen from the table of squares, square roots, and reciprocals in the appendix, $\sqrt{230} = 15.166$ and $\sqrt{231} = 15.199$. So $\sqrt{230.5} \cong 15.2$, to the nearest tenth.

of the mean, or within 30 of the mean of 461. This means that 95 percent of all results would lie between 431 and 491. Thus if the results were greater than 491 or less than 431, we would reject the null hypothesis that both parties have the same strength. (This involves an assumption of a significance level, α, of 0.05.)

Now because of the very close similarity between the binomial distribution and normal distribution, the chance of a binomial result of 492, 493, 494, etc., up to 922, or 430, 429, 428, etc., down to 0, is just about 0.05, too. (These are the numbers more extreme than 2 standard deviations from the mean of the binomial.)

We can now reach a conclusion: Since the observed count of 497 votes out of 922 for the Democrats is greater than the critical value of 491, we decide to reject the null hypothesis that each party has equal strength and to predict that the Democrats have greater strength and will win. If in fact the two parties were exactly equal in strength, there would be less than 1 chance in 20 that we would see so extreme a result in either direction.

What we have just done is logically a sign test: We tested a null hypothesis that each of two alternatives (Republican and Democratic) is equally likely. But we could not use the procedure introduced earlier in this book for calculating the probabilities of the 923 different possible events: 0/922, 1/922, 2/922, etc., 921/922, 922/922. That would have taken months. Instead, we took advantage of the fact there is a strong similarity between the sign-test probability distribution (symmetric binomial distribution) and a normal distribution with the same mean and variance. We are able to quickly find the mean and variance of the symmetric binomial distribution using the formulas $M = N/2$ and $SD^2 = N/4$.

In terms of the five-steps outlined for the Z test, we can summarize what we just did as follows:

Step A: State a null hypothesis. H_0 : For every person who votes for one of the two major parties, there is a 50 : 50 chance that he will vote Democratic, and a 50 : 50 chance that he will vote Republican. (Note that this is precisely the sort of null hypothesis we are used to seeing in a sign test.)

Step B: Choose a significance level. We decided $\alpha = 0.05$.

Step C: Determine the rejection region for the test statistic, Z. With a 0.05 significance level, we reject the null hypothesis whenever we observe a Z score smaller than -2 or larger than $+2$. (As usual, the rejection region is a *pair* of intervals.)

Step D: Compute the particular value of the test statistic, Z, which results from the data we've gathered. We observed 497 Democratic votes out of 922 votes for the two major parties. If the null hypothesis is true, the events "number of Democratic votes out of 922" will be distributed according to a binomial distribution with both alternatives equally likely. This distribution is very similar to a normal distribution with the same mean and variance. The mean of the binomial distribution is $N/2 = 461$. The variance is $N/4 = 230.5$.

We would like to know: What is the chance of seeing an observation as extreme as 497/922 in the *binomial* distribution with mean 461 and variance 230.5? Since the binomial distribution is very similar to the normal distribution, this question will have almost exactly the same answer as the following one: What is the chance of seeing an observation as extreme as 497 in a normal distribution with a mean of 461 and a variance of 230.5? It is this latter question that we can answer by computing a Z score. The Z score corresponding to our observation of 497 ($X_i = 497$), in a *normal* distribution with mean $M = 461$ and standard deviation $SD = \sqrt{230.5} \cong 15.2$, is given by

$$Z = \frac{X_i - M}{SD} = \frac{497 - 461}{15.2} = \frac{36}{15.2} = 2.4$$

This is the value of our test statistic.

Step E: Determine whether the computed test statistic lies in the rejection region. Since +2.4 is indeed greater than +2, we conclude that it is unlikely that we would have observed as many as 497 Democratic votes out of 922, if the two alternatives were really equally likely. We reject the null hypothesis and conclude that people are more likely to vote Democratic than Republican, and that the Democrats will very probably win.

USING THE NORMAL DISTRIBUTION AS AN APPROXIMATION TO THE BINOMIAL: A SECOND EXAMPLE

In the last chapter we looked at the exact probabilities for each of the 21 possible events in a 20-trial coin-tossing experiment, computed by the standard sign-test procedure. They are displayed again here for reference.

Event	Probability of event (to four decimal places)
0/20	0.0000
1/20	0.0000
2/20	0.0000
3/20	0.0011
4/20	0.0046
5/20	0.0148
6/20	0.0370
7/20	0.0739
8/20	0.1201
9/20	0.1602
10/20	0.1762
11/20	0.1602
12/20	0.1201
13/20	0.0739
14/20	0.0370
15/20	0.0148
16/20	0.0046
17/20	0.0011
18/20	0.0000
19/20	0.0000
20/20	0.0000

These are the exact probabilities. This is the binomial distribution for 20 trials, with the parameter, Probability of heads = 1/2, for each trial.

Now suppose, in running a sign test with a 0.05 significance level, you wanted to determine an appropriate rejection region. This would include the following events: 0, 1, 2, 3, 4, 5, 15, 16, 17, 18, 19, and 20. From the table, we can see that the combined probability of these 12 events is 0 + 0 + 0 + 0.0011 + 0.0046 + 0.0148 + 0.0148 + 0.0046 + 0.0011 + 0 + 0 + 0 = 0.410. So this is the rejection region we would obtain using the exact probabilities for the binomial.

Now, suppose that we didn't have time to compute these probabilities, and we wished to use the normal distribution as an approximation to the binomial. In this case there are 20 trials, so $N = 20$. The mean of the symmetric binomial is $N/2$ or $20/2$ or 10. The variance is $N/4$ or $20/4$ or 5. So the $SD = \sqrt{5} \cong 2.2$. The normal distribution with the same mean and variance will closely resemble the binomial. In the normal distribution, there is a 95 percent chance that the results will lie within 2 standard deviations of the mean, or within 2 times 2.2 of 10, or between 5.6 and 14.4. In the binomial distribution, only exact numbers are possible. The exact numbers smaller than 5.6 or greater than 14.4 are 0, 1, 2, 3, 4, 5, 15, 16, 17, 18, 19, and 20. These, then, are our rejection region, using the normal approximation. As you can see, they are precisely the same events we decided upon using the exact possibilities.

The normal distribution is a good enough approximation to the binomial distribution to be very useful.

THE NORMAL DISTRIBUTION, A TABLE OF AREAS FROM THAT DISTRIBUTION, AND SOME DEDUCTIONS FROM THAT TABLE

Although there has been extended discussion of the normal distribution in the past chapters, you have needed to use only one important numerical fact about that distribution: that 95 percent of the area of any normal distribution lies within 2 standard deviations of the mean. It is often useful to know a bit more than that, and the following table and graph are sources of most of the information you are ever likely to need.

The Unit Normal Distribution

Suppose you start with any normal distribution whatsoever, and sample numbers from it, over and over. Suppose further that you compute the Z score for each observation, X_i, using the familiar formula $Z = (X_i - M)/SD$. The Z scores which result will be normally distributed with mean $M = 0$ and standard deviation $SD = 1$. This particular normal distribution, with $M = 0$ and $SD = 1$, is known as the *unit normal distribution*. Any question about any other normal distribution can be reformulated into a question about the unit normal distribution. And it is the unit normal distribution (or Z distribution) which usually appears in tables. A picture of the unit normal distribution is presented in Figure 17.1. This has been drawn so that the total area under the curve is 100 squares. Each square thus represents a probability of 1/100, or 0.01. By using this picture, you can find the probability of any interval of interest.

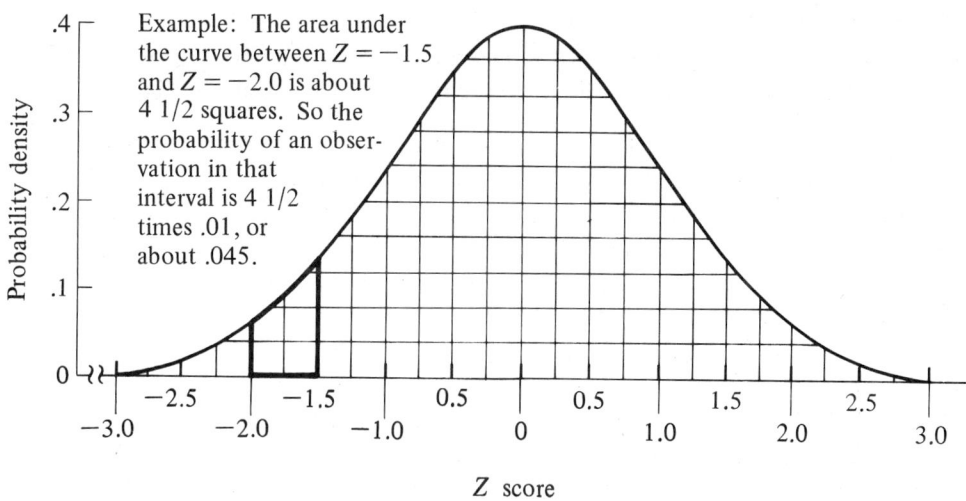

Figure 17.1   The total area under the curve corresponds to a probability of 1.00. This area is 100 squares. Thus, each square corresponds to a probability of .01.

Suppose you want to know the probability of seeing a Z score between 0 and +1 (see Figure 17.2). By counting the squares, you can quickly see that there are about 34 squares under the curve and over the interval between $Z = 0$ and $Z = 1$. Since each square corresponds to a probability of 0.01, the probability of a Z score between 0 and 1 must be about 0.34. This agrees with a statement made earlier that about 68 percent of all observations in a normal distribution are within 1 standard deviation of the mean. The probability of an observation between $Z = 0$

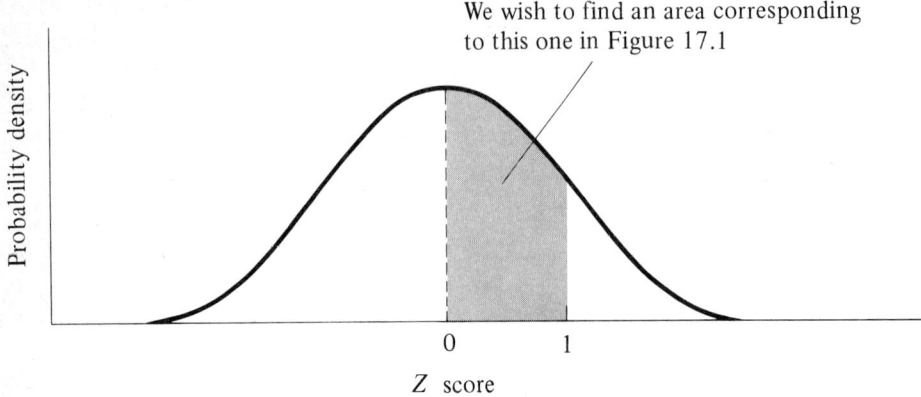

Figure 17.2

and $Z = 1$ is 0.34. Since the curve is symmetric (and you can check by counting squares), the probability of an observation between $Z = -1$ and $Z = 0$ is also about 0.34. The total probability of an observation within 1 standard deviation of the mean, then, is $0.34 + 0.34 = 0.68$, which is just about the same as $2/3 = 0.67$. You will have further opportunities to use this graph in the problems.

Counting squares is rather inexact, because there are so many fractional squares, and you have to estimate their contribution to the total area. To make things easier, someone else has already done the square counting for most intervals of interest, and summarized the results in a table. For any particular Z score of interest, this table gives the probability of seeing an observation between zero and +Z: For a Z score of 1, the corresponding probability is 0.3413 (see Figure 17.3). This agrees with the estimate, 0.34, which we made by counting squares.

Using the table to check the rule of thumb   Now suppose that we want to check the rule of thumb and find the probability of a Z score greater than +2 or less than −2. How do we proceed, using the table in Figure 17.3? We want to find the areas of the two shaded sections in Figure 17.4.

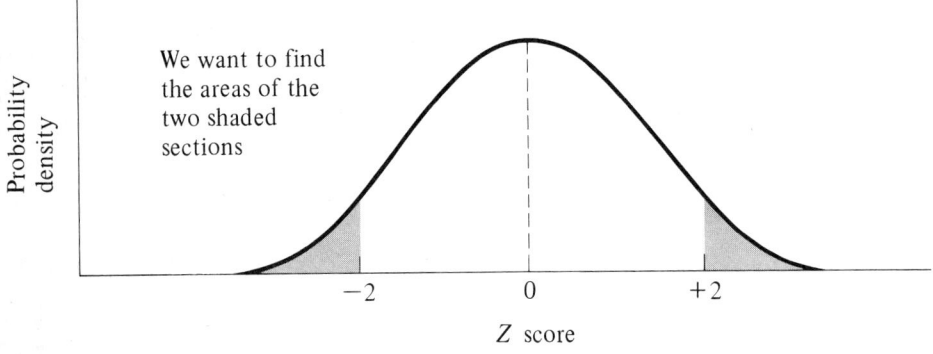

Figure 17.4

THE NORMAL APPROXIMATION TO THE BINOMIAL

TABLE OF AREAS OF THE NORMAL PROBABILITY CURVE (Z TABLE)

Z	0.00	0.01	0.02	0.03	0.04	0.05	0.06	0.07	0.08	0.09
0.0	.0000	.0040	.0080	.0120	.0160	.0199	.0239	.0279	.0319	.0359
0.1	.0398	.0438	.0478	.0517	.0557	.0596	.0636	.0675	.0714	.0753
0.2	.0793	.0832	.0871	.0910	.0948	.0987	.1026	.1064	.1103	.1141
0.3	.1179	.1217	.1255	.1293	.1331	.1368	.1406	.1443	.1480	.1517
0.4	.1554	.1591	.1628	.1664	.1700	.1736	.1772	.1808	.1844	.1879
0.5	.1915	.1950	.1985	.2019	.2054	.2088	.2123	.2157	.2190	.2224
0.6	.2257	.2291	.2324	.2357	.2389	.2422	.2454	.2486	.2517	.2549
0.7	.2580	.2611	.2642	.2673	.2704	.2734	.2764	.2794	.2823	.2852
0.8	.2881	.2910	.2939	.2967	.2995	.3023	.3051	.3078	.3106	.3133
0.9	.3159	.3186	.3212	.3238	.3264	.3289	.3315	.3340	.3365	.3389
1.0	.3413	.3438	.3461	.3485	.3508	.3531	.3554	.3577	.3599	.3621
1.1	.3643	.3665	.3686	.3708	.3729	.3749	.3770	.3790	.3810	.3830
1.2	.3849	.3869	.3888	.3907	.3925	.3944	.3962	.3980	.3997	.4015
1.3	.4032	.4049	.4066	.4082	.4099	.4115	.4131	.4147	.4162	.4177
1.4	.4192	.4207	.4222	.4236	.4251	.4265	.4279	.4292	.4306	.4319
1.5	.4332	.4345	.4357	.4370	.4382	.4394	.4406	.4418	.4429	.4441
1.6	.4452	.4463	.4474	.4484	.4495	.4505	.4515	.4525	.4535	.4545
1.7	.4554	.4564	.4573	.4582	.4591	.4599	.4608	.4616	.4625	.4633
1.8	.4641	.4649	.4656	.4664	.4671	.4678	.4686	.4693	.4699	.4706
1.9	.4713	.4719	.4726	.4732	.4738	.4744	.4750	.4756	.4761	.4767
2.0	.4773	.4778	.4783	.4788	.4793	.4798	.4803	.4808	.4812	.4817
2.1	.4821	.4826	.4830	.4834	.4838	.4842	.4846	.4850	.4854	.4857
2.2	.4861	.4864	.4868	.4871	.4875	.4878	.4881	.4884	.4887	.4890
2.3	.4893	.4896	.4898	.4901	.4904	.4906	.4909	.4911	.4913	.4916
2.4	.4918	.4920	.4922	.4925	.4927	.4929	.4931	.4932	.4934	.4936
2.5	.4938	.4940	.4941	.4943	.4945	.4946	.4948	.4949	.4951	.4952
2.6	.4953	.4955	.4956	.4957	.4959	.4960	.4961	.4962	.4963	.4964
2.7	.4965	.4966	.4967	.4968	.4969	.4970	.4971	.4972	.4973	.4974
2.8	.4974	.4975	.4976	.4977	.4977	.4978	.4979	.4979	.4980	.4981
2.9	.4981	.4982	.4983	.4983	.4984	.4984	.4985	.4985	.4986	.4986
3.0	.4987	.4987	.4987	.4988	.4988	.4989	.4989	.4989	.4989	.4990
3.1	.4990	.4991	.4991	.4991	.4992	.4992	.4992	.4992	.4993	.4993
3.2	.4993	.4993	.4994	.4994	.4994	.4994	.4994	.4995	.4995	.4995
3.3	.4995	.4995	.4996	.4996	.4996	.4996	.4996	.4996	.4996	.4997
3.4	.4997	.4997	.4997	.4997	.4997	.4997	.4997	.4997	.4997	.4998
3.5	.4998	.4998	.4998	.4998	.4998	.4998	.4998	.4998	.4998	.4998
3.6	.4998	.4998	.4999	.4999	.4999	.4999	.4999	.4999	.4999	.4999
3.7	.4999	.4999	.4999	.4999	.4999	.4999	.4999	.4999	.4999	.4999
3.8	.4999	.4999	.4999	.4999	.4999	.4999	.4999	.4999	.4999	.5000
3.9	.5000	.5000	.5000	.5000	.5000	.5000	.5000	.5000	.5000	.5000
Z	0.00	0.01	0.02	0.03	0.04	0.05	0.06	0.07	0.08	0.09

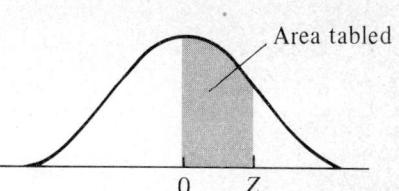

Figure 17.3

Example: The area corresponding to a Z score of 1.96 is .4750. Thus the probability of a Z score between 0 and 1.96 is .4750.

Note: This table is also reproduced in the Appendix for easy reference.

From Table II of Ronald A. Fisher and Frank Yates, *Statistical Tables for Biological, Agricultural and Medical Research*, 6th ed., published by Oliver and Boyd, Edinburgh, 1963, p. 45, and by permission of the authors and publishers.

Let's first find the probability of a Z score greater than +2 (see Figure 17.5). From Figure 17.5 we can see that the area of the shaded portion of the first curve is 0.50 − 0.4773 = 0.0227.

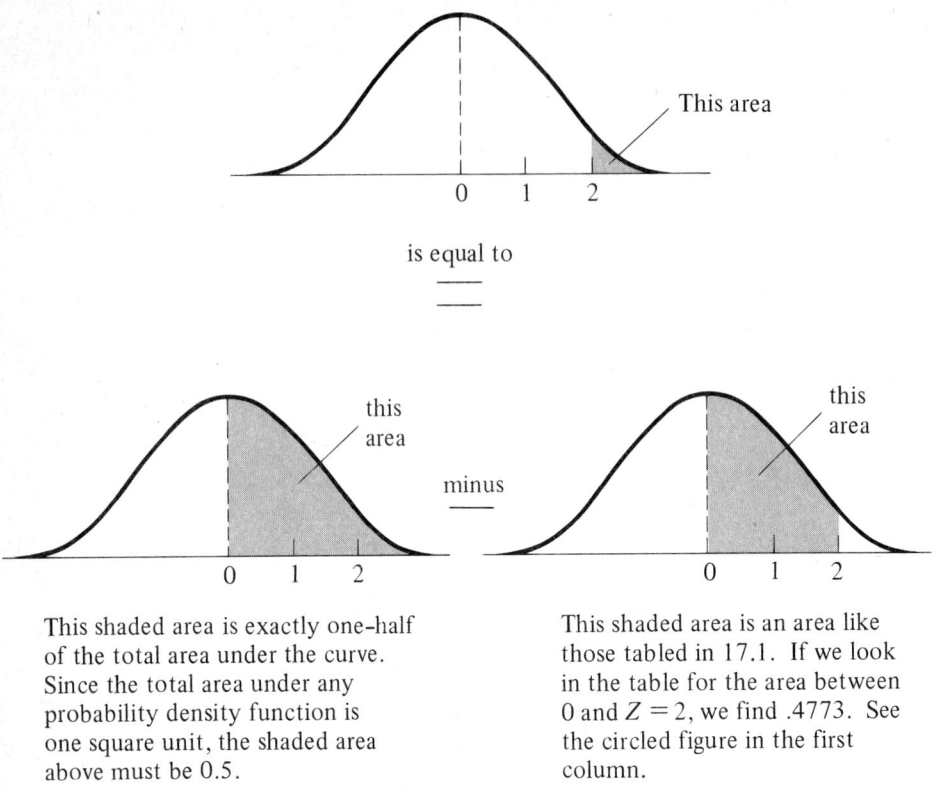

Figure 17.5 How to find the probability of a Z score greater than +2.

Thus the probability, in the unit normal distribution, of seeing an observation greater than 2 is 0.0227. This is the probability of a Z score greater than +2. Looking back at Figure 17.4, you can see that the figure is symmetric, so the area in the left-hand shaded area should be the same: 0.0227. The probability of a Z score less than −2 is also 0.0227. Putting these two together, the probability of a Z score more than +2 or smaller than −2 is just 0.0227 + 0.0227 = 0.0454. About 4 1/2 percent of the time you will see a Z score more than +2 or less than −2. And the rest of the time, or 95 1/2 percent of the time, you will see Z scores greater than −2 but less than +2: scores less than 2 standard deviations from the mean. The rule of thumb looks good.

Example of the use of the normal table   Using Figure 17.3, find the Z value which is exceeded only 1 percent of the time in both directions.

We are looking for a value which we shall call $Z_{0.01}$. The probability of a Z score greater than $+Z_{0.01}$ or less than $-Z_{0.01}$ is to be exactly 1 in 100, or 0.01. Using an argument exactly parallel to the previous one, we expect half of the 0.01 area to be in each of the two tails. The area of one tail should thus be 0.01/2, or 0.005. The area to look for in the table, then, is the difference between the entire right-hand side of the distribution (0.50) and the area of the right tail (0.005). Thus the area in the table corresponding to $Z_{0.01}$ should be 0.50 − 0.005, or 0.495. Looking in the table, we find the following two entries:

.4949   corresponds to   Z = 2.57   and   .4951   corresponds to   Z = 2.58

When more accurate tables are used, 2.58 is the appropriate value for $Z_{0.01}$. *There is a probability of 0.01 that you will observe a Z score greater than +2.58 or less than −2.58.*

This is of more than academic interest. If we are running a normal test (Z test) and wish to use a 0.01 significance level, our rejection region will be all Z values greater than +2.58 or less than −2.58. The probability of such a value when the null hypothesis is in fact true is just 1 in 100.

If you go back to the table and find a value for $Z_{0.05}$ by proceeding in exactly the same way, you will find that this critical value is 1.96. There is a 0.05 probability of a Z score greater than +1.96 or smaller than −1.96 when the null hypothesis is true. This is very close to the rule of thumb value of 2.00. Where great accuracy is required, 1.96 is the more precise number. It's harder to remember, though, and the rule of thumb is adequate for most purposes.

QUESTION AND ANSWER

Question: Just how does the Z table work? What do the numbers along the top mean?

Answer: To find an area corresponding to a Z score of 0.0, 0.1, 0.2, 0.3, etc., you can just look at the first entry of the corresponding row. (Look at the table to see what is meant.) But suppose you wish to find the area between the mean and a Z score of 0.04? You look in the top row, since the score starts 0.0. But then you look at the fourth column, since the last digit of the score 0.04 is 4. This gives an area of .0160. So the probability of a Z score between 0 and 0.04 is 0.0160.

Question: What is the probability of seeing a Z score *between* 0 and 2?

Answer: Looking in the 2.0 row in the table, we find .4773. This is the area between 0 (the mean) and the Z score of 2. So it is the probability of seeing an observation in that interval.

Question: What is the probability of an observation between +1.0 and +2.0 in a standard normal distribution?

Answer: This involves a subtraction. See Figure 17.6.
  If you have followed through so far, you should be ready for the next question.

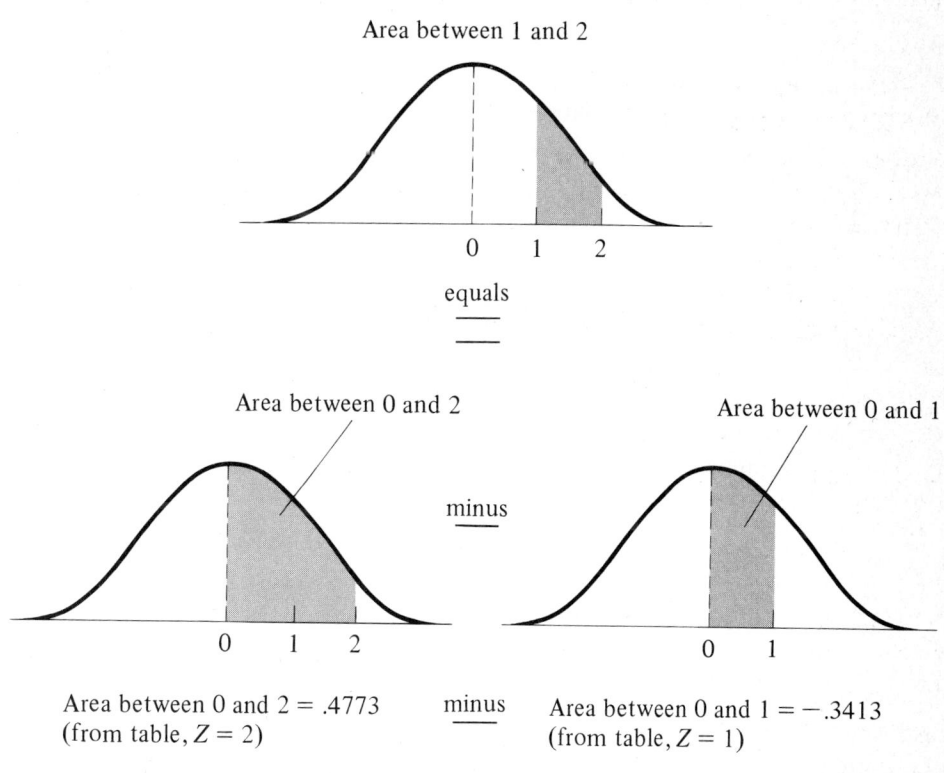

Figure 17.6 Finding the probability of a Z score between +1.0 and +2.0.

146   THE NORMAL DISTRIBUTION AND THE Z TEST

Question: What is the probability of observing a Z score between −1.43 and +0.78?

Answer: Again, this takes a two-stage analysis. See Figure 17.7.

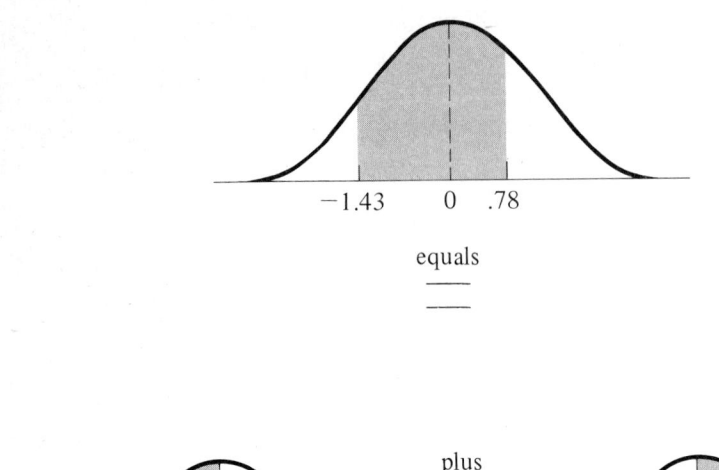

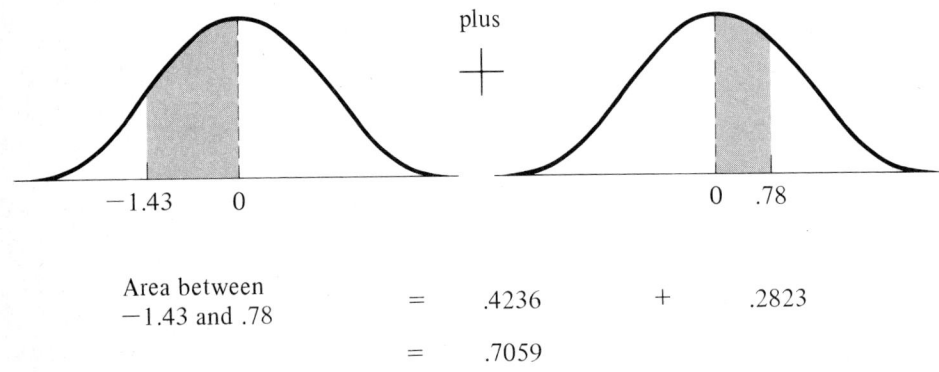

Figure 17.7   Finding the probability of a Z score between −1.43 and +0.78.

At this point you are really ready for anything: see the problems.

Summary

Discussion continued on the use of the normal distribution to approximate the binomial (sign-test) probability distribution. The null hypothesis for the sign test says that both of two alternatives are equally likely. For such a situation, when there are N trials or cases, the mean of the binomial distribution is $N/2$ and the variance is $N/4$. A normal distribution with the same mean and variance is very similar. Instead of asking, "Is this observation too extreme to have come from a binomial distribution with both alternatives equally likely and N trials?" we can ask, "Is this observation too extreme to have come from a normal distribution with mean $N/2$ and variance $N/4$? *The answers to both questions are usually the same.*

We looked at the 20-coin-toss situation (20-trial sign test), and we concluded that the rejection region, based on exact probabilities, for a 0.05 significance level included the events 0/20, 1/20, 2/20, 3/20, 4/20, 5/20, 15/20, 16/20, 17/20, 18/20, 19/20, and 20/20. When we used the normal approximation to the binomial, we reached exactly the same conclusion.

A graph was presented of the unit normal distribution (or Z distribution). This normal distribution has a mean $M = 0$ and standard deviation $SD = 1$. On the basis of the graph it is possible to determine the area under the curve corresponding to any interval of interest. All you need do is count squares to find probabilities for the normal distribution. A table was presented, which can be used for much the same purposes. Using that table, we found that the probability of a Z score greater than +2.58 or less than −2.58 was exactly 0.01, or 1 in 100. This means that for a Z test with $\alpha = 0.01$, the rejection region will be all Z scores greater than +2.58 or less than −2.58

Problems

1. Consider the election example discussed at the beginning of this chapter. How extreme a result (in terms of number of votes out of 922) would you need to observe to be able to conclude at the 0.01 significance level that the Republicans had a higher strength than the Democrats? (*Hint:* On a normal distribution, observations more than 2.58 standard deviations away from the mean occur exactly 1 percent of the time.)

2. Fifteen classes see a new film, "Highway Safety." For the next 24 months, a record is kept of the number of moving traffic violations for each class, and the violation rate is compared with the national average. Of the 15 classes, 11 do better than the national average and 4 do worse. Is this a significant result? Use a 0.05 significance level, and be specific about your logic.

3. Problems involving the normal distribution table:

   a. What is the probability of a Z score between 0 and 1.65?

   b. Find a critical value, $Z_{0.10}$, such that the probability of an observation less than $-Z_{0.10}$ or greater than $+Z_{0.10}$ is just 1 in 10, or 0.10. Follow a procedure analogous to that in the text for $Z_{0.01}$.

   c. Find a Z score such that 80 percent of all observations are smaller than Z_{80}. (*Hint:* What is the area which you should look for in the table? 0.20? 0.80? 0.30? 0.40? Something else?) *Draw a sketch!*

   d. IQ scores are approximately normally distributed with mean 100 and standard deviation 15. What is the score, $score_{80}$, such that 80 percent of the people taking the test score lower than $score_{80}$? (Use your answer to part c.)

   e. A large company selling pizza parlor franchises has determined that the cash capitalization required for a new franchise to break even is approximately normally distributed, with a mean of $18,000 and a standard deviation of $5,000. How much capitalization should John Walthey have if he wants to be sure of a 90 percent chance that he will start to make money before he goes broke?

4. At a wine tasting, 100 college professors were asked to state their preference between two wines in unmarked glasses. One wine was Chateau Haut Brion, 1959, an extremely expensive imported wine. The other was an inexpensive domestic wine, Mountain Red. Of the 100 people, 65 preferred the Haut Brion. Is this a statistically significant preference? Use a 0.01 significance level. What conclusion(s) would you draw? *Think about it.*

5. Look back at practice problem 2c, page 52, in Chapter 8. Perform a sign test to determine whether the observed preference for live presentation of classroom material is significant. Use a 0.05 significance level.

   Discuss what action you would take if you were a college administrator.

6. Using the normal table, find the probability of observing a score between zero and the following Z values (use Figure 17.3).

Z	Area
2.50	0.4938
3.42	0.4997
0.74	0.2704
1.53	
2.97	
3.99	
4.00	
8.00	

7. What is the Z score which is exceeded exactly 0.025 (2.5 percent) of the time? What is the negative Z score such that exactly 2.5 percent of all observations are less than that score?

8. Suppose that grades in a large course are normally distributed with mean 70 and standard deviation 10. Grades between 90 and 100 get an A, between 80 and 90 get a B, between 70 and 80 get a C, between 60 and 70 get a D, and below 60 the grade is F.

   a. Find the Z score associated with each of the cut-points: 60, 70, 80, 90.

   b. Find the probability of observing each of the five grades. (Remember that the probability of observing a grade between, say, 60 and 70 is the area under the curve between these two numbers.)

9. Suppose one wanted to run a Z test with a 0.08 significance level. One should reject the null hypothesis for any observed Z score greater than _____ or less than _____.

10. The odometers on 200 new cars are checked for accuracy. On 142 of them, they read more than 10 miles when the car has traveled exactly 10 miles. On 58 they read less. Is this statistically significant evidence of a consistent bias?

The *t* Test, Correlation, and Chi-Square

The *t* Test for the Mean of a Sample

THE *Z* TEST AND BEYOND

We have so far seen three different cases where the *Z* test can be used, directly or indirectly:

1. When we are asking whether one particular observation, call it X_i, came from a particular, known normal distribution with known mean and variance. For example, remember the problem of the chlorine concentration at the water works.
2. A more common case is that in which we are dealing with the mean of a sample, M_{samp}. Under various conditions which you should know by now, the sampling distribution of the mean is a normal distribution with known mean and variance. We ask, using a *Z* test, whether it is plausible to assume that the sample mean observed was indeed drawn from that normally distributed sampling distribution of the mean.
3. In a sign-test situation with a large number of trials, or scores, we have seen how to use the normal distribution as an approximation to the binomial distribution. The use of the normal distribution here is quite close to a *Z* test. In the following example, we can't quite run a *Z* test, but we can do something similar.

Example Leading Up to the *t* Test

Suppose that over the past ten years the grade point averages for upper division psychology majors at one university have been calculated, and an average was found each year. The overall grand average GPA, based on many thousands of individual GPAs, is 2.10. In the winter quarter this year, the GPAs for 160 students were found. The average of these 160 GPAs, M_{samp}, was found to be 2.24. No record has been kept of the variability of the scores over the past ten years. That is to say, we don't know anything about the population variance. But we do know the population mean, 2.10. *Question:* Is the mean GPA observed this winter, 2.24, significantly higher than the GPAs observed over the past ten years? Have we evidence leading to a conclusion that either the students have improved or the grading standards have changed? The question is whether these 160 numbers were drawn from the same population as the thousands of GPAs over the past ten years. The mean of that population is (almost exactly) known, and is 2.10. But we don't know the variance.

If we knew the population variance, we could answer the question easily. From the population variance we could compute the variance of the sampling distribution of the mean for samples of size 160. The population mean would tell us the mean of that sampling distribution. And we could then run a simple Z test to find out if the observed mean, 2.24, could plausibly be said to have come from that sampling distribution of the mean. *But we don't know the population variance.*

Subgoal question: Is there any way we can get any information about the population variance? Is there any way to get any idea about it? There is, for we do know the 160 individual numbers which were used to compute the sample mean, M_{samp}. On the basis of these numbers, using the relationship first presented in Chapter 10, we can compute an *estimate* of the population variance. This can then be used to estimate the variance of the sampling distribution of the mean. And from there we can decide whether the observed mean is plausible as a number from that sampling distribution.

From Chapter 10

$$\text{Best estimate of } SD_{pop}^2 = \frac{\Sigma (X_i - M_{samp})^2}{n - 1}$$

Here $n = 160$ and $M_{samp} = 2.24$, so

$$\text{Best estimate of } SD_{pop}^2 = \frac{\Sigma (X_i - 2.24)^2}{159}$$

Now after this computation was carried out, based on the 160 GPAs for this winter, the result was Estimated $SD_{pop}^2 = 0.64$.

With this estimate in hand, how do we proceed to decide if the observed mean of 2.24 is significantly different from the population mean, which is 2.10? We must use the estimated population variance to compute an estimated variance of the sampling distribution of the mean of samples of size 160. Remember from the central limit theorem that the variance of the sampling distribution of the mean is given by

$$SD_{samp\ dist\ of\ the\ mean}^2 = \frac{SD_{pop}^2}{n} \quad \text{where } n \text{ is the sample size}$$

In this particular case, we have estimated that $SD_{pop}^2 = 0.64$, and we know that $n = 160$. Substituting in the above equation, we find

$$\text{Est. } SD_{samp\ dist\ of\ the\ mean}^2 = \frac{0.64}{160} = 0.004$$

Taking the square root of this number, we can find the estimated SD of the sampling distribution of the mean; SD_M:

$$\text{Est. } SD_{samp\ dist\ of\ the\ mean} = \sqrt{0.004} = \sqrt{\frac{4}{1000}} = \sqrt{\frac{40}{10,000}} = \frac{\sqrt{40}}{\sqrt{10,000}} = \frac{6.3}{100}$$

$$\text{Est. } SD_{samp\ dist\ of\ the\ mean} = 0.063$$

You can turn to the table of squares, square roots, and reciprocals in the appendix to verify that $\sqrt{40} \cong 6.3$.

Now we are in a position to return to the original question: We know that the mean of the sampling distribution of the mean must be the same as the population mean, which is 2.10. And we have arrived at an estimate of the variance of the sampling distribution of the mean, and hence the standard deviation, which is 0.063. In this distribution, would an observation of 2.24 be surprisingly high? We can find how many standard deviations 2.24 is from M_M:

$$\frac{2.24 - 2.10}{0.063} = \frac{0.14}{0.063} = \frac{140}{63} = \frac{20}{9}$$

$$= 2.22$$

On the basis of everything we have learned, there is little chance of so extreme a Z score. (The rule of thumb said that there was only 1 chance in 20 of seeing an observation as far as 2 standard deviations away from the mean.) So we would probably conclude, finally, that this year's mean GPA *is* significantly different from the population of GPAs over the last decade.

There is one small flaw in this entire argument. We were forced to accept an *estimate* of the population variance, instead of knowing exactly what that variance is. This necessitates a systematic difference in how we interpret the scores obtained.

Recapitulation: We started with the following problem. We know the population mean ($M_{pop} = 2.10$) and the sample mean ($M_{samp} = 2.24$). We don't know the population variance, but we can estimate it on the basis of the variability of the numbers in the sample. The steps we went through were:

1. Estimate the population variance on the basis of the sample variability, using the relationship

$$\text{Est. } SD^2_{pop} = \frac{\Sigma (X_i - M_{samp})^2}{n - 1}$$

2. Use this estimate to compute the (estimated) variance of the sampling distribution of the mean:

$$\text{Variance of the sampling distribution of the mean} = SD^2_M = \frac{SD^2_{pop}}{n}$$

3. Compute a score just like a Z score (except that it uses an *estimate* instead of a known variance of the sampling distribution of the mean):

$$\text{``Z''} = \frac{M_{samp} - M_{samp\ dist\ of\ the\ mean}}{\text{Est. } SD_{samp\ dist\ of\ the\ mean}} = \frac{M_{samp} - M_M}{\text{Est. } SD_M}$$

4. Decide whether the resulting score is so high that you conclude the sample couldn't have been taken from the population in question.

The logic of this situation, which confuses most people at first, can be presented as a type of flow chart (see page 154). Putting the information from the chart together, we can specify an estimate of the sampling distribution of the mean for samples of size 160 from this population. This is different from other sampling distributions of the mean which we have seen, since we were forced to use an estimate of the population variance: Est. $SD^2_{pop} = 0.64$. On the basis of this estimate, we can estimate the variance of the sampling distribution of the mean, finding Est. $SD^2_M = 0.063$.

We observed a sample mean of 2.24. We are interested in deciding if it could plausibly have come from this sampling distribution of the mean. If not, then we reject the null hypothesis. Three numbers are shown in Figure 18.1: 2.10, the mean of the sampling distribution; 2.24, the single sample mean actually observed; and 0.063, the estimate of the standard deviation of the sampling distribution, Est. SD_M. On the basis of these three numbers, we compute a statistic like a Z score. It would *be* a Z score if we actually knew the variance of the population.

154 THE t TEST, CORRELATION, AND CHI-SQUARE

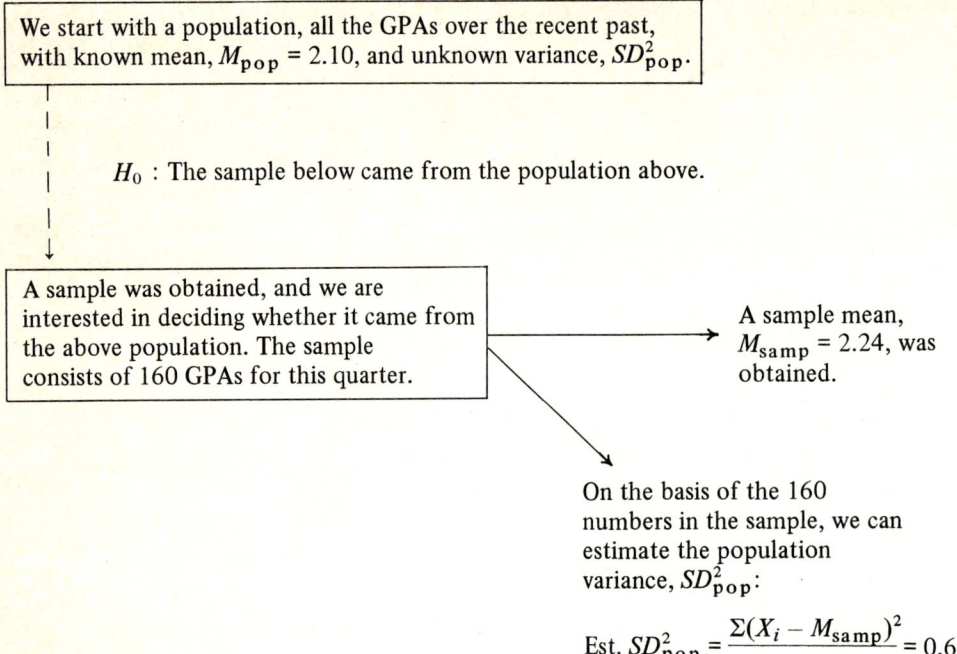

We start with a population, all the GPAs over the recent past, with known mean, $M_{pop} = 2.10$, and unknown variance, SD^2_{pop}.

H_0 : The sample below came from the population above.

A sample was obtained, and we are interested in deciding whether it came from the above population. The sample consists of 160 GPAs for this quarter.

A sample mean, $M_{samp} = 2.24$, was obtained.

On the basis of the 160 numbers in the sample, we can estimate the population variance, SD^2_{pop}:

$$\text{Est. } SD^2_{pop} = \frac{\Sigma(X_i - M_{samp})^2}{n-1} = 0.64$$

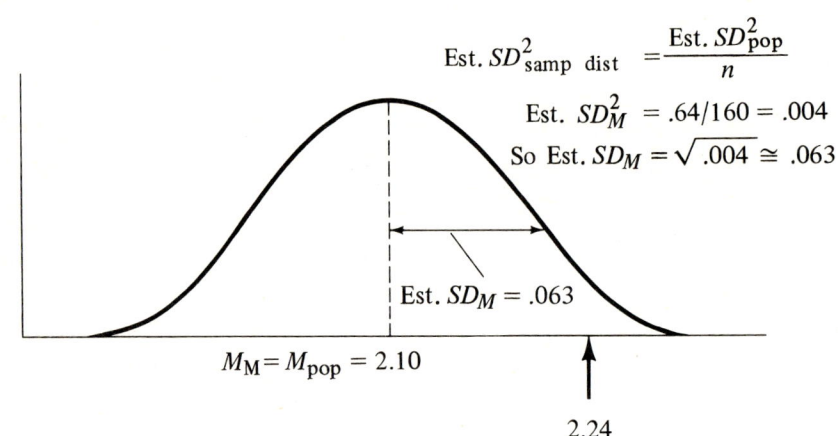

$$\text{Est. } SD^2_{samp\ dist} = \frac{\text{Est. } SD^2_{pop}}{n}$$

$\text{Est. } SD^2_M = .64/160 = .004$

So $\text{Est. } SD_M = \sqrt{.004} \cong .063$

Figure 18.1 Estimated sampling distribution of the mean for samples of size 160 from the parent population of all GPAs over the past years.

In general, $Z = (X_i - M)/SD$. Our score is

$$\frac{X_i - M}{\text{Est. } SD} = \frac{2.24 - 2.10}{0.063}$$

$$= \frac{0.14}{0.063} = \frac{140}{63} = \frac{20}{9}$$

$$= 2.22$$

This is called a t score or t statistic.

After you have worked through the following self-quiz, we will present some more material about the t score.

Self-Quiz to Review the Logic Behind the t Statistic

People in the general population are known to learn foreign-language vocabulary words at a rate which is normally distributed with mean $M = 7$ words per hour. The variance of the normal distribution of rates is not known. A sample of six Wilson College students is selected at random from the entire set, and they are given a language-learning test. They learn 9, 13, 8, 6, 9, and 15 words per hour. *Question:* Try to decide whether these numbers are sufficiently high to lead you to conclude that Wilson College students learn faster than the general population. Put differently: Do you think that the mean of all Wilson College students is higher than the mean, 7, for the general population?

Here is a suggested plan of attack:

1. Find the mean, M_{samp}, of the sample of six scores.
2. Make the best estimate you can of the variability in the general population by studying the variability of the sample and using the relationship (presented in Chapter 10) that

$$\text{Best estimate of } SD^2_{pop} = \frac{\Sigma (X_i - M_{samp})^2}{n - 1}$$

3. Use this estimate of SD^2_{pop} to generate an estimate of the variability of means of samples of size 6 from the general population. That is, find the (estimated) variance of the sampling distribution of the mean for samples of size 6: Est. $SD^2_{samp\ dist\ of\ the\ mean}$, which we denote Est. SD^2_M.
4. Figure out what the mean of the sampling distribution of the mean should be. Denote the mean of the means by M_M.
5. Compute a statistic like a Z score, to help decide whether the observed sample mean, M_{samp} (from step 1), is sufficiently high to lead us to conclude that Wilson College students are different from the general population.

If you have not already done so, I urge you to take the time now to work through the above quiz. It will substantially enhance your understanding of the material which is to come.

Solution to the quiz problem: The null hypothesis we are considering is that the six scores, 9, 13, 8, 6, 9, and 15, were drawn from the same normal distribution, with a mean of 7, which characterizes the general population. We follow the steps above.

1. Find the sample mean:

$$M_{samp} = \frac{\Sigma X_i}{n} = \frac{\Sigma X_i}{6} = \frac{9 + 13 + 8 + 6 + 9 + 15}{6} = \frac{60}{6}$$

$$M_{samp} = 10$$

2. Estimate the SD^2_{pop} on the basis of the sample variability:

$$\text{Best estimate of } SD^2_{pop} = \frac{\Sigma (X_i - M_{samp})^2}{n - 1} = \frac{\Sigma (X_i - M_{samp})^2}{5}$$

$$= \frac{(9 - 10)^2 + (13 - 10)^2 + (8 - 10)^2 + (6 - 10)^2 + (9 - 10)^2 + (15 - 10)^2}{5}$$

$$= \frac{1 + 9 + 4 + 16 + 1 + 25}{5}$$

$$= \frac{56}{5}$$

Best estimate of $SD^2_{pop} = 11.2$

3. Now we wish to deduce the variance of the sampling distribution of the mean, SD_M^2, on the basis of the SD_{pop}^2. Remember from the central limit theorem that

$$SD_M^2 = \frac{SD_{pop}^2}{n} \quad \text{where } n \text{ is the sample size}$$

Since we've estimated $SD_{pop}^2 = 11.2$, and we know that the sample size is $n = 6$, $SD_M^2 = SD_{pop}^2/n = 11.2/6 = 1.87$. Since $\sqrt{1.87} \cong 1.4$, we know that our estimate of SD_M is 1.4.

4. Since the mean of the sampling distribution of the mean is always equal to the population mean, M_{pop}, we know that in this case $M_M = M_{pop} = 7$. This says that if you take sample after sample from the parent population, and compute the mean for each sample, the average value of the sample mean will be exactly 7, the population mean.

5. Compute a Z-like statistic: Remember that a Z statistic or Z score for a particular number from a specified distribution is just the difference of that number from the distribution mean, divided by the standard deviation of the distribution. In this case, the observation is the mean of the six numbers, $M_{samp} = 10$. The distribution is the sampling distribution of the mean of samples of size 6, which we have deduced has mean $M_M = 7$ and standard deviation $SD_M = 1.4$ (see steps 3 and 4).

Our Z-like statistic, then, is

$$\frac{X_i - M_M}{\text{Est. } SD_M} = \frac{10 - 7}{1.4} = \frac{3}{1.4} \cong 2.14$$

It is essential, here, to be clear that the mean of our sample, $M = 10$, is a *single number* in the sampling distribution of the means. Having computed this score, do we decide to reject the null hypothesis and conclude that Wilson College students are superior? On the basis of everything you have learned up to this point, you probably think *Yes*, assuming a 0.05 significance level. Your decision would, however, be incorrect, as we shall shortly show.

The crux of the problem is that the statistic we just computed was not exactly a Z statistic. The Z statistic is defined for a situation in which we *know* both the mean, M, and the standard deviation, SD, of a distribution, and we want to decide whether or not a given number, X_i, came from that distribution. In our case we don't know the exact value of the standard deviation of the sampling distribution of the mean. We estimated it, on the basis of an estimate of the population variance. But as you can well imagine, that estimate is probably not exactly correct. Indeed, about half the time the estimate we get in this way will be too high and about half the time it will be too low. This means that the statistic we have just computed will not have exactly the same distribution as the Z statistic.

THE t STATISTIC CONTRASTED WITH THE Z STATISTIC

The statistic we just computed, which is a Z-like statistic based on a known distribution *mean* and an estimated distribution *standard deviation*, is called the t statistic. It turns out to be the case that large values of the t statistic are somewhat more likely than large values of the Z statistic when the null hypothesis is true. This, of course, influences our decision.

When you take observations at random from a normally distributed population with *known* mean and standard deviation, and then compute a Z score on the basis of each observation, the resulting scores will have the Z distribution, which is to say they will have a normal distribution with mean 0 and standard deviation 1. We know that for such a situation, scores greater than $+2$ or less than -2 will happen only about 5 percent of the time. In this situation, however, we didn't know the actual standard deviation of the normally distributed population.[1] We had to estimate it. When we do this, and then proceed as we did above, the resulting statistic is called a t statistic, and is said to have a t distribution.

[1] The normally distributed population, you recall, is the sampling distribution of the mean for samples of size 6.

In Figure 18.2 both distributions are similar in shape, and they are symmetric. But the t distribution is flatter and wider than the Z distribution. Whereas only about 5 percent of all Z scores are greater than +2 or less than −2, it turns out that fully 10 percent of the t scores are greater than +2 or less than −2 (for samples of size 6). *The important thing to remember is that extreme values of the t statistic are more likely to happen than extreme values of the Z statistic, because of the extra variability introduced when you have to estimate the standard deviation of the sampling distribution of the means.*

As you might expect, this is not much of a problem when you have very large samples: when the sample is large, say with 100 numbers in it, then it will yield a very precise estimate of SD^2_{pop}, and the net result will be just about the same as if you had exact knowledge of the variance of the population. On the other hand, when you are working with very small samples, the estimates of SD^2_{pop} will be very unstable, and the result will be great variability in the t statistic.

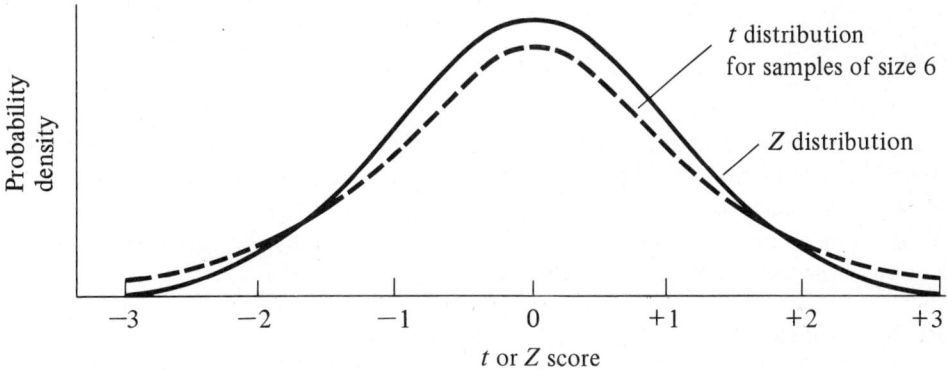

Figure 18.2 The Z distribution, or unit normal distribution, compared with a t distribution for sample size 6. Note that extreme values of the t statistic are more likely (and so less surprising) than extreme values of the Z statistic.

If we were to draw the distribution of the t statistic for samples of size 1000, it would be indistinguishable from the curve for the Z statistic or Z score. On the other hand, the t distribution for samples of size 2 would be even wider than the distribution in Figure 18.2 for samples of size 6.

A table for the various t distributions is presented at the end of this chapter. There is a *different* t distribution for each possible sample size. To give as much information about each t distribution as you have about the Z distribution (see Chapter 17, Figure 17.3), it would take 47 pages of separate tables. Each row in the table for t corresponds to a complete table. Naturally, the information provided is more limited. What you have in this table is a set of critical values for various significance levels. Consider the line which gives values for a sample of size 6: This is given opposite the number 5 in the Degrees of Freedom column. (For samples of size 23, you have 22 degrees of freedom, sometimes written 22 df; for samples of size n, these are $(n - 1)$ degrees of freedom.) Look down the column headed 0.05. This is the column which contains the critical values for a 0.05 significance level. In this case (5 degrees of freedom) the table entry is 2.57. This tells us that with samples of size 6, there is a probability of 0.05 that the t statistic will be greater than +2.57 or less than −2.57.

Put differently, this tells us that in order to be justified in rejecting the null hypothesis, with a significance level of 0.05, we would need to observe a t statistic of at least +2.57 or less than −2.57. Recall that in our earlier case we computed a value of 2.14. Accordingly, we cannot reject the null hypothesis at the 0.05 significance level.

QUESTION AND ANSWER

Question: Using the t table, how large a value of the t statistic would we need to observe to reject the null hypothesis if the sample size were 2 and the significance level was 0.05?

Answer: Looking at the very first row, for 1 degree of freedom (since the sample size is 2, and degrees of freedom is the sample size minus 1), we see a critical value of 12.71 in the $\alpha = 0.05$ column. This means that with very small samples (of size 2) we would need to see an observation 12.71 times the estimated standard deviation away from the mean, in order to reject the null hypothesis. Can you see why this is so?

When you take samples of size 2, it naturally sometimes happens by pure chance that both numbers in a sample are very close to each other. That means that when you estimate the SD^2_{pop} on the basis of the sample, you end up with a number which is far too low. This, in turn, gives rise to an underestimate of the standard deviation of the sampling distribution of the mean. And as a result, the t statistic has a very high value. Because of the great variability in the SD^2 estimates, the distribution of the t statistic when the null hypothesis is true is very flat. So we need to see a very extreme value in order to conclude that the null hypothesis is not true.

Summary

We have seen how to use a Z score to help decide whether a particular sample was taken from a specified population, with known mean, M_{pop}, and variance, SD^2_{pop}. We consider the sampling distribution of the mean for samples of the given size from the specified population. We compute the mean, M_M, and the variance, $SD^2_M = SD^2_{pop}/n$, of *this sampling distribution*. We then treat the observed sample mean as a single observation, X_i, in the sampling distribution of the mean, and compute a Z score:

$$Z = \frac{X_i - M_M}{SD_M} \quad \text{where } X_i \text{ is the sample mean}$$

When the null hypothesis is true, this score has a unit normal distribution.

Unfortunately, it is often the case that we would like to proceed as outlined above, but we do not know the variance of the hypothesized parent population, SD^2_{pop}. In this case, we estimate it on the basis of the numbers in the sample:

$$\text{Est. } SD^2_{pop} = \frac{\Sigma(X_i - M_{samp})^2}{n - 1}$$

We then proceed very much as we did for the Z test: We next find the (estimated) variance of the sampling distribution of the mean for samples from the parent population:

$$\text{Est. } SD^2_M = \frac{\text{Est. } SD^2_{pop}}{n}$$

We know the mean of the sampling distribution of the mean: $M_M = M_{pop}$. And we compute a score like a Z score to describe the position of the sample mean in the sampling distribution of the mean. Since the sample mean is a single number in the sampling distribution, we denote it X_i. This score is a t statistic:

$$t = \frac{X_i - M_M}{\text{Est. } SD_M} \quad \text{where } X_i \text{ is the sample mean}$$

When the null hypothesis is true, this t statistic has a distribution like the unit normal distribution, but it is flatter and wider. This means that extreme values of the t score are more likely to happen than extreme values of the Z score. And this, in turn, means that in order to demonstrate a significant result, we need to see a higher value for the t statistic than for the Z score.

The larger the sample size, the closer is the t distribution to a Z distribution.

Problems

1. To show that you understand the *t* tables at the end of the chapter answer the following questions:

    a. Supposing you observe a *t* statistic of −2.14 based on a sample of size 14. Is this sufficient evidence to reject the null hypothesis when your significance level is 0.05?

    b. How big a *t* score do you need in order to reject the null hypothesis at the 0.10 significance level for a sample of size 31?

    c. What is the probability of a *t* score *less than* −1.86, when your sample contains nine numbers?

    d. The bottom line of the table, with the symbol ∞, which stands for "infinity," in the degrees of freedom column, should look familiar. Where have you seen these numbers before? Does this make sense? Why?

2. A manufacturer offers an Accuracy-Increasing Attachment for target-shooting rifles, which he claims will increase precision for anybody as soon as it is added to the rifle. Five marksmen try it, and they observe the following *changes* in score: −1, +6, +3, +7, and +5. Using a *t* test, decide if the attachment had an effect. Use $\alpha = 0.10$. *Hint:* Follow a procedure similar to that used in the self-quiz example. Here, the null hypothesis is that the changes in score are drawn from a normal distribution with $M = 0$. Show all your steps clearly.

3. Return to the self-quiz example at the start of Chapter 8. Analyze the difference scores using a *t* test, to decide if the mean is significantly different from zero. Compare your conclusion to the one reached in Chapter 8. Use the same significance level, 0.05.

†4. Why doesn't it make sense to talk about a *t* distribution for samples of size 1? We mentioned the *t* distribution for samples of size 2 in the text.

TABLE 18.1 CRITICAL VALUES FOR THE *t* STATISTICS

Sample size for a test of a hypothesis about the mean of a single sample	Degrees of freedom	Probability				
		0.50	0.10	0.05	0.02	0.01
2 = m	(m–1) = 1	1.000	6.34	12.71	31.82	63.66
3	2	0.816	2.92	4.30	6.96	9.92
4	3	.765	2.35	3.18	4.54	5.84
5	4	.741	2.13	2.78	3.75	4.60
6	5	.727	2.02	2.57	3.36	4.03
7	6	.718	1.94	2.45	3.14	3.71
8	7	.711	1.90	2.36	3.00	3.50
9	8	.706	1.86	2.31	2.90	3.36
10	9	.703	1.83	2.26	2.82	3.25
11	10	.700	1.81	2.23	2.76	3.17
12	11	.697	1.80	2.20	2.72	3.11
13	12	.695	1.78	2.18	2.68	3.06
14	13	.694	1.77	2.16	2.65	3.01
15	14	.692	1.76	2.14	2.62	2.98
16	15	.691	1.75	2.13	2.60	2.95
17	16	.690	1.75	2.12	2.58	2.92
18	17	.689	1.74	2.11	2.57	2.90
19	18	.688	1.73	2.10	2.55	2.88
20	19	.688	1.73	2.09	2.54	2.86
21	20	.687	1.72	2.09	2.53	2.84
22	21	.686	1.72	2.08	2.52	2.83
23	22	.686	1.72	2.07	2.51	2.82
24	23	.685	1.71	2.07	2.50	2.81
25	24	.685	1.71	2.06	2.49	2.80
26	25	.684	1.71	2.06	2.48	2.79
27	26	.684	1.71	2.06	2.48	2.78
28	27	.684	1.70	2.05	2.47	2.77
29	28	.683	1.70	2.05	2.47	2.76
30	29	.683	1.70	2.04	2.46	2.76
31	30	.683	1.70	2.04	2.46	2.75
36	35	.682	1.69	2.03	2.44	2.72
41	40	.681	1.68	2.02	2.42	2.71
46	45	.680	1.68	2.02	2.41	2.69
51	50	.679	1.68	2.01	2.40	2.68
61	60	.678	1.67	2.00	2.39	2.66
71	70	.678	1.67	2.00	2.38	2.65
81	80	.677	1.66	1.99	2.38	2.64
91	90	.677	1.66	1.99	2.37	2.63
101	100	.677	1.66	1.98	2.36	2.63
126	125	.676	1.66	1.98	2.36	2.62
151	150	.676	1.66	1.98	2.35	2.61
201	200	.675	1.65	1.97	2.35	2.60
301	300	.675	1.65	1.97	2.34	2.59
401	400	.675	1.65	1.97	2.34	2.59
501	500	.674	1.65	1.96	2.33	2.59
1001	1000	.674	1.65	1.96	2.33	2.58
Infinity	∞	.674	1.64	1.96	2.33	2.58

Example
If a sample of size 8 is taken from a normal distribution with known mean and unknown variance, and a *t* statistic is computed, there is a 0.05 probability that the *t* score would be greater than +2.36 or smaller than −2.36.

Note: This table is also reproduced in the Appendix for easy reference.
From Table III of Ronald A. Fisher and Frank Yates, *Statistical Tables for Biological, Agricultural and Medical Research*, 6th ed., published by Oliver and Boyd, Edinburgh, 1963, p. 46, and by permission of authors and publishers.

The *t* Test for a Single Mean, Concluded; The *t* Test for the Mean of a Set of Difference Scores

We have seen that if a sample is hypothesized to have come from a parent population with known mean and variance, it is often possible to specify exactly the sampling distribution of the mean for samples of this size. Then it is easy to make a judgment whether an observed sample mean could plausibly have come from the hypothetical parent population. The test we use here is a *Z* test. We know both the mean and the variance.

In the last chapter we considered the problem of making a decision of this sort when the *variance* of the hypothetical parent population is *not known*, and must be estimated on the basis of the variability of the sample observed. In this situation the logic was like this:

1. Compute the mean of the sample, M_{samp}.
2. Estimate the variance of the population on the basis of the numbers in the sample, using the relationship introduced in Chapter 10: Est. $SD^2_{\text{pop}} = \Sigma(X_i - M_{\text{samp}})^2/(n-1)$.
3. Deduce the best estimate available of the variance of the sampling distribution of the sample mean, SD^2_M, by dividing the estimated population variance SD^2_{pop}, by the sample size, n: Est. SD^2_M = Est. SD^2_{pop}/n.
4. Specify the mean of the sampling distribution of the mean (which will, you recall, be equal to the mean of the population): $M_M = M_{\text{pop}}$.
5. Compute the *t* statistic: $t = (M_{\text{samp}} - M_M)/\text{Est. } SD_M$ where M_M is the mean of the sampling distribution and Est. SD_M is the estimated standard deviation of that distribution.

Having computed the *t* statistic, and assuming you have already decided on a significance level, you can check in a table of critical values of *t*, like the one in Chapter 18.

Note the close resemblance between the *t* statistic and the corresponding *Z* statistic, which is used when we know the variance of the parent population:

$$Z = \frac{M_{\text{samp}} - M_M}{SD_M}$$

where M_{samp} = the observed sample mean
M_M = the mean of the sampling distribution, and
SD_M = the standard deviation of the sampling distribution of the mean. The only difference is that here we *know* SD_M, and we don't have to estimate it.

The essential difference between the Z statistic and the t statistic is the following: When the null hypothesis is true, the numbers you observe by computing the t statistic tend to be larger than the numbers you observe by computing the Z statistic. The t table tells you how much larger.

A SAMPLING EXPERIMENT TO COMPUTE t AND Z SCORES

In an experimental demonstration, some students looked at how t statistics and Z statistics compare in one situation. They took a series of samples of size 4 from the standard normal distribution, which has mean 0 and variance 1. This is the distribution which is listed in the Z tables. For each sample they computed the sample mean. They also computed a Z statistic for each sample mean as follows:

$$Z = \frac{M_{samp} - M_M}{SD_M}$$

We know how to find SD_M, since

$$SD_M^2 = \frac{SD_{pop}^2}{n} = \frac{SD_{pop}^2}{4}$$

since sample size is 4, so

$$SD_M^2 = \frac{1}{4}$$

since $SD_{pop} = 1$. We know that $M_M = M_{pop} = 0$, since the mean of the sampling distribution of the mean is equal to the mean of the population, which is zero for the unit normal distribution. Thus the Z score for a specific mean, M_{samp}, is just

$$Z = \frac{M_{samp} - 0}{\sqrt{1/4}} = \frac{M_{samp} - 0}{1/2} = 2 \times M_{samp}$$

(Note that since $1/2 \times 1/2 = 1/4$, $\sqrt{1/4} = 1/2$.) In a directly parallel way, the students also computed a t statistic on the basis of each sample. But to do this, they pretended they didn't know the population variance, and so had to estimate it on the basis of the numbers in the sample. For each sample they computed a t statistic as follows:

$$t = \frac{M_{samp} - M_M}{\text{Est. } SD_M} = \frac{M - 0}{\text{Est. } SD_M}$$

$$\text{Est. } SD_M^2 = \frac{\text{Est. } SD_{pop}^2}{n} = \frac{\text{Est. } SD_{pop}^2}{4} \quad \text{since } n = 4$$

$$\text{Est. } SD_{pop}^2 = \frac{\Sigma (X_i - M_{samp})^2}{n - 1} = \frac{\Sigma (X_i - M_{samp})^2}{3}$$

Now for each of 50 samples they computed a t statistic and a Z statistic, as indicated. Note that this is a little artificial: if you know the population variance, you ordinarily never would use a

t statistic. But the students went through this to demonstrate that t statistics are not quite the same as Z statistics, but on the average are larger.

Figure 19.1 presents a pair of frequency distributions, showing how often various values of each statistic were found. Several features of this frequency distribution are important: Note that almost all of the observed Z scores were between -2 and $+2$. Out of 50 observations only two

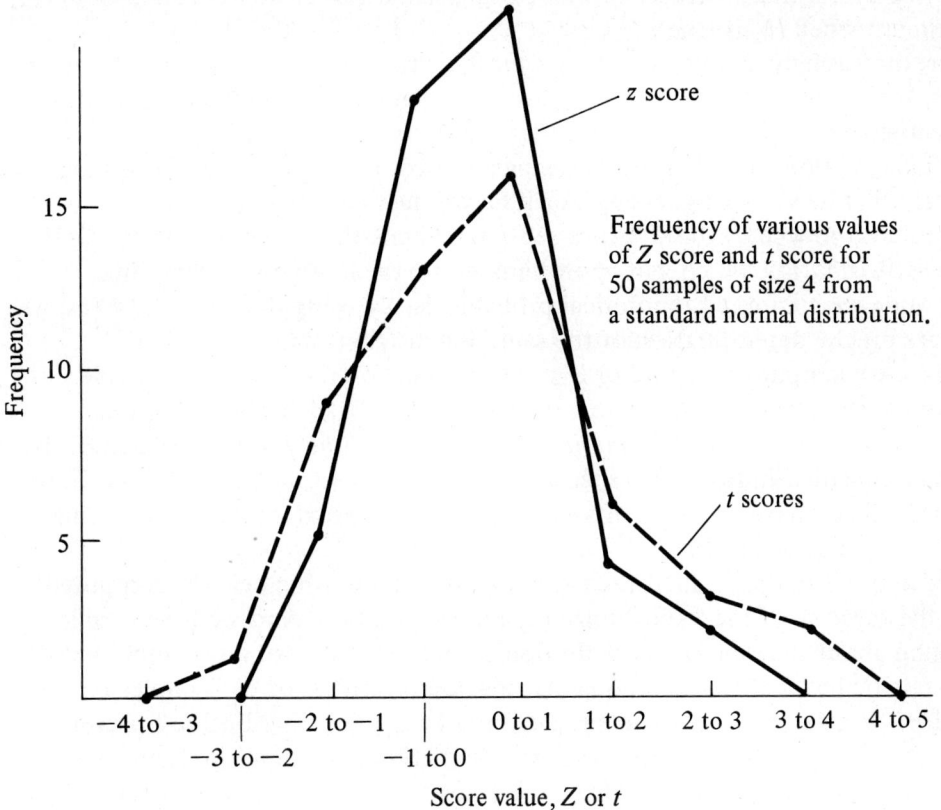

Figure 19.1 The frequency with which various t scores and Z scores were observed when 50 samples of size 4 were taken from a population with a unit normal distribution. For each sample mean, a Z score was computed. Then, pretending the population variance was unknown, a t score was computed. Note that extreme t scores are more likely than extreme Z scores.

scores were observed between $+2$ and $+3$. No larger values occurred. Values of the t statistic were more likely to be large: there were six samples yielding t scores less than -2 or greater than $+2$. The t distribution is lower and broader than the Z distribution. You should remember this when making decisions on the basis of a value of the t statistic. Compare the curves in Figure 18.2.

All this illustrates the need for special t tables. When H_0 is true, we observe Z statistics greater than 2 or less than -2 about 5 percent of the time. But we observe such extreme values of the t statistic much more often: in this case, with samples of size 4 (3 degrees of freedom), the corresponding critical value for t statistics is greater than $+3.18$ or less than -3.18 (see the t table). It also can be seen in the table that with a 0.10 significance level ($\alpha = 0.10$), the critical value for t is 2.35 (with sample size 4). This means that 10 percent of the time, by pure chance, we would see t scores more extreme than 2.35, when H_0 is true.

One other bit of evidence. In the set of 50 samples of size 4, the t and Z values were compared. In 42 out of the 50 cases, the t statistic was more extreme than the Z statistic.

You may have noted that the frequency distributions in Figure 19.1 are approximations to two *sampling distributions* for samples of size 4 from a standard normal distribution: the sampling distribution for the Z score and the sampling distribution for the *t* score. If we continued such a sampling process for a great many samples, we could come as close as we wished to the distribution which underlies the Z table (unit normal distribution) and to the distribution which underlies the third line of the *t* table (the line for samples of size 4). These tables provide us with information about the sampling distributions for our test statistics when H_0 is true, just as the marble box provided information about the sampling distribution of the score (8/10, 9/10, etc.) in the ESP experiment when H_0 is true.

It should be getting increasingly clear to you that *sampling distributions* are at the very heart of inferential statistics. We used the marble box, long ago, to estimate the likelihood of various scores (like 9/10 correct guesses) *when the null hypothesis was true*. Armed with this information, we were in a position to decide, upon seeing a subject guess correctly 10/10 times, that it was most unlikely that he was just guessing, for if he was just guessing, it was very improbable that he would do so well. It didn't make sense to assume that the observation 10/10 came from the *sampling distribution* which was approximated by the marble box sampling experiment. Later on when we computed theoretical probabilities by using the counting rules, we were just making a more precise determination of the sampling distribution.

Then we learned how to compute a Z score or Z statistic, based on the mean of a sample. And when the null hypothesis is true, we know how likely it is that we shall observe various values of the Z statistic. That is, we know the *sampling distribution of the Z statistic* when H_0 is true. This particular sampling distribution has a special name: the *unit normal distribution*. In the sampling experiment described above, the students experimentally approximated this sampling distribution (see the solid line in Figure 19.1).

Finally, we learned about a statistic called the *t* statistic, or *t* score, which can be computed when we don't know the variance of the hypothesized parent population. And we have a table summarizing information about the sampling distribution of the *t* statistic for any sample size of interest. While the Z statistic has the same sampling distribution regardless of the sample size (when the null hypothesis is true), the *t* statistic has a different sampling distribution for every different sample size. The dotted curve in Figure 19.1 is an experimental approximation to the *sampling distribution for the* t *statistic, for sample size* 4. It gives a rough idea of the probability of seeing various values of the *t* statistic when the null hypothesis is true.

Now that you thoroughly understand the *t* statistic, you can run *t* tests without limit. The five steps involved are just about the same as the five steps for the Z test. For the *t* test, however, the null hypothesis is of the following form: H_0 : The sample was taken from a specified population that is approximately normally distributed and has a given sample mean, M_{pop}. (The variance of the parent population is not known.)

Some Situations in Which a *t* Test Would be Helpful

1. In one college the percentage of students with high grades (A or B) in large lecture courses averaged 35 percent over the last 15 years, and it has been very stable. In the ten large lecture courses given last quarter, the following percentages were observed: 38, 31, 42, 50, 25, 43, 39, 37, 45, and 34. Does this provide strong evidence that grades are getting higher than they were in the past?

    This is a typical situation for the *t* test. We are asking whether it is plausible to assume that these ten numbers came from a population with a mean of 35. The variance of the parent population (percentages of grades A or B in large lecture courses over the past 15 years) is not known. The null hypothesis is that the numbers *were* taken from that population. H_0 : The ten scores given were taken from a population with mean 35. In order to compute a *t* statistic we go through the same steps as above. After computing the sample mean, M_{samp}, on the basis of these ten numbers, we estimate the population variance. Then we derive an estimate of the variance of the sampling distribution of the mean for samples of size 10. Then we compute the *t* statistic. After choosing a significance level, we can look in the *t* table to determine a rejection region. (The rejection region includes all values more extreme than the entry in the table.) We can then find out whether our computed t score lies in the rejection region. If it does, we reject the null hypothesis and conclude that something has changed: either grading is easier, or people are learning more this quarter.

2. Students at Weatherly Teachers College have Scholastic Aptitude Test scores which are approximately normally distributed, with mean $M_{pop} = 610$. Sixteen students were arrested in a demonstration. It was discovered that their SAT grades were 657, 604, 691, 785, 510, 654, 597, 680, 615, 800, 549, 593, 517, 724, 698, and 610. Do these scores give evidence that the students arrested were significantly different from the general student body in aptitude?

   Again, this is a good situation for a t test. The null hypothesis is that the scores were taken from a population with mean 610; that is, as far as SAT scores go, the students were typical of the general student body.

Two examples involving difference scores

3. Protein molecules in the brain apparently have something to do with learning. A psychologist has studied senile patients who have memory problems to see if providing them with a high-protein diet would help their memory. Each subject was tested twice in a standard memory test, once before the high-protein diet started and once after it had been in effect for several weeks. The results are presented here:

Subject	1	2	3	4	5	6	7	8
Score after diet	61	80	67	72	72	79	42	52
Score before diet	56	73	49	74	63	71	33	60

Did the high-protein diet have a significant effect on memory? This question, it turns out, can be answered with a t test. Try to figure out how before reading on.

The t test is good for answering a question of the following form: Did a single sample of numbers come from a population with known mean and unknown variance? Here, however, we have two sets of numbers, and no mention of a parent population. What do we do?

Here, just as we did in the case of the sign test, we look at the difference scores: What is the difference for each subject between his score after diet and his score before diet? Those differences are shown below:

Subject	1	2	3	4	5	6	7	8
Score after diet	61	80	67	72	72	79	42	52
Score before diet	56	73	49	74	63	71	33	60
Difference score After − Before	+5	+7	+18	−2	+9	+8	+9	−8

Now we can summarize the results of the experiment by a sample of eight numbers from the population of all possible difference scores. If the diet has no influence on memory, then the average difference score should be zero: a person should do about as well before the diet as after the diet. So we ask the following question: Could the above sample of difference scores (+5, +7, +18, etc.) plausibly have come from a population with a mean of zero? The null hypothesis says yes: H_0 : The eight difference scores were taken from a population with a zero mean. We compute a t statistic, and see if this null hypothesis should be rejected.

4. On alternate days in an 18-day period, the state police in Arizona maintain "high visibility" by patrolling a test section of highway with conspicuous police cars, or "low visibility" by patrolling it with unmarked cars. For each pair of days, two scores are obtained: number of accidents with high visibility police and number of accidents with low visibility police. The results are presented below:

Number of accidents under two conditions, for nine day-pairs

High visibility	31	16	29	38	24	22	30	18	32
Low visibility	42	35	27	36	28	45	36	25	26

Did the visibility of the state police make a difference?

A COMMENT ON THE POWER OF TWO TESTS

Recall problem 2 in Chapter 18, which dealt with an accuracy-increasing attachment for target-shooting rifles. Five marksmen tried it and observed changes of -1, $+6$, $+3$, $+7$, and $+5$ when they compared their before and after scores. If you used a t test to decide whether these five numbers could plausibly have come from a normal distribution with a zero mean (which would mean the attachment had no effect), you found that you should reject the null hypothesis and conclude that the attachment did indeed help accuracy. (Significance level = 0.10)

Now suppose that instead of running a t test you had decided to run a sign test. With a significance level of 0.10, as before, you would have analyzed this five-trial situation and decided that only two events were in the rejection region: 0/5 and 5/5. If the scores of all five men changed in the same direction, you would conclude the attachment had a significant effect. However, since only 4 out of 5 improved, you could not reject the null hypothesis. Using two different tests to analyze the same data, you reached two different conclusions: With the t test you were able to reject the null hypothesis. With the sign test, you were not. *What gives?* Is statistics a crazy black art, full of internal inconsistencies? The essential difference is that the t test considers not only the *direction* but also the *magnitude* of change. The sign test looks only at the direction (+ or −) of the change. As a result, the t test is more *powerful* than the sign test. That is, with the t test *you have a greater chance of finding an effect when there is one* than you do with the sign test. If we denote the probability of a Type Two error by β, the Greek letter beta, then we can write the following definition.

The power of a statistical test is $1 - \beta$, where β is the probability of a Type Two error. The smaller the chance of missing an effect when one is really there, the more powerful a statistical test is.

We have just seen an example in which the t test is more powerful than a sign test. Since it usually is, why bother to use the sign test? There are several reasons. You can run a sign test without referring to statistical tables. The computations are simple, and the theory is very clear. Often, as you have seen, the sign test is adequate to find significant effects. And when it is, it is the simplest way.

You can probably run a sign test in your head by now, using the normal approximation to the binomial. It is almost never possible for most people to run a t test mentally, because of the more extensive computations required.

And the sign test can be applied in situations where the t test cannot: where we observe only the direction of change and cannot attach a number to the magnitude of the change; or where the parent population is violently different from a normal distribution.

QUESTION AND ANSWER

Question: It was mentioned in this chapter that some students "took samples of size 4 from a unit normal distribution." How do you do that?

Answer: You have to find something which has exactly or approximately a normal distribution with $M = 0$ and $SD = 1$. In the above demonstration, the students used a computer which had been programmed to sample in such a distribution. But you could do the same thing without a computer if you were locked in a dungeon and had pencil, paper, and a penny. Here's how you might proceed. Suppose you were to toss the penny 100 times. The event "number of heads" has a binomial distribution. It has mean $N/2 = 100/2 = 50$. And it has variance $N/4 = 100/4 = 25$. So it has $SD = \sqrt{25} = 5$. Now recall that this binomial distribution and the normal distribution are very similar. So you could closely approximate a normal distribution as follows: Call a single sequence of 100 coin tosses *one experiment*. Now you write down the result of the experiment, for example, 58 heads. Then you perform the experiment again, and write down the result, for example, 47 heads. You could continue this indefinitely if you wanted to. Suppose you conducted the experiment 1000 times. The 1000 observations would very closely approximate a normal distribution with $M = 50$ and $SD = 5$. Now if you computed a Z score for each observation, by finding $(X_i - 50)/5$ for each score, X_i, you would have a close approximation to the unit normal distribution.

To return to the students, they could have performed the 100-coin-toss experiment four times, and computed a Z score each time. The four resulting numbers, then, would be four numbers taken from a normal distribution with $M = 0$ and $SD = 1$. These four numbers could be used to make up 1 of the 50 samples mentioned. It would have taken a while, however.

The point is to understand the ideas, not to do the entire experiment that way.

Question: In running a t test, we go through a long series of steps: Given the mean of the population, find the mean of the sampling distribution of the mean, M_M; given the numbers in the sample, find the sample mean, M_{samp}; given the numbers in the sample, estimate the variance of the population from which the sample was drawn:

$$\text{Est. } SD^2_{\text{pop}} = \frac{\Sigma(X_i - M_{\text{samp}})^2}{n-1}$$

Then, given the estimate of the population variance, use it to compute an estimate of the variance of the sampling distribution of the mean, SD_M^2:

$$\text{Est. } SD^2_M = \frac{\text{Est. } SD^2_{\text{pop}}}{n}$$

Finally, compute a t statistic:

$$t = \frac{M_{\text{samp}} - M_M}{\text{Est. } SD_M}$$

Isn't there an easier way?

Answer: There is, but if you start off by doing things the easier way, you lose track of the similarity between the t statistic and the Z statistic. Using all the steps above, and combining them, you can come up with the following formula or computation rule:

$$t = \frac{M_{\text{samp}} - M_M}{\text{Est. } SD_M} = \frac{M_{\text{samp}} - M_{\text{pop}}}{\text{Est. } SD_M} \qquad \text{since } M_M = M_{\text{pop}}$$

$$t = \frac{M_{\text{samp}} - M_{\text{pop}}}{\sqrt{\text{Est. } SD^2_M}} \qquad \text{since Est. } SD_M = \sqrt{\text{Est. } SD^2_M}$$

$$t = \frac{M_{\text{samp}} - M_{\text{pop}}}{\sqrt{\dfrac{\text{Est. } SD^2_{\text{pop}}}{n}}} \qquad \text{since Est. } SD^2_M = \frac{\text{Est. } SD^2_{\text{pop}}}{n}$$

$$t = \frac{M_{\text{samp}} - M_{\text{pop}}}{\sqrt{\dfrac{\dfrac{\Sigma(X_i - M_{\text{samp}})^2}{n-1}}{n}}} \qquad \text{since Est. } SD^2_{\text{pop}} = \frac{\Sigma(X_i - M_{\text{samp}})^2}{n-1}$$

and finally, simplifying the expression under the $\sqrt{}$ we have

$$t = \frac{M_{samp} - M_{pop}}{\sqrt{\dfrac{\Sigma(X_i - M_{samp})^2}{n(n-1)}}}$$

In the last expression, M_{samp} is the sample mean; M_{pop} is the mean of the hypothesized parent population; n is the sample size; and the summation in the denominator is over all the different values, X_i, in the sample. If you have a lot of t scores to compute, this formula will save you time. But if you are mostly interested in *understanding* what is going on, you may do better by going through all the steps. By so doing, you review several important properties of sampling distributions, and you review the rule for estimating population variance on the basis of a sample.

A final note on simplifying the computation of the t statistic: The following formula for t was presented above:

$$t = \frac{M_{samp} - M_{pop}}{\sqrt{\dfrac{\Sigma(X_i - M_{samp})^2}{n(n-1)}}}$$

The quantity in the denominator (which, you recall, is the estimated standard deviation of the sampling distribution of the mean, Est SD_M) is difficult to compute if the sample mean, M_{samp}, doesn't happen to be an integer. The following version of the formula often simplifies the number manipulation a bit further:

$$t = \frac{M_{samp} - M_{pop}}{\sqrt{\dfrac{\Sigma(X_i)^2 - nM_{samp}^2}{n(n-1)}}}$$

Here M_{samp} is the sample mean; M_{pop} is the hypothesized population mean; n is the size of the sample; and $\Sigma(X_i)^2$ is the sum of the squares of all the numbers in the sample.

Summary

The logic of the computation of the t statistic was reviewed. We asked whether a particular sample mean is so extreme that it could not plausibly have come from the sampling distribution of the mean for samples of the given size from the parent population specified in a null hypothesis.

The essential difference between the t statistic and the Z statistic is that in the t situation we have to *estimate* the variance in the parent population on the basis of the numbers in the sample. In order to run a Z test we need to *know* the variance of the parent population, as well as its mean.

A pair of frequency distributions was presented, showing how t scores and Z scores compare. Fifty samples of size 4 were taken from the standard normal distribution. For each sample, two scores were computed: a Z score and a t score. For the latter, we estimated the variance of the parent population from the numbers in the sample. It was found that extreme values of the t score are more likely than extreme values of the Z score. This is consistent with the fact that, for a given significance level, you need to observe a more extreme t score to reject the null hypothesis than you would if you were computing a Z score.

Several examples were presented in which a t test would be helpful. For two of these examples it was necessary to compute a set of difference scores, and then test whether the mean of the difference scores was significantly different from zero. If not, the two sets of numbers are not significantly different from each other.

Problems

1. The length of time that a cold will last is normally distributed with a mean of 6 days and an unknown variance. A group of people decide to try Vitamin C therapy and keep track of how long their colds last. They observe the following durations: 5, 2, 4, 9, 6, 1, 2, 3, and 3 days. Are these numbers sufficiently low, on the average, to convince you that they couldn't have come from a population with a mean of 6? Run a t test to make this decision. Use a significance level of 0.05. Show all five steps, just as you did for the Z test (see page 127). State your null hypothesis. Be sure you know what probability distribution you are talking about at all times. Any time you talk about a mean, be sure it's clear what it is the mean of. Likewise be certain for any variance or standard deviation.

    What would you have decided with a significance level of 0.10?

2. Run a t test for example 1 of page 164. Use a significance level $\alpha = 0.10$.
3. Run a test to make a decision about example 2 on page 165. Use a significance level of 0.05.
4. Run a test to decide whether the protein diet had any effect (example 3, page 165). Use a 0.01 significance level.
5. Run a test to determine whether visibility had an important effect on accidents in the Arizona test (example 4, page 165). Use a significance level of 0.10.
6. For each of the problems above, state what decision you would have made, in order to risk a Type One error.
7. Could you have run a sign test instead of a t test for problem 4 or 5 above? What would have happened?
†8. In the sampling experiment reported in this chapter, 50 samples were taken. For each sample a t statistic was computed, as if the population variance were unknown. Then a Z score was computed, using the true population variance. In 42 of 50 cases, the t statistic was more extreme than the Z score. Why wasn't it more extreme for 50 out of 50 cases?
†9. In the section on Power in this chapter we discussed an example in which the powerful t test detected a significant difference whereas the comparatively weak sign test failed to detect it. This is typical of the general pattern. But do you think there could *ever* be a case in which the opposite result would obtain: when a sign test would lead you to conclude a significant effect, and a t test would not? Try to write down a set of numbers, some positive and some negative, so that a sign test would convince you that positive numbers are more likely than negative numbers, but a t test would *not* convince you that the mean was different from zero.

    Although this may appear to be a paradox, it is not. Try to write an explanation of the situation.

Correlation

20

The word correlation is a familiar part of the everyday English of most of us. When two events are correlated, knowing about one of them will help you to predict something about the other. The idea is important in statistics, too. Here is a simple example.

For each of ten students in a college math course, two scores are obtained: a pretest score, based on a test taken at the start of the year to assess how much the student knows then; and a final exam grade, obtained at the end of the year. One might well expect such scores to be related: people who do well on the pretest might be expected to do well, in general, on the final. And those who had trouble with the pretest might well tend to have trouble with the final, too. The scores, and a *scatterplot:* a graph where each point shows the two scores for one student, are presented in Figure 20.1. Please turn to that figure.

SCATTERPLOTS

Note that the points in the scatterplot are concentrated around the diagonal from the lower left to the upper right side of the graph. This scatterplot has ten points, one to represent each pair of scores. The point for the first student, who scored 39 on the pretest and 65 on the final, is shown by a pointer. If we had data for 100 students, there would be 100 points on the scatterplot.

The reason for drawing such a plot is that it makes it easy to see that the scores are related. In general, a low score on one exam is paired with a low score on the other. And people who did well on the pretest are likely to do well on the final.

As you probably know, a correlation coefficient is a measure of such a relationship. It is symbolized r, and is always between -1 and $+1$. A correlation of $+1.00$ means a perfect relationship. Knowing one score, you can exactly predict the other. A correlation of -1.00 also means a perfect relationship, but an inverse one: high scores on one measure predict low scores on the other, and conversely. A correlation of 0.00 means that the two variables are unrelated.

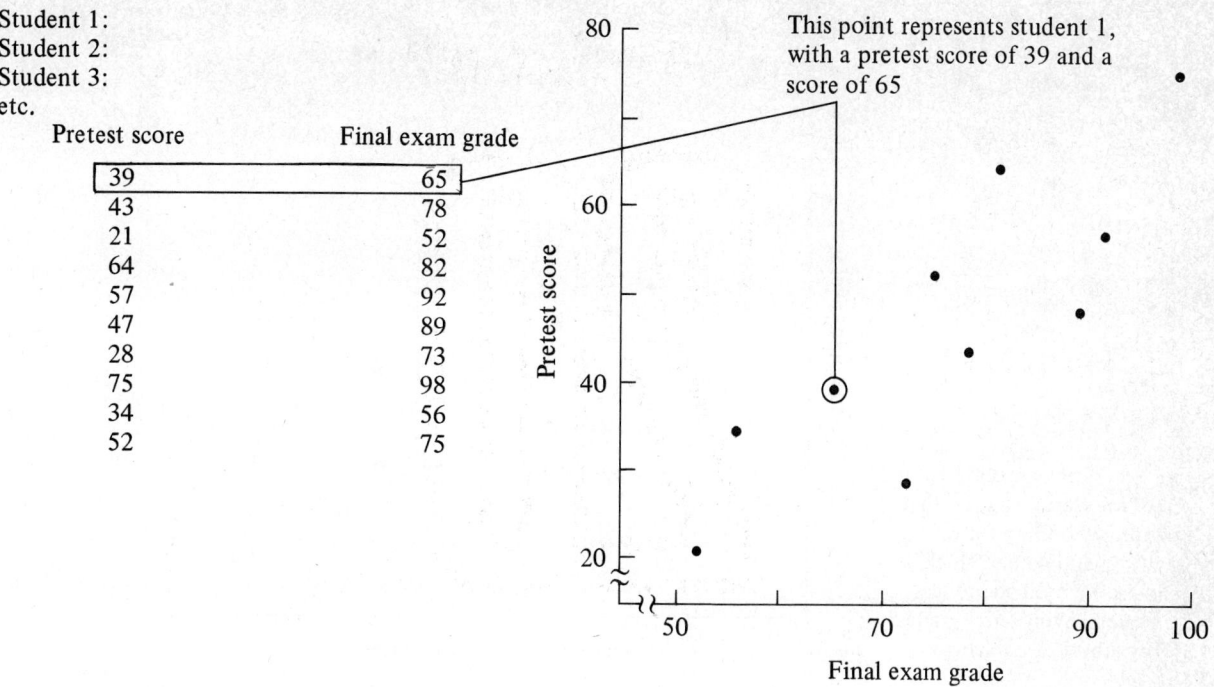

Figure 20.1 A scatterplot showing ten pairs of scores.

For the scatterplot in Figure 20.1, the correlation is +0.84. This is a strong positive relationship.

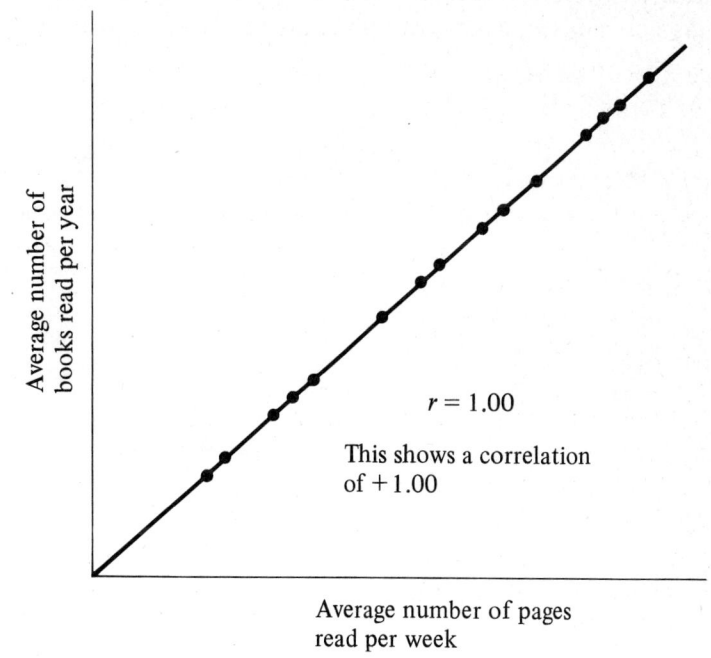

Figure 20.2 A perfect positive correlation, $r = +1.00$.

A correlation of +1.00 means that all the points lie along a straight line in the scatterplot, as in Figure 20.2. If you took a random sample of surfers and found two measures for each, you might find a negative correlation (Figure 20.3). A scatterplot like this would suggest that people who did a lot of surfing (and so had thick calluses) would spend less time improving their grades.

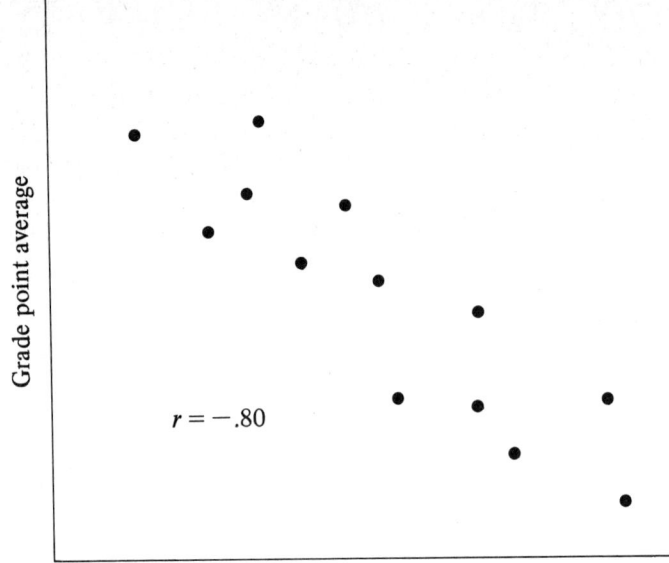

Figure 20.3 A scatterplot like this would suggest that people who did a lot of surfing (and so had thick calluses) would spend less time improving their grades. (Hypothetical data.)

A scatterplot for a situation in which two variables are totally unrelated will usually look something like Figure 20.4. In Figure 20.4, $r = 0.00$. For example, if you counted the number of letters in each person's entire name, for a group of 40 people, and then measured each person's height in inches, there would be no relationship between the two measures. We say they are uncorrelated.

Note that the correlation tells how well a *straight line* can summarize the data in a scatterplot. As you can see, a straight line could easily be used to roughly summarize the first

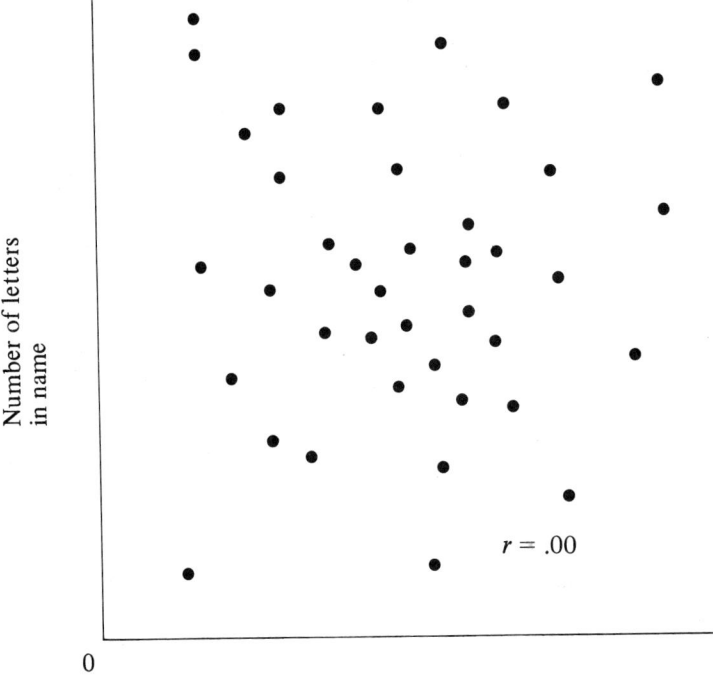

Figure 20.4 These two variables are uncorrelated.

three scatterplots in this chapter, but it would be very hard to choose a straight line which best fits the set of points in Figure 20.4.

The next scatterplot, Figure 20.5, presents a situation in which there is a relationship, but no straight line could be used to summarize it. In such a situation we say there is a *curvilinear* relationship. If you calculated the correlation coefficient, r, it would be about 0.00. But obviously there's a relationship here. It's just not a simple straight-line relationship.

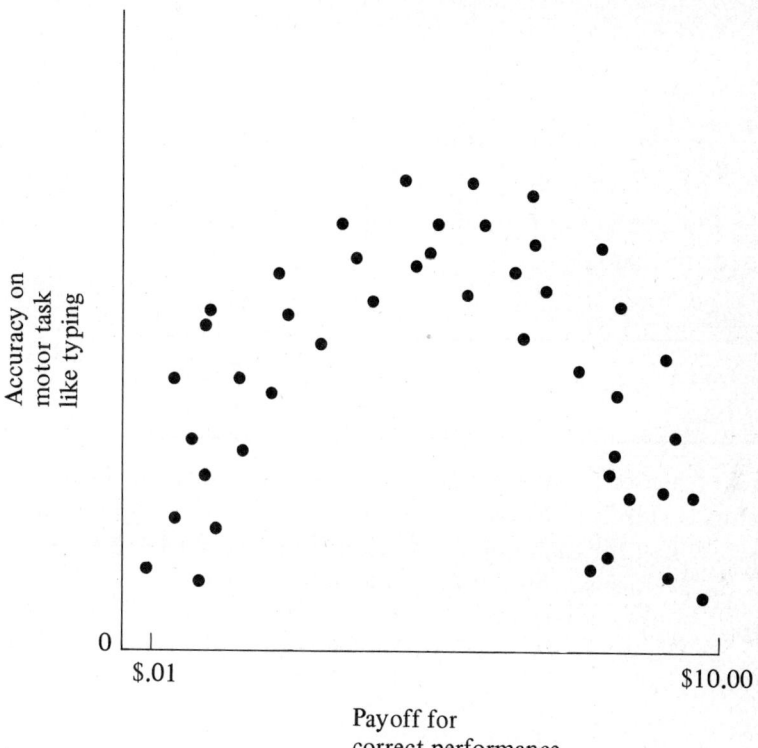

Figure 20.5 An example of a curvilinear relationship between two variables.

Since you immediately recognize a situation like this once you have drawn a scatterplot, it is a very good idea to draw a scatterplot before computing a correlation coefficient. The scatterplot may tell you more. (*Note*: The relationship indicated in Figure 20.5 is known in psychology as the law of Yerkes and Dodson. It says that as motivation increases, performance increases, up to a certain point. But after that point, performance starts to markedly deteriorate.)

HOW DO YOU CALCULATE THE CORRELATION COEFFICIENT?

There are a variety of approaches to this problem, but one of them is conceptually much simpler than the others. This is the one you should learn, since it is the easiest to remember.

For each of the scatterplots above, there were two sets of scores. In the first case, they were pretest scores and final exam grades. In the third case, they were grades of surfers and thickness of knee calluses of surfers. Now call one of these measures the X scores, and compute the mean and the standard deviation of that set of scores. (For example, you might start with the ten pretest scores.) Then compute a Z score, Z_{X_i}, corresponding to each data value, X_i. All data values lower than the mean will have negative Z scores, and those greater than the mean will have positive Z scores.

After you have found the Z scores for each of these data values, you repeat the entire calculation for the numbers in the second set. If the grades on the pretest were the X scores, then the second set, the grades on the final exam, could be called the Y scores. You must compute the mean and the variance of the final exam scores, and then compute the Z score for each of these scores, Z_{Y_i}.

You will finish with two new sets of scores, which are transformations of the initial two sets. Then you can calculate the correlation coefficient as follows:

$$r = \frac{Z_{X_1} \times Z_{Y_1} + Z_{X_2} \times Z_{Y_2} + Z_{X_3} \times Z_{Y_3} + \text{etc.}}{n}$$

$$= \frac{\Sigma Z_{X_i} \times Z_{Y_i}}{n} \quad \text{where } n \text{ is the number of pairs}$$

Example of calculation: Consider Table 20.1, which lists five pairs of scores on two separate quizzes for five people.

TABLE 20.1

| Score on Quiz 1 | 4 | 2 | 5 | 1 | 3 | X scores |
| Score on Quiz 2 | 4 | 8 | 2 | 10 | 6 | Y scores |

It would be a good, easy exercise to draw a scatterplot for these five pairs before reading on.

The next step is transforming the X scores. For the population 4, 2, 5, 1, 3 we wish to find the Z score, Z_X, for each number. The mean is $M_X = (4 + 2 + 5 + 1 + 3)/5 = 3$. The variance, SD^2, is the average squared deviation from the mean,

$$SD_X^2 = \frac{\Sigma(X_i - M)^2}{5} = \frac{(4-3)^2 + (2-3)^2 + (5-3)^2 + (1-3)^2 + (3-3)^2}{5}$$

$$= \frac{1+1+4+4+0}{5}$$

$$= 2$$

So $SD_X = \sqrt{2}$. Now we can find each value of Z_X:

$$Z_{X_i} = \frac{X_i - M_X}{SD_X} = \frac{X_i - 3}{\sqrt{2}}$$

TABLE 20.2

X_i	4	2	5	1	3
Z_{X_i}	$\frac{4-3}{\sqrt{2}} = \frac{1}{\sqrt{2}}$	$\frac{-1}{\sqrt{2}}$	$\frac{2}{\sqrt{2}}$	$\frac{-2}{\sqrt{2}}$	$\frac{0}{\sqrt{2}}$

Now we can similarly transform the Y scores into Z scores. For the Y population, 4, 8, 2, 10, 6, we must find the mean, M_Y, and the standard deviation, SD_Y:

$$M_Y = \frac{4+8+2+10+6}{5} = 6$$

$$SD_Y{}^2 = \frac{\Sigma(Y_i - M_Y)^2}{n} = \frac{(4-6)^2 + (8-6)^2 + (2-6)^2 + (10-6)^2 + (6-6)^2}{5}$$

$$= \frac{4 + 4 + 16 + 16 + 0}{5}$$

$$= 8$$

So $SD_Y = \sqrt{8}$.

Now, just as above, we can find a Z score, Z_Y, for each of the Y values:

$$Z_{Y_i} = \frac{Y_i - M_Y}{SD_Y} = \frac{Y_i - 6}{\sqrt{8}}$$

TABLE 20.3

Y_i	4	8	2	10	6
Z_{Y_i}	$\frac{4-6}{\sqrt{8}} = \frac{-2}{\sqrt{8}}$	$\frac{2}{\sqrt{8}}$	$\frac{-4}{\sqrt{8}}$	$\frac{4}{\sqrt{8}}$	$\frac{0}{\sqrt{8}}$

The next step is to compute a new number for each of the five students. Each student had two scores: X_i and Y_i. We have found a Z score corresponding to each of these, as indicated in Tables 20.2 and 20.3. Now we will find the product of the two scores for each student: $Z_{X_i} \times Z_{Y_i}$. We will then find the average value of these five products, averaged over all the students. That is the correlation coefficient. Now we will put all the information we have obtained together into Table 20.4.

TABLE 20.4 COMPUTING THE CORRELATION BETWEEN THE SCORES ON TWO QUIZZES

Student	X score Quiz 1	Corresponding Z_X score	Y score Quiz 2	Corresponding Z_Y score	Product $Z_{X_i} \times Z_{Y_i}$
1	4	$\frac{1}{\sqrt{2}}$	4	$\frac{-2}{\sqrt{8}}$	$\frac{1}{\sqrt{2}} \times \frac{-2}{\sqrt{8}} = \frac{-2}{\sqrt{16}}$
2	2	$\frac{-1}{\sqrt{2}}$	8	$\frac{2}{\sqrt{8}}$	$\frac{-1}{\sqrt{2}} \times \frac{2}{\sqrt{8}} = \frac{-2}{\sqrt{16}}$
3	5	$\frac{2}{\sqrt{2}}$	2	$\frac{-4}{\sqrt{8}}$	$\frac{2}{\sqrt{2}} \times \frac{-4}{\sqrt{8}} = \frac{-8}{\sqrt{16}}$
4	1	$\frac{-2}{\sqrt{2}}$	10	$\frac{4}{\sqrt{8}}$	$\frac{-2}{\sqrt{2}} \times \frac{4}{\sqrt{8}} = \frac{-8}{\sqrt{16}}$
5	3	$\frac{0}{\sqrt{2}}$	6	$\frac{0}{\sqrt{8}}$	$0 \times 0 = 0$
				Sum	$\frac{-20}{\sqrt{16}} = -5$

If this looks like an incomprehensible mass of nasty numbers, please take a couple of minutes to see where each part comes from. The only new numbers in this table are those in the right-hand column. The numbers in the second and third columns are from Table 20.2. The numbers in the next two columns are from Table 20.3. And the numbers in the last column are obtained by multiplying together the two Z scores for each student.

In doing this, we have used the rule that $\sqrt{2} \times \sqrt{8} = \sqrt{16}$. Now we can find the correlation coefficient by averaging the numbers in the last column, according to the formula for r presented above.

$$r = \frac{\Sigma Z_{X_i} \times Z_{Y_i}}{5} = \frac{(-2/\sqrt{16}) + (-2/\sqrt{16}) + (-8/\sqrt{16}) + (-8/\sqrt{16}) + 0}{5}$$

$$= \frac{-2/4 - 2/4 - 8/4 - 8/4 + 0}{5}$$

$$= \frac{-1/2 - 1/2 - 2 - 2 + 0}{5}$$

$$= \frac{-5}{5}$$

So

$$r = -1.00$$

Now if you are going to *remember* the formula for calculating the correlation coefficient, you had better understand it pretty well. It says that the correlation coefficient is the average of the products of the Z scores within each pair, averaged over all pairs. In this case we have calculated that $r = -1.00$. This means a perfect negative relationship between the two scores. A person who did well on one quiz did poorly on the other, and vice versa.

Remember that the Z scores are negative when a person did worse than the mean, and they are positive when he did better than the mean. In this example, whenever one of the scores is positive (above the mean), the other one is negative (below the mean). You can see this in Table 20.4. This is the essence of what is meant by a negative correlation! And the relationship is captured in the products of the Z scores, since for this example none of the products is a positive number.

If we were looking at a case where there was a strong positive correlation, then a person who scores high on one quiz would score high on the other. The product $Z_X \times Z_Y$ would be positive. On the other hand, a person who did poorly on one quiz would probably do poorly on the other: both Z_X and Z_Y would thus be *negative* numbers. Remember that the product of two negative numbers is a positive number, so his product $Z_X \times Z_Y$ will be a positive number again! Thus the correlation coefficient will be the average of a series of positive numbers, and it will be high and positive.

Consider, finally, a case in which there is no relationship between the two variables. The Z scores will be sometimes negative and sometimes positive. The products, $Z_X \times Z_Y$, will also be sometimes positive and sometimes negative. On the average, they will average out to around 0.00 – the correlation which says there is no relationship.

If you understand the logic of this calculation, you should be able to remember how to compute a correlation coefficient years from now. If you do not, you will forget the formula in a few weeks. *So why not understand it?*

Another correlation example: scatterplots with raw scores and with Z scores. Ten students kept track of how many hours they spent working in a math course, including time in class. The final grade which each obtained is shown in Table 20.5, along with the time spent.

TABLE 20.5

Student	1	2	3	4	5	6	7	8	9	10
X, Final grade	42	18	84	81	100	28	38	69	57	93
Y, Hours spent	90	70	100	82	104	66	78	95	85	90

Is there any correlation between the final grade and the number of hours spent? Just looking at the table, it's not easy to tell. It helps to make a scatterplot. In Figure 20.6, the scores are

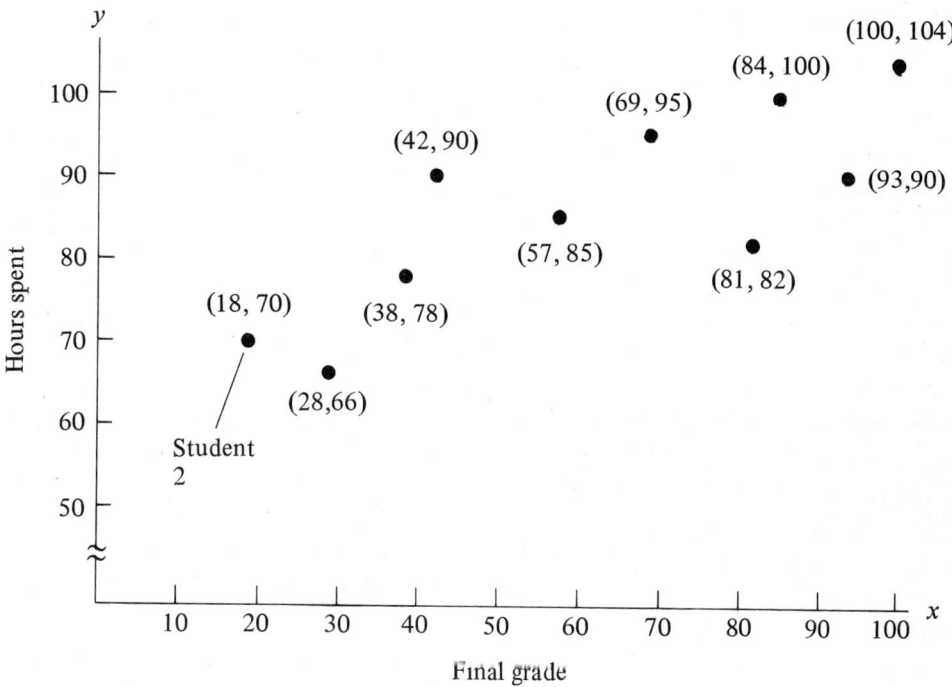

Figure 20.6 A scatterplot showing the relationship between hours spent in a course and final grade in the course for 10 students.

written beside the point for each student. For example, the numbers 18, 70 beside the point for student 2 indicate that his final grade was 18, and that he spent 70 hours working on the course. Having the graphic display should help you see that there is a positive relationship: The higher the number of hours, the higher the final grade. But it is hard to guess, from Figure 20.6, just what the correlation coefficient will be. In Figure 20.7 the same scatterplot is drawn, but this time the scores have been transformed into Z scores, as one might do in computing the correlation coefficient.

Figure 20.7 is a rather formidable figure. But if you spend about five minutes looking it over, it should help you to understand why the correlation coefficient, the average value of $Z_X \times Z_Y$, is a positive number for this scatterplot. The scatterplot has been divided into four quarters, or *quadrants*, by a vertical line at 61, the mean of the final grades (M_X), and by a horizontal line at 86, the mean of the hours spent scores (M_Y). Now consider just the lower left quadrant: any point in this quadrant has an X score less than M_X, which means that the corresponding Z score, Z_X, will be negative. Any point in this quadrant also has a Y score less than M_Y, which means

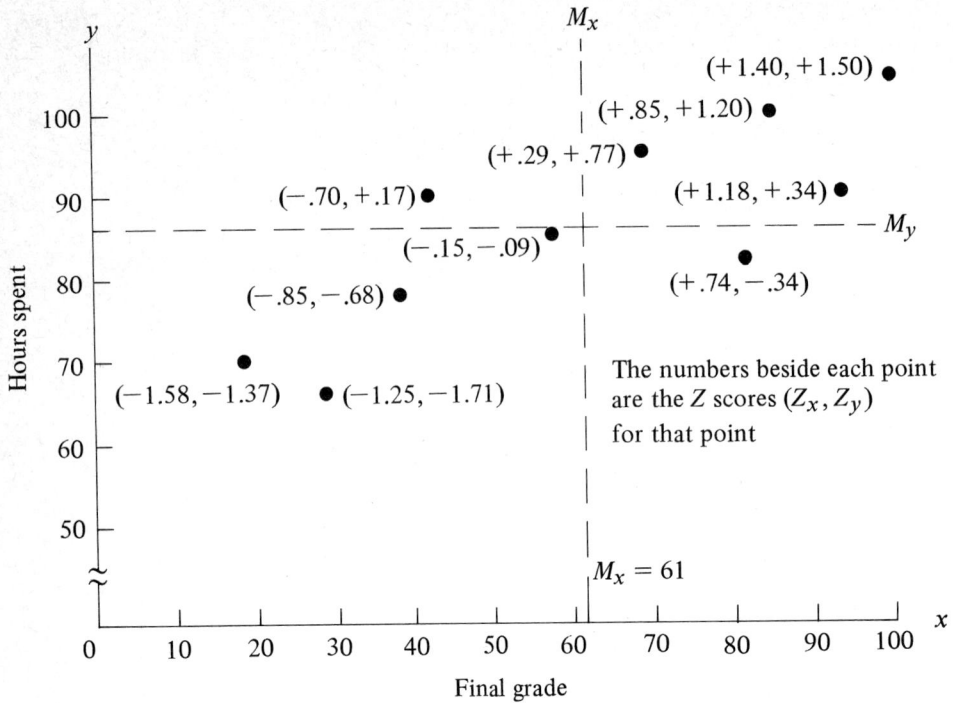

Figure 20.7 The data from Figure 20.6 are here replotted in terms of Z scores. Beside each point is a pair of numbers, Z_X, Z_Y, the Z scores for that point.

that the corresponding Z_Y score is negative too. Since both the Z scores are negative, the product, $Z_X \times Z_Y$, will be a *positive* number, and will tend to increase the correlation coefficient.

A similar argument can be made for the points in the upper right quadrant: in that area, both Z scores are positive, and again the products, $Z_X \times Z_Y$, will be positive. Note that these two quadrants contain most of the points in this scatterplot.

One point, with $Z_X = -0.70$ and $Z_Y = +0.17$, lies in the upper left quadrant. And another point lies in the lower right quadrant. Points in these two quadrants have one positive Z score and one negative Z score, so for such points the products $Z_X \times Z_Y$ will be negative.

If you have understood this discussion, you should anticipate that for this scatterplot there is a strong positive correlation. In fact, the correlation is +0.87, which is not far from a perfect positive correlation of 1.00.

A COMPUTATIONAL FORMULA FOR r

The rule that $r = (\Sigma Z_X \times Z_Y)/n$ will always work, and it helps many people understand the correlation coefficient. If you have many correlations to compute, however, the numbers get difficult. A more efficient computation system is shown below.

$$r = \frac{n\Sigma(X \times Y) - (\Sigma X) \times (\Sigma Y)}{\sqrt{[n\Sigma(X^2) - (\Sigma X)^2] \times [n\Sigma(Y^2) - (\Sigma Y)^2]}}$$

where n is the number of points in the scatterplot, and $\Sigma(X \times Y)$ is the sum of the products $(X \times Y)$, one for each point in the scatterplot. (For example, if a person studied $Y = 70$ hours, and received a final grade $X = 18$, his product $X \times Y = 18 \times 70 = 1260$.) $\Sigma(X \times Y)$ is the sum of many such products.

$\Sigma(X^2)$ represents the number you obtain if you square each number in the X list, and add up the squares. $(\Sigma X)^2$ looks similar, but is different: to get it you add up all the numbers in the X list, ΣX, and then square that total.

Sample computation. Table 20.1, repeated here for convenience, lists the X scores and the Y scores for five people who took two quizzes.

| Score on Quiz 1 | 4 | 2 | 5 | 1 | 3 | (X scores) |
| Score on Quiz 2 | 4 | 8 | 2 | 10 | 6 | (Y scores) |

Let's quickly use the computational formula to check our earlier result:

$n = 5$ (the number of pairs)

$\Sigma(X \times Y) = (4 \times 4) + (2 \times 8) + (5 \times 2) + (1 \times 10) + (3 \times 6)$
$= 16 + 16 + 10 + 10 + 18$
$= 70$

$\Sigma X = 4 + 2 + 5 + 1 + 3 = 15$

$\Sigma Y = 4 + 8 + 2 + 10 + 6 = 30$

$\Sigma(X^2) = 4^2 + 2^2 + 5^2 + 1^2 + 3^2 = 16 + 4 + 25 + 1 + 9 = 55$

$\Sigma(Y^2) = 4^2 + 8^2 + 2^2 + 10^2 + 6^2 = 16 + 64 + 4 + 100 + 36 = 220$

$$r = \frac{n\Sigma(X \times Y) - (\Sigma X) \times (\Sigma Y)}{\sqrt{[n\Sigma(X^2) - (\Sigma X)^2] \times [n\Sigma(Y^2) - (\Sigma Y)^2]}}$$

Substituting

$$r = \frac{(5 \times 70) - (15 \times 30)}{\sqrt{[5 \times 55 - (15)^2] \times [5 \times 220 - (30)^2]}}$$

$$= \frac{350 - 450}{\sqrt{[275 - 225] \times [1100 - 900]}}$$

$$= \frac{-100}{\sqrt{50 \times 200}} = \frac{-100}{\sqrt{10000}} = \frac{-100}{100}$$

$$= -1.00$$

just as we calculated above.

Which way is better for calculating a correlation coefficient? It depends. If you want to understand how the formula relates to the scatterplot, the first method, $r = (\Sigma Z_X \times Z_Y)/n$, is best. It has the great advantage that it is easy to remember. With luck, you will still remember it a year or two after reading this book. But I'll bet you couldn't remember the computational formula right now, without checking above.

If you have several correlation coefficients to compute, the computational formula will save you some time. If you have more than about half a dozen, you'll probably use a computer to help you. So my personal preference is for the form which is simple to understand and remember, $r = (\Sigma Z_X \times Z_Y)/n$.

Summary

We say that two variables are *correlated* if knowing the value of one will help us to predict the value of the other. The correlation coefficient, r, is equal to zero if the two variables are unrelated. It is equal to +1.00 if they are

180 THE t TEST, CORRELATION, AND CHI-SQUARE

perfectly correlated. And it is equal to −1.00 if they are perfectly negatively correlated: if the highest values of one are paired with the lowest values of the other, second highest and second lowest, and so on.

A *scatterplot* results when each pair of numbers in a correlation situation is represented as a single point on a graph. Several typical patterns were presented.

A conceptually simple way to calculate the correlation coefficient was presented: Call the first variable the X score, and the second the Y score. Transform the scores in the first list into standard scores, Z_X, using the mean and the standard deviation of that set of scores, and do the same for the Y scores. Then the correlation coefficient, r, is given by

$$r = \frac{\Sigma Z_X \times Z_Y}{n}$$

the average value of the product $Z_X \times Z_Y$, averaged over all the pairs.

A computational formula for the correlation, which is harder to remember and understand, but somewhat easier to apply, was also presented. It always gives the same answer

Problems

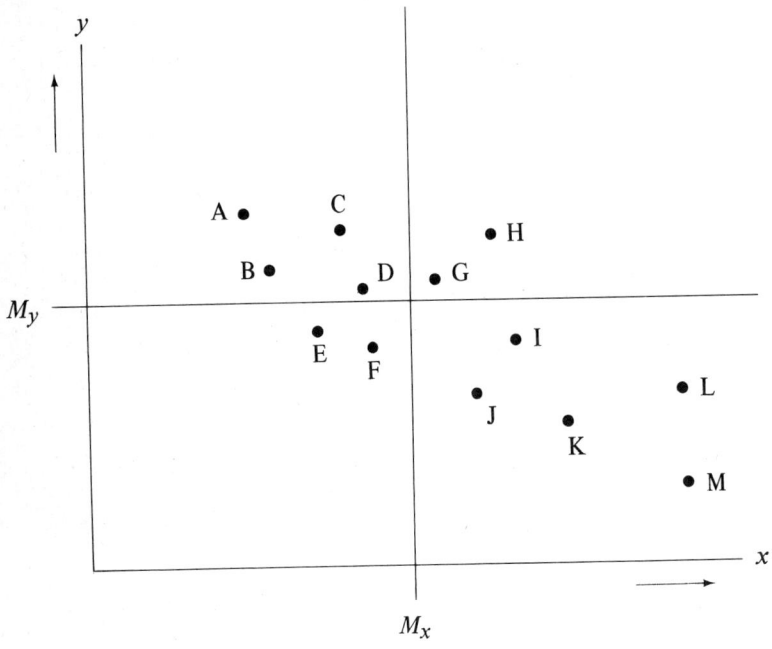

1. In the accompanying scatterplot each point is identified with a letter of the alphabet, from A to M. Which points will have a positive product $Z_X \times Z_Y$, and so tend to increase the correlation coefficient? Which points will have a negative product $Z_X \times Z_Y$, and so tend to decrease the correlation coefficient? Would you guess this scatterplot represents a positive or a negative correlation? (If you find this difficult, start by deciding which points will have a positive Z_X score.)

2. Have you ever noticed that when you ask for "half a pound" butchers seem to give you about 10 ounces? I watched mine during 15 transactions. In 13 of the 15 he gave the customer *more* than requested, and in 2 he gave less. Does this constitute convincing evidence that he is consistently biased in the direction that sells more meat? Use a 0.05 significance level.

3. A Chamber of Commerce, interested in the source of frugality, investigated the relationship between the amount of money a man had saved at age 40, and his parents' average income while he was a child (age 5 to 12). They found the following for ten people:

Parents' income, ($/yr, in thousands)	8	16	6	12	12	20	22	12	12	10
Savings at age 40 ($, in thousands)	9	5	10	5	3	5	4	9	6	6

a. Draw a scatterplot for these data.

b. Compute the correlation coefficient.

4. For the correlation example on page 177, the set of X scores has mean $M_X = 61$ and standard deviation $SD_X = 27.2$. The set of Y scores has $M_Y = 86$ and $SD_Y = 11.7$. Knowing this information, find Z_X and Z_Y for student 2, who had $X = 18$ and $Y = 70$. Do your results agree with those in Figure 20.7? Could you, if asked, compute the rest of the numbers on that figure?

5. Draw a scatterplot for the quiz data in Table 20.1. Does your figure make sense, given the correlation computed for these data?

†6. Does it matter which set of data is called the X set and which set is called the Y set? Can you prove it?

†7. It was mentioned in this chapter that when two variables are perfectly correlated, the scatterplot will be a straight line. This means that there are two constants, a and b, such that all the points lie along the line with the equation $Y = aX + b$.

a. Find out what these constants are for the quiz data in Table 20.1.

b. Make up two new data lists which contain at least ten data values, and which, you know, must be perfectly correlated since they lie along a straight line.

c. If you are feeling strong, compute the correlation coefficient for these new lists, to check your work.

Note: The best-fitting straight line through a scatterplot is called a *regression line*.

Correlation, Concluded

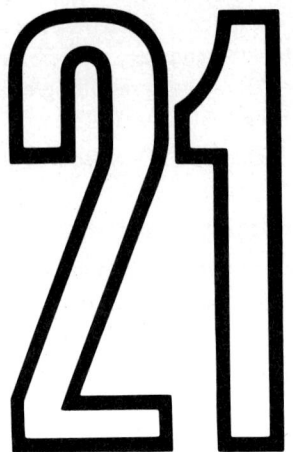

In Chapter 20 the fundamental rule for computing a correlation coefficient was introduced. Suppose you are calculating the correlation coefficient between pretest scores (called the X numbers) and final grades (called the Y numbers). You first consider all the X numbers as a population, and find the mean and variance. Then you transform them into Z scores, Z_X. Similarly, you transform all the Y numbers into Z scores, using the mean and variance of the Y numbers considered as a population. Finally, you multiply together the two scores for each person, and find the average, the *correlation coefficient*, r:

$$r = \frac{\Sigma Z_{X_i} \times Z_{Y_i}}{n}$$

We shall now work through a small example of this computation, which will serve to introduce the next topic: How large a correlation coefficient do you need to see before you conclude that it is significant? That is, it couldn't plausibly be said to be the result of chance variations?

TESTING THE SIGNIFICANCE OF r

The following demonstration can easily be done with any four numbers. To prove this to yourself, write any four numbers you like into the grid (Figure 21.1). Then, after you have read the following example, find the correlation between the two pairs of numbers.

The following four numbers were chosen at random by four different students on one occasion: 68, 54, 10, 90. Suppose that these are taken to be the scores of John and Mary on two different tests they took, quantitative and verbal. *Question:* What is the correlation between the quantitative scores and the verbal scores? In Figure 21.2 call the quantitative scores the

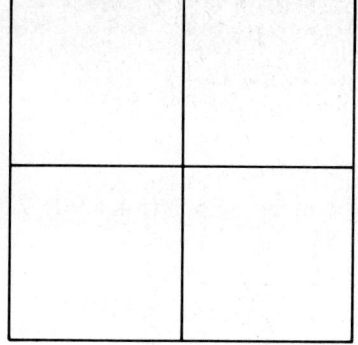

Figure 21.1 Write one number in each of the four boxes.

	Quantitative	Verbal
John	68	54
Mary	10	90
	X scores	Y scores

Figure 21.2

X numbers, and the verbal scores the Y numbers. First we must transform them into Z scores.

Consider the X population, 68, 10. The mean, M_X, is 39. The variance, the average squared deviation from the mean, is

$$SD_X^2 = \frac{(68 - M_X)^2 + (10 - M_X)^2}{2} = \frac{(29)^2 + (-29)^2}{2} = (29)^2 = 841$$

$$SD_X = \sqrt{SD_X^2} = \sqrt{841} = 29$$

Knowing the mean, M_X, and the standard deviation, SD_X, it is easy to find the Z scores corresponding to the two X numbers, 68 and 10.

When $X = 68$:

$$Z_X = \frac{68 - M_X}{SD_X} = \frac{68 - 39}{29} = 1$$

When $X = 10$:

$$Z_X = \frac{10 - M_X}{SD_X} = \frac{10 - 39}{29} = -1$$

We record the following results:

	Z_X	X	Y
John	1	68	54
Mary	−1	10	90

Similarly, the Y population is 54, 90. The mean, $M_Y = (54 + 90)/2 = 72$. The standard deviation may be found:

$$SD_Y{}^2 = \frac{\Sigma(Y_i - M_Y)^2}{n} = \frac{(54-72)^2 + (90-72)^2}{2} = \frac{(-18)^2 + (18)^2}{2} = 18^2$$

So

$$SD_Y = \sqrt{SD_Y{}^2} = \sqrt{18^2} = 18$$

Now, knowing $M_Y = 72$ and $SD_Y = 18$, we can transform the Y scores into Z scores.

When $Y = 54$:

$$Z_Y = \frac{54 - M_Y}{SD_Y} = \frac{54 - 72}{18} = -1$$

When $Y = 90$:

$$Z_Y = \frac{90 - 72}{18} = 1$$

Adding these two numbers to the above table we get:

	Z_X	X	Y	Z_Y
John	1	68	54	−1
Mary	−1	10	90	1

Using the new table we can find r:

$$r = \frac{\Sigma Z_{X_i} \times Z_{Y_i}}{n}$$

$$n = \frac{Z_{X_1} \times Z_{Y_1} + Z_{X_2} \times Z_{Y_2}}{2}$$

$$= \frac{(1) \times (-1) + (-1) \times (1)}{2}$$

$$= \frac{-2}{2}$$

$$= -1.00$$

After picking four numbers completely at random, and calculating the correlation coefficient based on the two pairs, we have obtained a perfect negative correlation: $r = -1.00$.

No matter what numbers you choose (as long as both people didn't have identical scores on either test), you would always obtain a perfect correlation. Half the time it will be +1.00, and half the time it will be −1.00. At this point, you should carry out a similar computation with the four numbers you wrote in the grid above.

Does this make sense? It does, for the following reason. Remember that a correlation coefficient represents a measure of how well the dots in a scatterplot can be represented by a straight line. If all the dots lie along one straight line, the correlation will be perfect: +1.00 or −1.00. *Question:* How well can you fit a *pair of dots* with a single straight line? Plainly, it is easy to draw a straight line which passes through both of the dots in Figure 21.3.

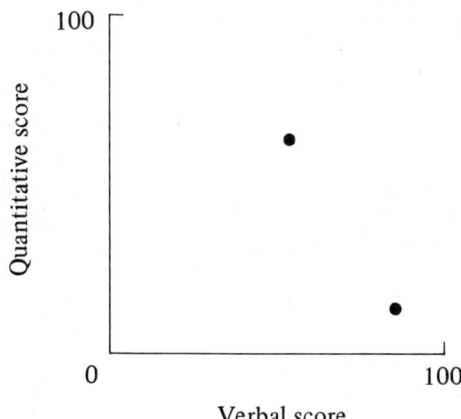

Figure 21.3 It is easy to draw a straight line that passes through all (both) points in this scatterplot.

Now suppose someone tells you that he has done an experiment to test the relationship between "Amount of time spent in an alpha rhythm state, during an experiment" and "Amount of learning during the experiment." He reports a correlation of +1.00. That's as high a correlation as you can ever observe. Should you be impressed? Maybe not. If he ran only two subjects, and so had only four numbers in his data, he would have found a correlation of +1.00 1 time out of 2, by pure chance, as we have just seen. In order to interpret a correlation coefficient, it is important to know something about the r values we would observe if in fact two variables were completely unrelated.

The Sampling Distribution of the r Statistic.

Consider the following sampling experiment. In a large population of students we know the IQ and the social security number of each person. A mad statistician samples ten students, and finds the last three digits of the social security number and the IQ for each. He calls the shortened social security number the X variable, and the IQ the Y variable, and computes a correlation coefficient. He finds $r = +0.17$. Then he takes ten more students and computes a correlation coefficient on the basis of the ten new pairs of numbers: this time he gets −0.06. And he continues this process, over and over, 1,000 times. At the end, he has a list of 1,000 correlation coefficients, each based on a sample of ten students from the population. They range from −0.71 to +0.67, and 90 percent of the correlations are between −0.55 and +0.55. How can he interpret such numbers? Does the correlation of +0.67, which he observed once, mean a positive relationship between social security number and IQ? Or what?

Finally, he does a grand computation: He takes all the thousands of students in the population, and computes the correlation coefficient between IQ and social security number for the entire group. He finds it to be 0.00. So the set of 1,000 correlation coefficients which he found in his sampling experiment is an approximation to the sampling distribution of the correlation coefficient, for samples of size 10 (10 pairs), from a population where the true correlation is zero.

The fundamental point is this: Even if two variables are completely unrelated, as are IQ and social security number, the correlation coefficient found on the basis of a small number of pairs will not always be zero. By pure chance, it will sometimes be positive and sometimes negative. Sometimes it will be quite far from zero; remember that the statistician once saw a value of +0.67. But knowing the sampling distribution helps us to interpret our own correlation coefficients.

Suppose, in an experiment, you measured spelling ability and skill in a ring-toss game for ten students. For each student you had a pair of scores. And you found the correlation coefficient between spelling and ring-toss scores to be +0.28. Should you conclude that you have firm evidence of a true relationship between these two variables? Probably not. For our mad statistician found values far more extreme than that, using samples of the same size, when in fact there was no relationship. Ten percent of the time he saw correlation coefficients greater than +0.55 or less than −0.55. So there's nothing very unusual about a score of +0.28. It could well be the result of chance fluctuation. If you were persistent, however, and repeated your experiment with 100 subjects, and again found a correlation of +0.28, that would be far more impressive. As you will be able to check below, such an extreme correlation would almost never arise by pure chance, if in fact the two variables, spelling score and ring-toss score, were unrelated.

The r Table: In Figure 21.4 there is a table of critical values for the correlation coefficient. Please refer to the table as you read the following discussion. In the left-hand column is the number of pairs of scores for an experiment. The various P values across the top of the table are significance levels. Suppose you have done an experiment with 22 pairs of scores (22 subjects, two scores each). In the second column, $P = 0.10$, we see that for 10 pairs, the number .549. This means that 10 percent of the time, if you have ten subjects and there is absolutely no relationship between the two measures you took, you will still observe a correlation greater than +0.549 or less than −0.549. Each row of the table summarizes information about one entire sampling distribution of the r statistic.

Example: Suppose you do an experiment with 42 subjects, and you calculate the correlation between the GPA of the subject and his "success score" in the experiment. How big a correlation coefficient do you need to observe in order to conclude the following: That couldn't possibly be due to chance fluctuations; so extreme a value of r would happen less than 1 time in 100 if there were really no relationship.

Solution: Looking across the row for 42 subjects in the 0.01 column, you find .393. This means that you would have to observe a correlation greater than +0.393 or less than −0.393 in order to confidently reject the idea that the correlation could have been the result of random fluctuations.

It helps, for the present discussion, to remember that when we compute a correlation coefficient we do so on the basis of a *sample* of pairs of numbers. But we usually are interested in making an inference to an entire *population* of such pairs. In the example at the start of the previous chapter, we considered the pretest scores and final exam grades of ten students in a college math course. But we were really interested in finding out if, for the general population of students in such courses, the final grade is highly correlated with the pretest score. Our inference problem is complicated, since the correlation coefficient computed for a sample will rarely be exactly the same as the coefficient you would obtain if you considered the entire population.

r TABLE: CRITICAL VALUES OF THE CORRELATION COEFFICIENT

Number of pairs (number of points in scatterplot)	Level of significance for two-tailed test (P)			
	0.10	0.05	0.02	0.01
3	.988	.997	.9995	.9999
4	.900	.950	.980	.990
5	.805	.878	.934	.959
6	.729	.811	.882	.917
7	.669	.754	.833	.874
8	.622	.707	.789	.834
9	.582	.666	.750	.798
10	.549	.632	.716	.765
11	.521	.602	.685	.735
12	.497	.576	.658	.708
13	.476	.553	.634	.684
14	.458	.532	.612	.661
15	.441	.514	.592	.641
16	.426	.497	.574	.623
17	.412	.482	.558	.606
18	.400	.468	.542	.590
19	.389	.456	.528	.575
20	.378	.444	.516	.561
21	.369	.433	.503	.549
22	.360	.423	.492	.537
23	.352	.413	.482	.526
24	.344	.404	.472	.515
25	.337	.396	.462	.505
26	.330	.388	.453	.496
27	.323	.381	.445	.487
28	.317	.374	.437	.479
29	.311	.367	.430	.471
30	.306	.361	.423	.463
31	.301	.355	.416	.456
32	.296	.349	.409	.449
37	.275	.325	.381	.418
42	.257	.304	.358	.393
47	.243	.288	.338	.372
52	.231	.273	.322	.354
62	.211	.250	.295	.325
72	.195	.232	.274	.303
82	.183	.217	.256	.283
92	.173	.205	.242	.267
102	.164	.195	.230	.254

Example
If two variables are really unrelated, and you observe 17 pairs of scores, there is a probability of 0.02 that you will observe a correlation greater than +0.558 or less than −0.558.

Note: This table is also reproduced in the Appendix for easy reference.

From Table VII of Ronald A. Fisher and Frank Yates, *Statistical Tables for Biological, Agricultural and Medical Research*, 6th ed., published by Oliver and Boyd, Edinburgh, 1963, p. 63, and by permission of the authors and publishers.

Figure 21.4

Suppose you started with a pair of variables which are totally uncorrelated, like zip code and height in inches, for each of 10,000 people. These are almost surely uncorrelated. But if you took a sample of five people, found the two scores for each person, and then computed the correlation coefficient, it would not always be zero. Indeed, if you took 1000 such samples (with replacement) and computed the r statistic each time, you would approximate the entire *sampling distribution of the r statistic* (for sample size 5). It is just such a sampling distribution which is involved in the corresponding row (for sample size 5) in the correlation table. On the basis of such a sampling distribution, it was determined just how often one would see various values of the correlation coefficient, based on samples of size 5 from a population of pairs in which the true (population) correlation is zero.

We saw a variety of sampling distributions in earlier chapters. For each case, we can find out, or calculate, how likely it is that we shall see various values of a statistic of interest: a score, like 8/10, in the ESP experiment; a statistic, like a t score or a Z score, under various conditions. Now we have a table which tells us how likely it is that we shall see various values of the r statistic as a function of the sample size. In every case, we know about a sampling distribution when the null hypothesis is true: For the ESP experiment, we calculated the probability of various scores *when the subject was just guessing*. Later we looked in a table to find the chance of various Z or t scores, when the null hypothesis was true, and the sample did indeed come from the hypothesized population. And now we look in the r table to find a value of the r statistic which is very unlikely to be exceeded when the null hypothesis is true and the two variables involved are in fact uncorrelated in the parent population.

PRACTICE JUDGING SCATTERPLOTS

This section is intended to help you learn to estimate scatterplots by eye. This is an extremely simple way to get a first idea about the relationship between two variables. Figure 21.5 shows a variety of correlations, ranging from −0.99 to +0.99. Each figure can be viewed in four ways, showing two scatterplots each for two correlations. For example, the scatterplot for a correlation of +0.99 can be rotated clockwise through 90, 180, or 270°, and shows a correlation of −0.99, +0.99, and −0.99 in those three other orientations. The 12 scatterplots, each viewed in four orientations, thus show you 4 x 12 = 48 different pictures. After looking them over, and trying to guess a correlation while you cover the numbers, you should become quite skillful at estimating a correlation on the basis of a scatterplot.

CORRELATION AND CAUSALITY

One of the most common and most serious of all statistical errors can occur when someone infers causality from a correlation. We already discussed this in Chapter 2; we saw there that just because there is a correlation between marijuana use and hard drug use (as there may well be) does not *necessarily* mean that the former *causes* the latter. There *may* be a causal relationship; but without further information, we cannot be sure simply because there is a correlation. Similarly, there is probably a correlation between the number of swallows flying low and millimeters of rain in the next three hours. But this does not mean that the swallows *caused* the storm.

This is a difficult situation, because when one variable causes another they are often strongly positively correlated. But the opposite is not true. If, for any large city, you were to plot the number of illegitimate children born in each of the past 30 years, and also the number of

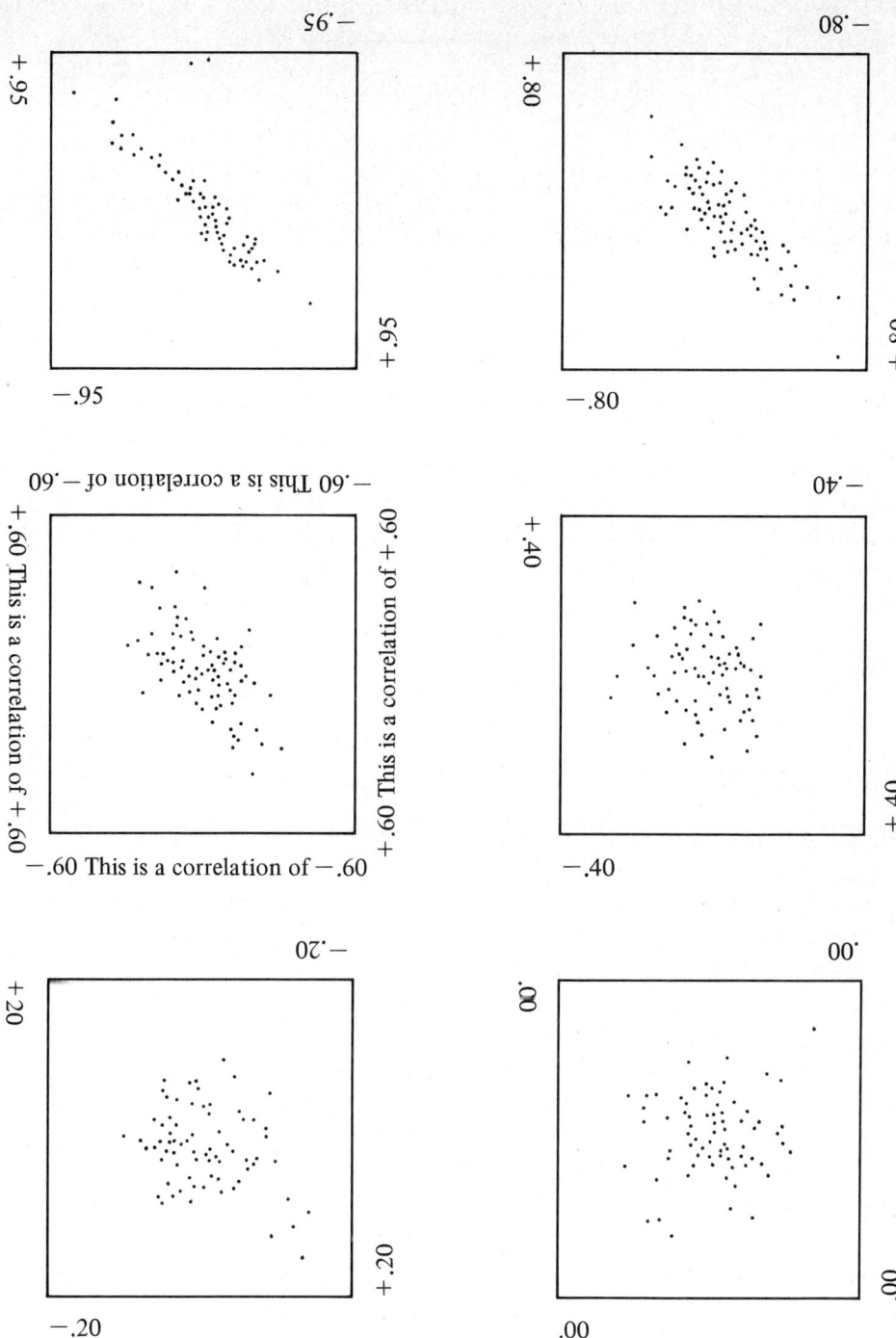

Figure 21.5 Each scatterplot can be viewed four ways. Rotate the book to see this. The correlation you see is noted in the lower left-hand corner for each possible orientation.

clergymen living in the city in each year, there would probably be a strong positive correlation. This would not, however, mean that there is a causal relationship: both variables are directly related to population, and so they co-vary.

The essential point is this: If someone tells you that two variables are highly correlated, and

190 THE t TEST, CORRELATION, AND CHI-SQUARE

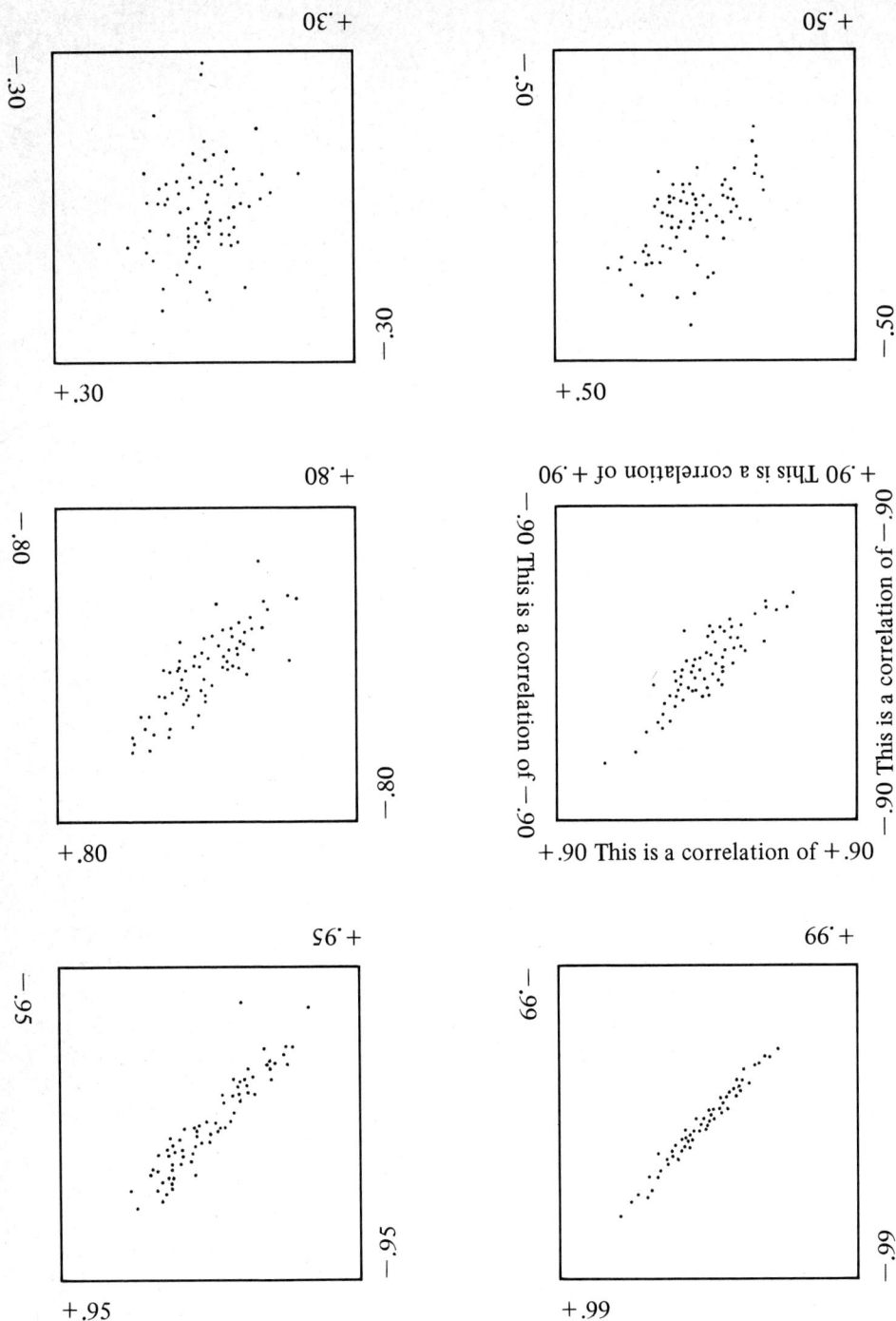

Fig. 21.5   *(Continued)*

proceeds to deduce that one of them causes the other, he *may* well be mistaken. On the basis of the correlation, he can't be sure.

In previous chapters you were introduced to a variety of statistical tests. In each case you should have learned when that test is appropriate and useful. But to give a general review, and interrelate what you have learned so far, it may be helpful to consider all six tests you have learned about, and to see where they are useful.

GENERAL REVIEW OF CONDITIONS
UNDER WHICH YOU APPLY EACH OF SIX STATISTICAL TESTS

Test	Conditions for Application
Sign test	The sign test is used when you are interested in evaluating the statement (null hypothesis) that each of two different and nonoverlapping results occurs just half the time. For example, you might be interested in finding out if it were true that "boy babies and girl babies are equally likely" under certain conditions. Or you might be interested in deciding whether to accept the following premise (null hypothesis): An increase in score is just exactly as likely as a decrease in score, as a result of this treatment. The procedures you have used for computing probabilities become quite difficult for situations in which there are more than about 12 events, or trials. A precise null hypothesis for the sign test can almost always be put in the following form: For each _____ there is a 0.5 chance that _____ will occur, and an equal chance that _____ will occur.
Sign test, using the normal distribution to approximate the true probabilities given by the binomial distribution	This test is appropriate in exactly the same sort of situation as the basic sign test, provided that there are so many trials or difference scores or results that the other procedure is too cumbersome. The normal approximation is not very accurate for less than about 30 events, though it can still be used. Remember that you use the fact that the mean of the symmetric binomial distribution is $N/2$ and the variance is $N/4$ to help you draw a normal distribution which will have the same shape.
The Z test for a single observation	This is the simplest possible Z test. It isn't often useful, but helps understand the more complex Z test, which is next in line. You use this test when your data consist of a *single observation* and you are interested in considering the hypothesis that this observation came from a specific normal distribution with known mean and variance. To decide whether the hypothesis is plausible, you compute a Z score based on the single observation. If it is very extreme, you decide your hypothesis is incorrect.
The Z test for the mean of a sample of size n	Quite often you are given the mean of a sample, and you are interested in knowing whether that sample came from a known, well-specified parent population. If the sample contained 30 or more numbers, the sampling distribution of the mean for such samples will be a normal distribution. It will have the same mean as the parent distribution, and a variance equal to SD^2_{pop} divided by n, the sample size. At this point, you can consider the observed sample mean as a single observation, and ask whether it came from the well-defined normal distribution which is the sampling distribution of the mean. From this point on, the procedure is identical to that for the previous test. This test is also useful for smaller samples, if they come from a known parent population which is a normal distribution. If so, the sampling distribution of the mean is a normal distribution, even for small samples.
The t test for the mean of a sample	This test is used in situations very like the Z test, except that it can be used (and is only used) when we don't know the variance of the hypothetical parent population. The theory behind the t test assumes that the hypothetical parent distribution is a normal distribution. If it is approximately like one, the t test may be appropriately used. If the parent distribution were very different from a normal distribution, the t test might get you into trouble. In short, the t test is used to test a hypothesis that an observed sample came from a normal distribution with a known, specified mean, but variance unknown. You use the variability of the sample to estimate the variance of the population.
The r test	When you have computed a correlation based on a sample of pairs, and wish to decide whether it could plausibly have resulted from random fluctuations alone, in a situation in which two variables are really totally unrelated, you use the r tables in Figure 21.4. The null hypothesis in this situation is that in the population of pairs from which you have taken a sample, there is no correlation between the two variables: the true correlation is zero.

192   THE t TEST, CORRELATION, AND CHI-SQUARE

QUESTION AND ANSWER

Question: Suppose that for each person in the United States we obtain two numbers: the number of pages he reads in a given year, and the amount of taxable income he reports in that year. If we computed a correlation coefficient based on the 200 million pairs of numbers, it might well be that it is 0.00. But I looked at the data for ten people on my street, and computed the correlation coefficient: it was $r = -0.34$. How can the correlation be 0.00 and -0.34 at the same time?

Answer: Although the *true* population correlation coefficient might be 0.00, this is not a fact that anyone is likely to know; it would be too difficult to get the information, or to do the calculation. All one can do is to take a sample of people and find out what the correlation is in the sample. Suppose you take a sample of ten people. If you do this over and over, you will get a different value almost every time. If you were to continue this for thousands of samples, and you made a frequency distribution of the various values of the correlation coefficient, you would have approximated the *sampling distribution* of the correlation coefficient for samples of size 10, from a population in which the two variables (pages read and dollars earned) are actually unrelated. This distribution would reveal that although values around 0.00 are more likely than others, there is still a broad spread.

This is important, since whenever we compute a correlation coefficient it is based on a *sample*. But we are almost always interested in the correlation which prevails in the *population*. Thus we are concerned with the variability of the statistic (the variance of the sampling distribution of the statistic).

Summary

The simple rule for computing the correlation coefficient, r, was reviewed. If we write down your weight and height and my weight and height, we will find that there is a perfect correlation between them. This is the case since any two pairs of numbers are perfectly correlated ($r = +1.00$ or $r = -1.00$). If two variables are uncorrelated, and a sample of pairs is obtained and the correlation coefficient computed on the basis of the sample, the correlation will not always be 0.00. Indeed, it will almost never be *exactly* zero. For a small number of pairs, the correlations will fluctuate widely. If you look at six pairs, there is 1 chance in 10 that you will see a correlation coefficient greater than $+0.729$ or less than -0.729 — even when the two variables are actually unrelated. The larger the number of pairs (the larger your sample), the greater the probability that your computed value of r will be close to zero, when that is the true correlation.

Tables exist to tell you critical values for the correlation coefficient, for each possible number of pairs. Such a table is included in this chapter, and reproduced in the appendix for easy reference.

Problems

1. A psychology student runs five subjects in an experiment. For each subject he obtains two measures: a measure of idealism, and a measure of anxiety. Find the correlation coefficient between the two scores (you probably will want to make a quick scatterplot):

   | Idealism | 4 | 5 | 1 | 3 | 2 |
   | Anxiety: | 8 | 6 | 10 | 4 | 2 |

2. In running a t test, the following may be involved: $M_{\text{samp dist of the mean}}$; M_{samp}; Est.$SD^2_{\text{samp dist of the mean}}$; M_{pop}; and Est.SD^2_{pop}. For each of these, state whether it is a parameter or a statistic.

3. A company claims that the average life of its batteries is 21.5 hours. A laboratory tests the accuracy of this statement. For a sample of five batteries, the following results were recorded: 19, 18, 22, 16, 25. Do these results indicate that batteries of this type have a shorter life than that claimed by the company? Use a significance level of 0.05.

4. How large a correlation would you need to observe, based on 17 pairs of numbers, to be convinced that a correlation couldn't be due to chance fluctuation? Use $\alpha = 0.01$.

†5. Suppose that you were locked in a dungeon with pencils, paper, index cards, and large hat. You were told that in order to escape, you must find a close approximation to the sampling distribution of the correlation coefficient, r, for samples of size 10, in a situation where the two variables are absolutely unrelated. How could you proceed? Assume that there are no statistics books in the room, but that you know how to calculate correlation coefficients.

In other words, you are to find answers to the following questions, when you have ten pairs of numbers from a situation in which the two variables are uncorrelated:

a. How likely is it that you'll observe a correlation between -1.00 and -0.95?

b. How likely is it that you'll observe a correlation between -0.95 and 0.90?

c. How likely is it that you'll observe a correlation between -0.90 and -0.85?

Etcetera up to, How likely is it that you'll observe a correlation between $+0.59$ and $+1.00$?

(*Hint*: Suppose that half the index cards are blue, and the other half white. Write the numbers 1, 2, 3, 4, etc., up to 100, on white cards. Then write the numbers 501, 502, 503, 504, etc., up to 600, on blue cards. Now mix the cards together in the hat, and draw a pair (one white and one blue) without looking at the numbers. Repeat the operation nine more times.).

The Chi-Square (χ^2) Test to Help Decide Whether Two Variables Are Related

THE MUSICAL EDUCATION OF THE LABORATORY RAT

An experimental psychologist was interested in finding out whether musical taste is influenced by early experience in a simple animal like the laboratory rat. He raised 48 animals from birth in a world in which classical music was played for 12 hours a day. For half of the rats the music was composed by Mozart. For the other half it was composed by Schoenberg. When the rats were 6 months old, he asked them which composer they preferred by giving them a choice of two cages: in one cage they heard music by Mozart, and in the other cage they heard music by Schoenberg. The Schoenberg-raised rats showed no consistent preference: 12 chose Schoenberg and 12 chose Mozart. For the Mozart-raised rats, however, there seemed to be a difference: 20 of the 24 rats chose to listen to Mozart, while only 4 chose Schoenberg. At this point it is time for a statistical decision. You might enjoy trying to state just what that decision is before reading on.

The psychologist had to decide whether the upbringing of the rats had any real relationship to their choices of music. For the Schoenberg rats it didn't seem to,[1] but for the Mozart-raised rats it did. The statistical question is whether, by pure chance, it is at all likely that one would see such a pattern of choices, if in fact the "education" had no influence on the preference.

There is a statistical procedure that is useful for making decisions like this. In order to use it, we first display the experimental results in what is known as a *contingency table* (Figure 22.1). This summarizes the results already presented. We can split our 48 rats into two groups: those raised with Mozart and those raised with Schoenberg. After the choice test, we can split them into a different pair of groups: those who later preferred Mozart and those who later preferred Schoenberg. We are asking whether there is any relationship between these two categorizations.

The next step in the analysis involves answering a new question: If there were absolutely no

---

[1] But maybe it did. Perhaps rats with no experience would have *avoided* Schoenberg, instead of showing a 12 : 12 neutral preference.

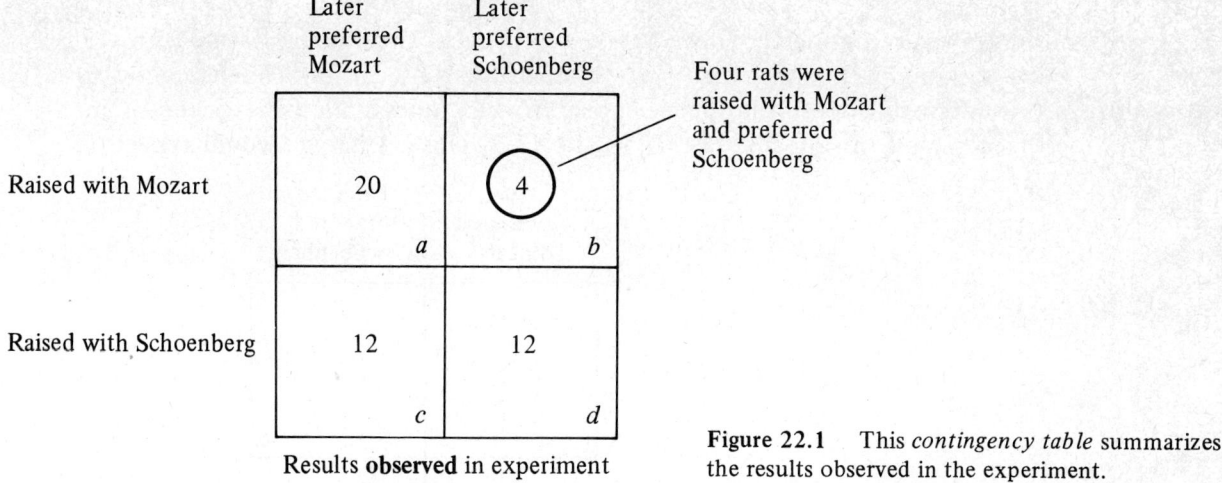

Figure 22.1 This *contingency table* summarizes the results observed in the experiment.

relationship between early experience and subsequent choice, what results would we expect to see? This is a difficult question, and if you enjoy challenges you might stop here and try to answer it. What would the contingency table look like if the two variables (early experience and musical preference) were absolutely unrelated? To answer this question, it helps to add some numbers to the above table: these numbers are called the *marginal totals*, and they are given in Figure 22.2. The marginal totals are obtained by adding up the appropriate cells in the

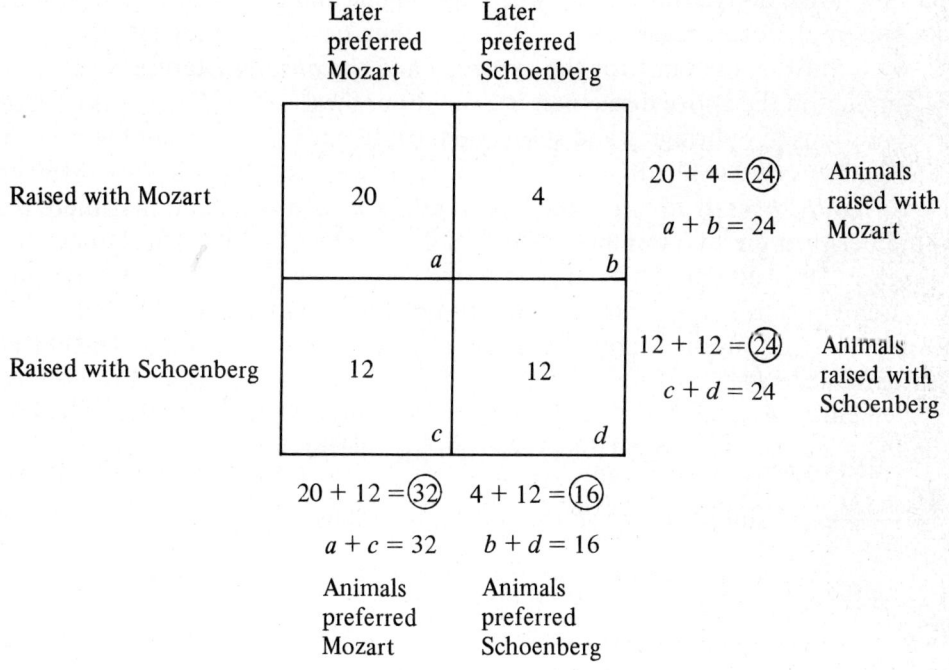

Figure 22.2 **Observed** results (with marginal totals added)

contingency table to find the total number of entries in each row (row totals) and in each column (column totals). They are the starting point in figuring out what would have happened if the two variables were unrelated. We can see that 24 of the 48 animals were raised with Mozart, and likewise half were raised with Schoenberg. We can also see that 32 of the 48 animals, or two-thirds of the animals, preferred Mozart, whereas only 16 out of 48, or one-third of the animals, preferred Schoenberg. Now if there were no relationship between these two variables, we would guess that two-thirds of rats generally prefer Mozart, and one-third Schoenberg, regardless of the conditions of their upbringing. So we would *expect* two-thirds of the Raised with Mozart rats to prefer Mozart, and we would *expect* two-thirds of the Raised with Schoenberg rats to prefer Mozart, if upbringing doesn't influence choice. We can summarize this expectation in another contingency table of *expected* results, Figure 22.3. This is what we would *expect* to

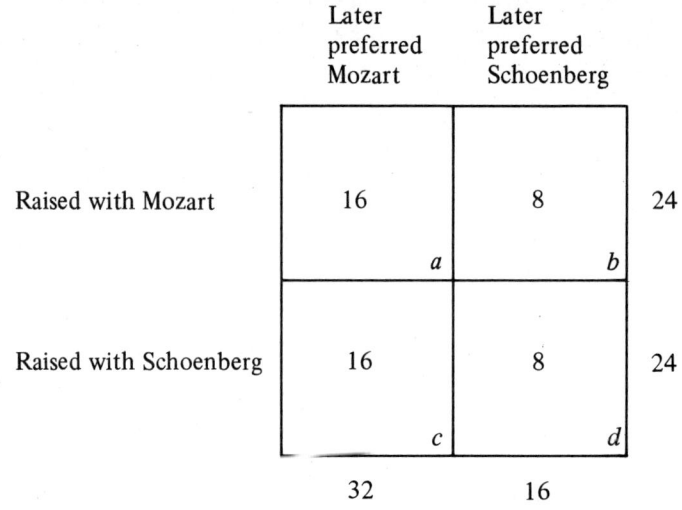

Figure 22.3 These are the results you would *expect* if upbringing has no influence on subsequent preference.

happen if there were no relationship between upbringing and preference; that is, if 32/48, or 2/3, of all rats preferred Mozart regardless of the music they heard as young rats. Note that the marginal totals must be the same for the *expected* and the *observed* tables.

Now, implicit in the above discussion is a null hypothesis: H_0 : There is *no* relationship between conditions of upbringing and subsequent preference. If this were the case, we would *expect* a contingency table like Figure 22.3. The table we *observed* is plainly different. But *is it so different that it forces us to reject the null hypothesis*, and conclude that there is a relationship between the two variables? This is a different sort of statistical question from others we have seen so far. And in order to answer it, we need a new procedure. We compute a test statistic, a number called chi-square (χ^2), which reflects the difference between the two tables. Each of the above contingency tables has four cells, labeled *a, b, c,* and *d*. The chi-square statistic for this situation is made up of four terms, one for each of the four cells. For each cell we denote the Expected entry by the letter *E*, and the Observed entry by the letter *O*. Thus for cell *a*, $E = 16$ and $O = 20$ (see the above tables). We can then write

$$\chi^2 = \sum \frac{(E-O)^2}{E} \quad \text{summed over all the cells in the table}$$

Reproducing the two tables for easy reference:

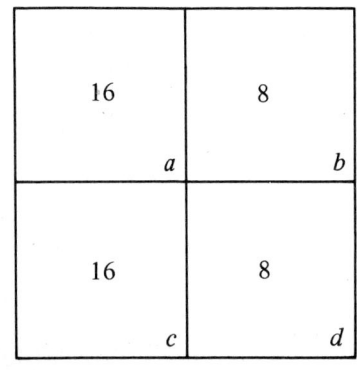

Figure 22.4

Expected

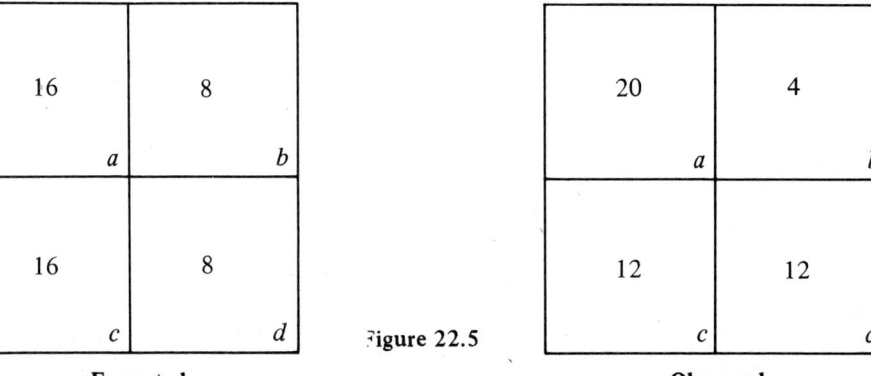

Figure 22.5

Observed

we can compute

$$\chi^2 = \underbrace{\frac{(16-20)^2}{16}}_{\text{cell } a} + \underbrace{\frac{(8-4)^2}{8}}_{\text{cell } b} + \underbrace{\frac{(16-12)^2}{16}}_{\text{cell } c} + \underbrace{\frac{(8-12)^2}{8}}_{\text{cell } d}$$

$$= \frac{(-4)^2}{16} + \frac{(4)^2}{8} + \frac{(4)^2}{16} + \frac{(-4)^2}{8}$$

$$= \frac{16}{16} + \frac{16}{8} + \frac{16}{16} + \frac{16}{8}$$

$$= 1 + 2 + 1 + 2$$

$$= 6$$

The *chi-square* number is a test statistic based on our data and the null hypothesis, which we can use, just as we use the *t* statistic and the *Z* statistic in evaluating the results. In order to do this we must know how likely it is that we would observe a chi-square value as large as 6 if there were absolutely no relationship between the two variables, upbringing and preference. The probability of seeing various values of chi-square in this situation has been tabulated, and is given in Table 22.1.

Before we can use Table 22.1, however, we must determine how many degrees of freedom are present in this situation. Here's a simple way to do that. If you write down the empty contingency table, with just the marginal totals noted, you obtain Figure 22.6. Now if you

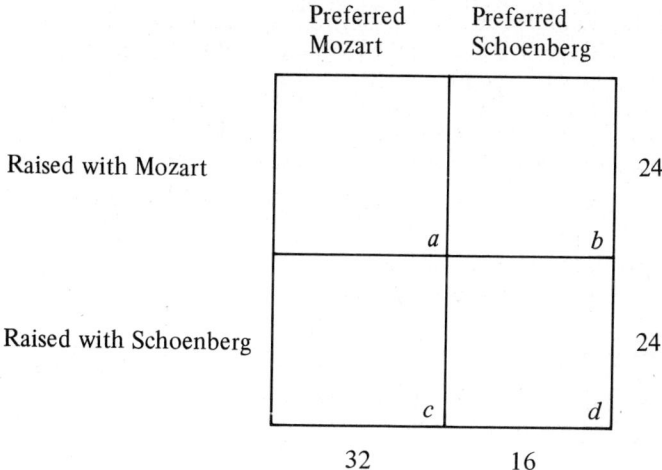

Figure 22.6

specify the observed frequency for just one of the cells, say cell *a*, then all the rest are determined. You can choose one of the frequencies freely, and then the rest are constrained. For example, suppose we enter just the frequency in cell *a*, as in Figure 22.7. We can determine the frequency which must go in cell *b*, since $a + b = 24$. It must be that $b = 24 - 20 = 4$. Similarly, since $a + c = 32$, we know that $c = 12$. And likewise *d* can be found once we know either *c* or *b*.

When only one entry in a contingency table can be freely chosen, and the rest are then fixed, we say the situation has *1 degree of freedom*. As will be made clear shortly, a table which has the form seen in Figure 22.8 will have 2 degrees of freedom. Once two of the entries of such a table have been specified, and the marginal totals given, all the other entries are fixed, too.

Like the *t* table, the chi-square table really summarizes information on many different

TABLE 22.1 CRITICAL VALUES[1] FOR THE χ^2 STATISTIC

	Significance level (α)								
Degrees of freedom	$P = 0.90$	0.70	0.50	0.30	0.20	0.10	0.05	0.02	0.01
1	0.02	0.15	0.45	1.07	1.64	2.71	3.84	5.41	6.63
2	0.21	0.71	1.39	2.41	3.22	4.60	5.99	7.82	9.21
3	0.58	1.42	2.37	3.66	4.64	6.25	7.81	9.84	11.34
4	1.06	2.19	3.36	4.88	5.99	7.78	9.49	11.67	13.28
5	1.61	3.00	4.35	6.06	7.29	9.24	11.07	13.39	15.09
6	2.20	3.83	5.35	7.23	8.56	10.64	12.59	15.03	16.81
7	2.83	4.67	6.35	8.38	9.80	12.02	14.07	16.62	18.47
8	3.49	5.53	7.34	9.52	11.03	13.36	15.51	18.17	20.09
9	4.17	6.39	8.34	10.66	12.24	14.68	16.92	19.68	21.67
10	4.86	7.27	9.34	11.78	13.44	15.99	18.31	21.16	23.21
11	5.58	8.15	10.34	12.90	14.63	17.27	19.67	22.62	24.72
12	6.30	9.03	11.34	14.01	15.81	18.55	21.03	24.05	26.22
13	7.04	9.93	12.34	15.12	16.98	19.81	22.36	25.47	27.69
14	7.79	10.82	13.34	16.22	18.15	21.06	23.68	26.87	29.14
15	8.55	11.72	14.34	17.32	19.31	22.31	25.00	28.26	30.58
16	9.31	12.62	15.34	18.42	20.46	23.54	26.30	29.63	32.00
17	10.08	13.53	16.34	19.51	21.61	24.77	27.59	30.99	33.41
18	10.86	14.44	17.34	20.60	22.76	25.99	28.87	32.35	34.80
19	11.65	15.35	18.34	21.69	23.90	27.20	30.14	33.69	36.19
20	12.44	16.27	19.34	22.77	25.04	28.41	31.41	35.02	37.57
21	13.24	17.18	20.34	23.86	26.17	29.61	32.67	36.34	38.93
22	14.04	18.10	21.34	24.94	27.30	30.81	33.92	37.66	40.29
23	14.85	19.02	22.34	26.02	28.43	32.01	35.17	38.97	41.64
24	15.66	19.94	23.34	27.10	29.55	33.20	36.41	40.27	42.98
25	16.47	20.87	24.34	28.17	30.67	34.38	37.65	41.57	44.31
26	17.29	21.79	25.34	29.25	31.79	35.56	38.88	42.86	45.64
27	18.11	22.72	26.34	30.32	32.91	36.74	40.11	44.14	46.96
28	18.94	23.65	27.34	31.39	34.03	37.92	41.34	45.42	48.28
29	19.77	24.58	28.24	32.46	35.14	39.09	42.56	46.69	49.59
30	20.60	25.51	29.34	33.53	36.25	40.26	43.77	47.96	50.89
Degrees of freedom	$P = 0.90$	0.70	0.50	0.30	0.20	0.10	0.05	0.02	0.01

EXAMPLE
There is 1 chance in 100 of seeing a χ^2 score of 6.63 or greater when the null hypothesis is true and you have 1 degree of freedom.

From Table IV of Ronald A. Fisher and Frank Yates, *Statistical Tables for Biological, Agricultural and Medical Research*, 6th ed., published by Oliver and Boyd, Edinburgh, 1963, p. 47, and by permission of the authors and publishers.

chi-square distributions; the distributions with 1, 2, 3, 4, 5, etc., up to 30 degrees of freedom. For the musical preference example, we could choose only one of the four numbers freely, so we say there is *1 degree of freedom*. Looking in the chi-square table, we find columns corresponding to a variety of significance levels. If we pick $\alpha = 0.05$, we see a critical value of 3.84 for the chi-square statistic. That means that with 1 degree of freedom there is just 1 chance in 20 that you will observe a χ^2 value greater than 3.84. In our example, we found a chi-square of 6. This is

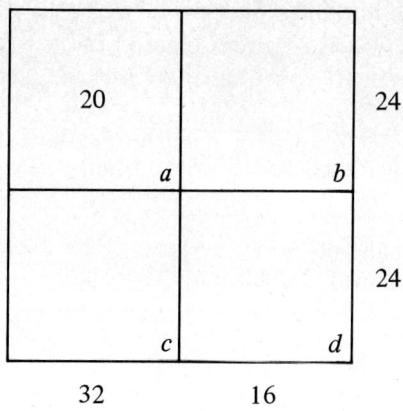

Figure 22.7

Figure 22.8

considerably greater than the critical value 3.84, and it indicates that it is a safe conclusion that the rats' upbringing *did* influence their subsequent musical preference.

Finding Expected Frequencies

There is just one part in the computation of a chi-square statistic which usually causes any difficulty for the student: finding the *expected* frequencies on the basis of the marginal totals. Suppose that a group of students is interviewed, some of them seniors and some freshmen. Each is asked to self-rate his political position as conservative, middle-of-the-road, liberal, or radical. Figure 22.9 is the resulting contingency table, with marginal totals.

	Conservative	Middle-of-the-road	Liberals	Radicals	
Freshmen	9	6	25	10	50
Seniors	21	24	15	40	100
	30	30	40	50	

Figure 22.9 **Observed**

THE *t* TEST, CORRELATION, AND CHI-SQUARE

The question of interest here is whether the freshmen and the seniors are really any different politically, or whether these two variables (class and politics) are unrelated. To test this we can use the chi-square statistic. But before doing that we must compute the *expected* frequencies.

Exercise: Try to find the *expected* frequencies which should go in Figure 22.10, on the basis of the marginal totals. Please don't read on until you have tried to fill in the blanks in the table.

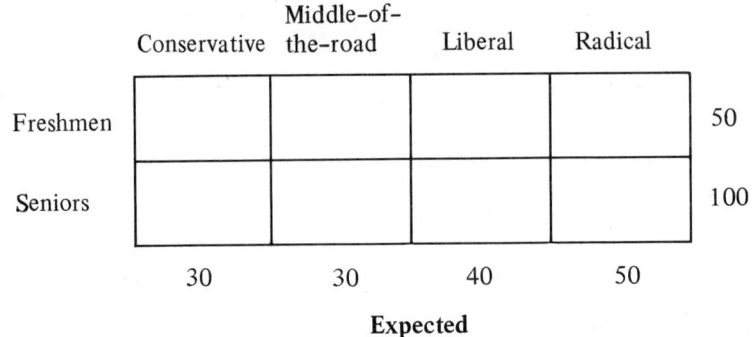

Figure 22.10 Try to fill in the expected frequencies before reading on.

Note that overall, 50 out of 150 students are freshmen. One-third of the students are freshmen. And 100 out of 150 students, or 2/3, are seniors. If there is *no relationship* between class status and political position, then we would expect 1/3 of the conservatives and 1/3 of the middle-of-the-roaders to be freshmen, and so on. That is, we would expect 1/3 of the 30 conservatives, or 10, to be freshmen. Similarly, we would expect 1/3 of the 40 liberals, or 13 1/3, to be freshmen, in the ideal case. Filling in the rest of the frequencies in this way, we get Figure 22.11.

	Conservative	Middle-of-the-road	Liberal	Radical	
Freshmen	10	10	13 1/3	16 2/3	50
Conservative	20	20	26 2/3	33 1/3	100
	30	30	40	50	

Figure 22.11   Expected

A formula for finding expected frequencies is presented in the Question and Answer section of this chapter.

QUESTION AND ANSWER

Question: You've given two examples of computations of *expected* contingency tables once the *observed* table (with marginal totals) is known. I followed both of those, but I'm not sure I've got the general principle: What is it?

Answer: The general principle is that the ratio between the entries in any given column is just the same as the ratio between the marginal totals beside the column. For example, in the first example we computed the *expected* table in Figure 22.12. As a second example, note the same pattern in the *expected* table in Figure 22.13, from the example on political positions of freshmen and seniors.

There is another way of putting this which is more helpful when you are faced with a problem of

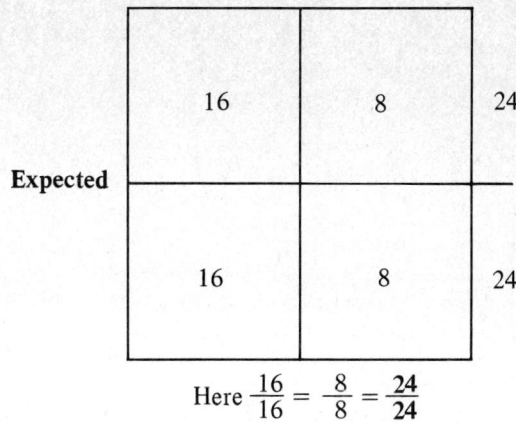

Figure 22.12 In each column the entries are in the ratio of 1 : 1.

Here $\frac{16}{16} = \frac{8}{8} = \frac{24}{24}$

	Conservative	Middle-of-the-road	Liberal	Radical	
Freshmen	10	10	13 1/3	16 2/3	50
Seniors	20	20	26 2/3	33 1/3	100
	30	30	40	50	

Here $\dfrac{10}{20} = \dfrac{10}{20} = \dfrac{13\tfrac{1}{3}}{26\tfrac{2}{3}} = \dfrac{16\tfrac{2}{3}}{33\tfrac{1}{3}} = \dfrac{50}{100}$

Figure 22.13 In each column the entries are in the ratio of 1 : 2.

computing *expected* entries: each entry will be the same fraction (or percentage) of its column's marginal total as the row total is of the total number of entries in the table. In the table in Figure 22.13, there are 150 students in all. According to the marginal totals, 50 of the 150, or 1/3, are freshmen. So it must be the case that 1/3 of the 30 conservatives, 1/3 of the middle-of-the-roaders, and 1/3 of the liberals are freshmen.

A Formula for Calculating Expected Frequency:

A slightly different way of remembering this is helpful to some people. The *expected frequency* for any cell in the contingency table is given by $(R \times C)/T$, where R is the row total for the row in which the cell lies, C is the column total for the column in which the cell lies, and T is the total number of entries in the entire table. For example, consider cell a of Figure 22.14. The row total, R, is 91: $R = 91$. The column total, C, is 56: $C = 56$. The total, T, is 171: $T = 171$. Thus the expected frequency for cell a is given by

$$E = \frac{R \times C}{T} = \frac{91 \times 56}{171}$$

This relationship can be used to find the expected frequencies for any contingency table. You might like to try it to find the entries for Figure 22.10.

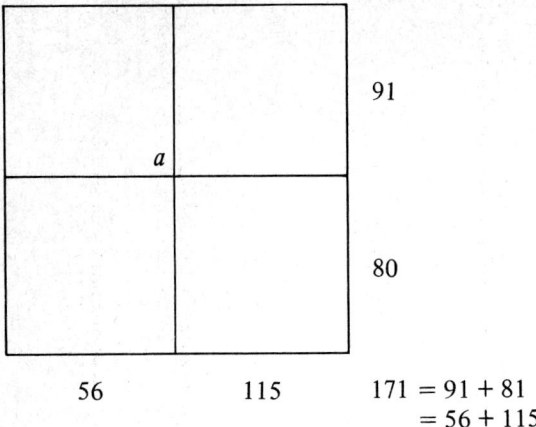

```
56      115       171 = 91 + 81
                      = 56 + 115
```

Figure 22.14

Question: In discussing the other tests we've seen, you always talked about a sampling distribution for a statistic. Why didn't you mention sampling distributions here?

Answer: The sampling distribution is still there, but it didn't quite rise to the surface. When the null hypothesis is true and the two variables are unrelated, the observed frequencies will rarely be exactly equal to the expected frequencies. They should be pretty close, but there is still random variation. The chi-square test is based on a sampling distribution for the chi-square statistic when the null hypothesis is true. For example, suppose that learning in a particular large course is unrelated to the sex of the student — a plausible hypothesis to a reasonable degree of accuracy. Now if 40 people are selected at random, they can be categorized as *male* or *female*, and they can be categorized as *above* or *below* the median score on the final exam. We could make up a contingency table and compute a chi-square statistic. Suppose another sample of 40 people is then taken, and another chi-square statistic computed. In principle, this process could be repeated over and over. We would end up with a sampling distribution of values for the chi-square statistic. And, as is indicated in Table 22.1, there is a 0.05 probability of seeing such a statistic greater than 3.84. So in our sampling experiment, we would probably see about 1 observation in 20 which was greater than 3.84. The main point is that there *is* a sampling distribution involved: the distribution of the chi-square statistic when the null hypothesis is true. Only a few special features, or critical values of that sampling distribution are summarized in the chi-square table for each value for degrees of freedom.

Summary

In this chapter the chi-square (χ^2) test was introduced. It can be used to help decide whether two different categorizations are related. For example, we looked at the question of whether the music a young laboratory rat heard in the experiment influenced his later preferences. In another example we looked to see if a student's political position was related to his class standing. The chi-square test is used to decide if an entire array, or *contingency table*, is significantly different from what one would expect if the two variables or categorizations were unrelated.

In conducting a chi-square test, the following steps are used:

1. Arrange the observed results into a *contingency table*.
2. State a null hypothesis, H_0, that the two variables involved are not related to each other.
3. Compute the marginal totals for the *observed* table.
4. Draw a new table, and find the *expected* cell entries if there were no relationship (if the null hypothesis is true.) For each cell you can compute $E = (R \times C)/T$, where R is the marginal total for the row in which the cell is located, C is the marginal total for the column in which the cell is located, and T is the total of all the cells in the table.

5. Figure out how many degrees of freedom you have. If the contingency table has R rows and C columns, then the number of degrees of freedom (df) is given by $df = (R - 1) \times (C - 1)$. For example a table with three rows and two columns has $(3 - 1) \times (2 - 1) = 2$ degrees of freedom.
6. Compute the chi-square statistic on the basis of the *expected* and *observed* tables, using the relationship

$$\chi^2 = \sum \frac{(E - O)^2}{E}$$

where the summation is carried out over all the cells in the contingency table. E denotes the entry in the *expected* table, and O denotes the entry in the corresponding cell of the *observed* table.
7. Decide on a significance level, and look in the chi-square table to determine the corresponding critical value.
8. Make your decision.

Problems

1. Compute the chi-square statistic for the example in Figure 22.9. In this case, three of the numbers can be chosen "freely" before the remainder are constrained by the marginal totals, so we have 3 degrees of freedom (3 df). Compute the test statistic and then use the chi-square table to decide if there is a significant relationship between class status and political position. Use a 0.01 significance level.

2. Prove to yourself that 3 df is appropriate for the previous example by filling in the empty cells in the following table, for which only three numbers are given:

4	1	8	?	20
?	?	?	?	30
13	14	12	11	50

3. Does the color of a car affect its resale value? A company buys a fleet of 44 cars, 24 red and 20 black. Two years later, it divides them into two groups on the basis of resale value. Below are the observed frequencies:

	High Resale	Low Resale
Red Cars	16	8
Black Cars	6	14

Observed

Run a chi-square test to decide whether there is a significant relationship between car color and resale value. Use a 0.05 significance level. State your null hypothesis.

a. Compute the marginal totals.

b. Draw a new table and find the expected frequencies if there is no relationship.

c. Figure out how many degrees of freedom you have here.

d. Compute the chi-square statistic, using the relationship

$$\chi^2 = \sum_{\text{all cells}} \frac{(E - O)^2}{E}$$

e. Consult the table and find a critical value.

f. Make your decision.

4. (Read the Question and Answer section for this chapter before proceeding.)
For each of the following *observed* contingency tables, find the marginal totals, and then find the corresponding *expected* tables:

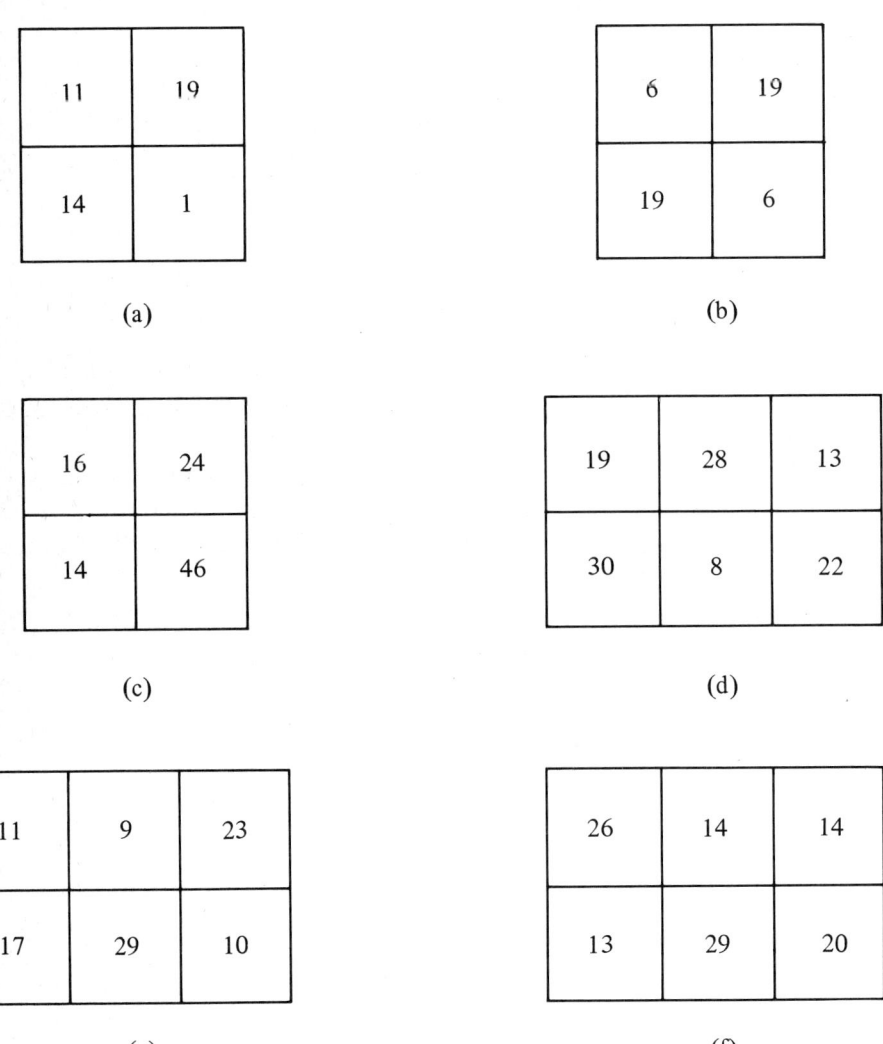

5. Can you ever observe a negative value for the χ^2 statistic? Why, or why not?

6. For each of the tables in problem 4, specify the number of degrees of freedom.

The Chi-Square (χ^2) Test, Concluded

23

ANOTHER CHI-SQUARE EXAMPLE

After studying the previous chapter you should be pretty good at using the chi-square statistic to test whether two variables are associated or related. Try the following problem to test your skills before reading ahead to the solution.

A study was performed to see if ecologists practice what they preach. A sample of 40 ecologists and a sample of 60 executives were asked whether they drove economy, standard, or luxury cars. Ecologists drove 35 economy cars and no standard cars. For both groups combined, the totals were 50 economy cars, 30 standard cars, and 20 luxury cars.

a. State a null hypothesis appropriate to this situation.
b. Fill in the remaining cells in Figure 23.1, and then compute a χ^2 statistic.

Figure 23.1

	Economy car	Standard car	Luxury car	
Ecologists	35	0		40
Executives				
	50	30	20	

205

206 THE t TEST, CORRELATION, AND CHI-SQUARE

Please try it before reading on.
Here is the solution to the problem posed above.

a. H_0 : There is no relationship between a man's profession and the kind of car he drives.
b. See Figure 23.2.

	Economy car	Standard car	Luxury car	
Ecologists	35	0	40 − 35 = 5	40
Executives	50 − 35 = 15	30 − 0 = 30	20 − 5 = 15	60
	50	30	20	

Observed

Figure 23.2 There is enough information in Figure 23.1 to enable us to fill in the missing cells.

Now we must compute the *expected* frequencies, if there is no relationship between the two variables (occupation and car choice). The basic rule to use here is as follows: *If there is no relationship between the two variables, then the ratios between the marginal totals will be the same as the ratios between corresponding cells within a category in the table.* The following examples will show what that means.

Note by looking at the totals along the right-hand margin in Figure 23.2 that there are 100 people in all, 40 ecologists and 60 executives. If there is no relationship between occupation and car choice, then we would expect about 40 percent of the economy car owners to be ecologists, since 40 percent of the entire sample are ecologists. In this case, since there are 50 economy car owners in all, we would expect 40 percent of 50, or 20 economy car owners, to be ecologists. So the entry in cell *a* should be 20 (see Figure 23.3). (You get the same result using the formula $E = (R \times C)/T = (40 \times 50)/100 = 2000/100 = 20$.)

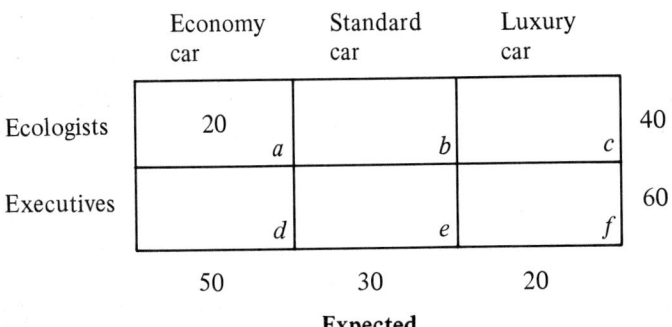

Figure 23.3

Expected

By an exactly parallel argument, since 40 percent of all the people are ecologists, we would expect 40 percent of the standard car owners to be ecologists if the two variables are unrelated. So the entry in cell b should be 40 percent of the total number of standard car owners, or 40 percent of 30, or 12. Similarly, if the two variables are unrelated, we would expect 40 percent of the 20 luxury car owners to be ecologists. Filling in the remaining gaps, we get Figure 23.4.

	Luxury car	Standard car	Economy car	
Ecologists	20	12	8	40
Executives	30	18	12	60
	50	30	20	

Figure 23.4   Expected

Now we can compute the chi-square statistic. To save having to look back, the two contingency tables are reproduced here in Figure 23.5.

a 35	b 0	c 5
d 15	e 30	f 15

20	12	8
30	18	12

Figure 23.5   Observed   Expected

$$\chi^2 = \sum_{\substack{\text{all cells} \\ \text{in table}}} \frac{(E-O)^2}{E} = \underbrace{\frac{(20-35)^2}{20}}_{\text{cell } a} + \underbrace{\frac{(12-0)^2}{12}}_{\text{cell } b} + \underbrace{\frac{(8-5)^2}{8}}_{\text{cell } c} + \underbrace{\frac{(30-15)^2}{30}}_{\text{cell } d} + \underbrace{\frac{(18-30)^2}{18}}_{\text{cell } e} + \underbrace{\frac{(12-15)^2}{12}}_{\text{cell } f}$$

$$= \frac{225}{20} + \frac{144}{12} + \frac{9}{8} + \frac{225}{30} + \frac{144}{18} + \frac{9}{12}$$

$$= 11.25 + 12 + 1.12 + 7.50 + 8 + 0.75$$

$$= 40.6$$

The number of degrees of freedom for any table like this is $(R-1) \times (C-1)$ where R is the number of rows and C is the number of columns. $df = (R-1) \times (C-1) = (2-1) \times (3-1) = 2$. If we pick a stringent significance level, $\alpha = 0.01$, we find a critical value for χ^2 with 2 df of 9.21 (see Table 22.1). The value we observed is far greater, so we can reject the null hypothesis, and conclude that there is a significant difference in the car-buying patterns of the two groups.

A RESTRICTION ON EXPECTED FREQUENCIES

There is a special restriction on the use of the chi-square test statistic for analyses like this: It is essential that all of the expected frequencies be 5 or greater. Otherwise the distribution of the test statistic, χ^2, is not the one given in the tables, and you will be trapped into errors. In any situation where you obtain an *expected* frequency of 4 or less, you must combine the category containing that frequency with some other category before continuing the analysis, or use another test. Suppose the *expected* table for a similar situation looked like Figure 23.6. This is not an acceptable table, since one of the cells contains the number 4, which is beneath our lower limit. What can we do? We can combine the first two columns: We can make a single category, economy cars, to combine the first two categories. Then we will have Figure 23.7. A similar strategy will work in many situations.

	Foreign economy cars	Domestic economy cars	Standard	Luxury
Ecologists	4	16	12	8
Executives	6	24	18	12
	10	40	30	20

Figure 23.6 **Expected**

	Economy car	Standard car	Luxury car
Ecologists	4 + 16 = 20	12	8
Executives	6 + 24 = 30	18	12

Figure 23.7

THE CHI-SQUARE TEST TO HELP DECIDE WHETHER SOME OBTAINED RESULTS ARE CONSISTENT WITH A THEORY

An officer of the Nevada State Gaming Commission, after watching a craps game, seized some dice he believed to be loaded. He rolled one of them 60 times and observed the six possible results with the following frequencies:

Result: No. of spots showing	1	2	3	4	5	6
Observed frequency, out of 60 throws.	12	9	17	4	8	10

Observed

Now if the die were perfectly fair, and each of the six outcomes were equally likely, we would expect to see each one about 1/6 of the time, or about 10 times out of 60 throws. The officer thought there were too many 3s, and not enough 4s, in his results. Do you think he has good evidence that the die is unfair?

Here is how one can test the null hypothesis that the die is fair: Just as in the examples we saw in the previous chapter, here we are concerned with a comparison between an *observed* array of numbers, 12, 9, 17, 4, 8, 10, and an *expected* array of numbers, 10, 10, 10, 10, 10, 10. And it turns out that a chi-square test statistic, computed in much the same way, can be used here, too. Computing the test statistic:

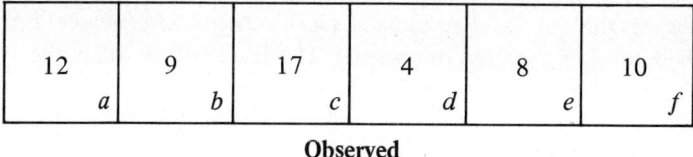

Observed

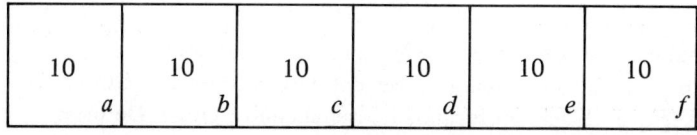

Expected

$$\chi^2 = \sum \frac{(E-O)^2}{E} = \underbrace{\frac{(10-12)^2}{10}}_{\text{cell } a} + \underbrace{\frac{(10-9)^2}{10}}_{\text{cell } b} + \underbrace{\frac{(10-17)^2}{10}}_{\text{cell } c} + \underbrace{\frac{(10-4)^2}{10}}_{\text{cell } d} + \underbrace{\frac{(10-8)^2}{10}}_{\text{cell } e} + \underbrace{\frac{(10-10)^2}{10}}_{\text{cell } f}$$

$$= \frac{4}{10} + \frac{1}{10} + \frac{49}{10} + \frac{36}{10} + \frac{4}{10} + \frac{0}{10}$$

$$= \frac{94}{10}$$

$$= 9.4$$

Before we can consult the chi-square table, we must figure out how many degrees of freedom (df) there are in this situation: How many entries can be placed in a table like the following one, before the rest are determined? Remember that the sum of all the numbers in the table must be 60, the number of throws. It will, I hope, be clear that we can *freely* specify five of the six

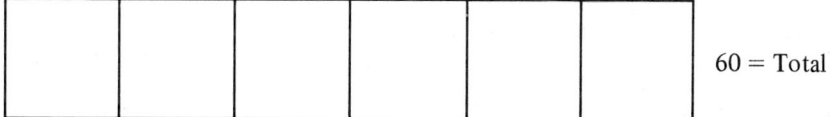

entries. Once any five are known, the sixth is determined, since the entries must add up to 60. In this situation there are 5 degrees of freedom. In general, in situations where we are comparing two lists of numbers, each n numbers long, there are $n - 1$ degrees of freedom.

Returning to the example, suppose we assume a significance level of 0.05. Is a chi-square value of 9.4 significantly large in this situation with 5 df? Looking at the chi-square table (Table 22.1) with 5 df and $\alpha = 0.05$, we find a critical value of 11.07. This means that even when the null hypothesis is true (the die is fair) we would expect chi-square values greater than 11.07 1 time in 20. So a chi-square value of 9.4 is not astonishing, and the officer had better get some more data before he takes the casino to court.

Summary

Another example was presented for the chi-square test to see if two variables are related. A special condition was noted for using the chi-square test statistic: all *expected* frequencies must be at least 5. Sometimes, if a smaller expected frequency is computed, it is possible to combine two adjacent rows (or columns) of a contingency table.

A second use of the chi-square test was presented: the test was used to decide if the observed frequencies for a set of categories differ significantly from the theoretical or expected frequencies. This is sometimes called the chi-square test for a single categorical variable.

Problems

1. In a certain metropolitan area, the voters can be categorized by ethnic background as follows: Anglo, 44 percent; Black, 36 percent; Chicano, 12 percent; Oriental, 5 percent; Other, 3 percent. Out of 100 people called for grand jury duty, 57 were Anglo; 26 Black; 9 Chicano; 7 Oriental; and 1 Other. Do these numbers provide convincing evidence that the selection system is unfair, and Anglos are overrepresented on grand juries? Run a chi-square test to help decide.

   a. Compute the *expected* frequencies, for 100 people, if every voter had an equal chance of being called.

   b. If any expected frequencies are less than 5, combine categories until the problem is eliminated (see Summary).

   c. Write out the table of *observed* frequencies.

   d. Compute the chi-square statistic.

   e. Find the number of degrees of freedom.

   f. Set a significance level.

   g. Check the chi-square table for a critical value.

   h. Make your decision

2. The question was raised, early in the book, whether the marble box was a good model for the ESP experiment when the null hypothesis was true. Now you have the necessary statistical tool to help answer that problem. The theoretical probabilities of the ten possible events are presented here:

Event (number of heads)	0/10	1/10	2/10	3/10	4/10	5/10	6/10	7/10	8/10	9/10	10/10
Probability	.001	.010	.044	.118	.205	.246	.205	.118	.044	.010	.001

The actually observed frequencies, in 50 trials, are the following:

Number of green marbles in paddle	0/10	1/10	2/10	3/10	4/10	5/10	6/10	7/10	8/10	9/10	10/10
Frequency	0	0	0	10	9	12	9	8	1	1	0

Do a chi-square test to see if the observed frequencies are consistent with the mathematical theory. To do this you must compute a table of *expected* frequencies in 50 trials. Start by multiplying each probability by the number of trials, 50. For example, the expected frequency of the observation 3/10 correct is 50 x 0.118 = 5.9. Similarly, compute the ten other expected frequencies. To keep all expected frequencies greater than 5, you will need to lump 0/10, 1/10, 2/10, 8/10, 9/10, and 10/10 into a single category *unlikely events*. Then compare the expected and observed frequency tables. Use a significance level $\alpha = 0.05$. Do you think the marble box data are consistent with the mathematical theory?

†3. Try to think how, given sufficient time and patience, you could *empirically* derive a chi-square distribution with 1 degree of freedom. That is, how could you figure out the probability of chi-square values between 0.00 and 0.10, between 0.10 and 0.20, etc. Suppose that you were locked in a dungeon with pencils, paper, coins, and dice, and told that your release depended on a close approximation to the true chi-square distribution with 1 degree of freedom.

The *t* Test for Means of Uncorrelated Samples

A commercial dog breeder was interested in comparing two different kinds of dog chow for pregnant bitches. He fed Wilson's chow to ten bitches and J.F.'s Superchow to ten others. Since the pups produced seemed of comparable quality, he decided to analyze the *number* produced in each litter. He observed the following:

Mothers fed Wilson's chow: 2, 3, 5, 6, 2, 4, 4, 3, 4, and 7 pups

Mothers fed J.F.'s chow: 7, 6, 5, 8, 4, 5, 6, 6, 7, and 6 pups

He then had to decide: Was there a significant difference between the average production with the two chows?

The first step is to compute the means of the two samples.

Sample 1 (Wilson's):

$$M_1 = \frac{2+3+5+6+2+4+4+3+4+7}{10}$$
$$= \frac{40}{10} = 4$$

Sample 2 (J.F.'s):

$$M_2 = \frac{7+6+5+8+4+5+6+6+7+6}{10}$$
$$= \frac{60}{10} = 6$$

Now in terms of mean production, J.F.'s seems superior. But does this represent a real, statistically significant difference? Or is it plausible that one could take two samples *from the same population* and see a comparable difference in the means?

There is clearly variability in pup production. Even if the two chows were identical, we would expect one sample of ten litters to be somewhat different from another. But how different? This is the question we must answer. The following statistical analysis will help. The null hypothesis states that there is no difference between the chows: the two samples were taken from the same population.

If you take two independent samples of size 10 from a normal distribution, you can compute the means of the two samples, M_1 and M_2. From these two you can compute a difference score, M_1 minus M_2. Sometimes this difference score will be a positive number, and sometimes it will be a negative number. Such difference scores will have a normal distribution, with mean zero. If we know the variance of the parent population from which the two samples were taken, we can deduce the variance of the sampling distribution of differences, M_1 minus M_2. Call this variance SD_{diff}^2. We can sketch the sampling distribution, which shows how likely are various values of M_1 minus M_2 (Figure 24.1).

Now let's return to the dog chow example. There we observed a difference score, $M_1 - M_2 = 4 - 6 = -2$. Now we would like to figure out if this is such a big difference score that it couldn't plausibly have come from the sampling distribution sketched in Figure 24.1. To figure this out we need to know the standard deviation, SD_{diff}, of the above distribution.

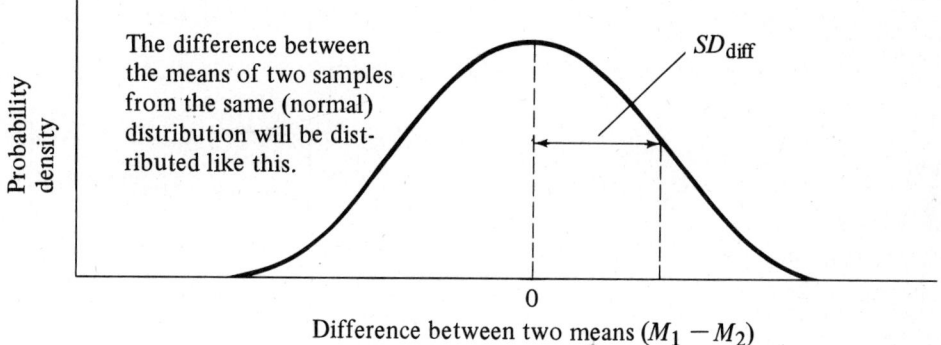

Figure 24.1 This is the sampling distribution for the *difference*, $M_1 - M_2$, when two samples are drawn from the *same* population, the mean of each sample is computed, and the difference, $M_1 - M_2$, is computed.

We are interested in deciding whether the two samples could have come from the same population. But we don't know the variance of that population, so we can't specify the exact value of SD_{diff}^2. But through a set of steps that are discussed in problem 4 at the end of this chapter, we can *estimate* that variance, using the following formula:

$$\text{Est. } SD_{\text{diff}}^2 = \frac{\overset{\text{summed over}}{\underset{\text{the first sample}}{\Sigma(X_1 - M_1)^2}} + \overset{\text{summed over}}{\underset{\text{the second sample}}{\Sigma(X_2 - M_2)^2}}}{n_1 + n_2 - 2} \times \left(\frac{1}{n_1} + \frac{1}{n_2}\right)$$

The first summation adds the squared deviations from the mean, M_1, for all the numbers in the first sample. The second summation similarly adds up the squared deviations from the mean, M_2, for the numbers in the second sample. n_1 is the size of the first sample (10, for Wilson's chow), and n_2 is the size of the second sample (10, for J.F.'s chow).

Sample Computation:

For the two samples above, here is the computation of the estimated variance, Est. SD^2_{diff}.

Sample 1 (Wilson's):   2, 3, 5, 6, 2, 4, 4, 3, 4, 7 $M_1 = 4$
Sample 2 (J.F.'s):   7, 6, 5, 8, 4, 5, 6, 6, 7, 6 $M_2 = 6$

Computing the first summation:

$$\Sigma(X_1 - M_1)^2 = (2-4)^2 + (3-4)^2 + (5-4)^2 + (6-4)^2 + (2-4)^2$$
$$+ (4-4)^2 + (4-4)^2 + (3-4)^2 + (4-4)^2 + (7-4)^2$$
$$= 4 + 1 + 1 + 4 + 4 + 0 + 0 + 1 + 0 + 9$$
$$= 24$$

Computing the second summation:

$$\Sigma(X_2 - M_2)^2 = (7-6)^2 + (6-6)^2 + (5-6)^2 + (8-6)^2 + (4-6)^2$$
$$+ (5-6)^2 + (6-6)^2 + (6-6)^2 + (7-6)^2 + (6-6)^2$$
$$= 1 + 0 + 1 + 4 + 4 + 1 + 0 + 0 + 1 + 0$$
$$= 12$$

Substituting these totals into the formula for Est. SD^2_{diff}, we get

$$\text{Est. } SD^2_{diff} = \frac{24 + 12}{n_1 + n_2 - 2} \times \left(\frac{1}{n_1} + \frac{1}{n_2}\right)$$

But we know that $n_1 = 10$ and $n_2 = 10$, so we can finish the calculation:

$$\text{Est. } SD^2_{diff} = \frac{24 + 12}{10 + 10 - 2} \times \left(\frac{1}{10} + \frac{1}{10}\right)$$

$$= \frac{36}{18} \times \left(\frac{2}{10}\right) = \frac{4}{10}$$

$$= 0.40$$

Taking the square root of both sides, we can get the standard deviation:

$$\text{Est. } SD_{diff} = \sqrt{\text{Est. } SD^2_{diff}}$$

$$= \sqrt{0.40} = \sqrt{\frac{40}{100}} = \frac{\sqrt{40}}{\sqrt{100}} = \frac{\sqrt{40}}{10} = \frac{6.32}{10}$$

$$= 0.632$$

Now we are almost done. We started asking whether two samples could plausibly have come from the same distribution. We have worked down to the final question: Could the difference score, $M_1 - M_2 = 4 - 6 = -2$, plausibly have come from the normal distribution in Figure 24.1? The mean of that normal distribution is zero, and we have estimated the standard deviation to be 0.632. We can compute a score which is like a Z score, but which is in fact a t statistic:

$$t = \frac{\text{observed difference} - \text{mean difference}}{\text{Est. } SD_{diff}}$$

$$= \frac{(-2) - 0}{0.632} = -3.16$$

We need to know the number of degrees of freedom involved before we can use the t table to evaluate this result. For a test of this sort, when we're comparing the means of two independent samples, the number of degrees of freedom is given by $df = n_1 + n_2 - 2$. For our sample computation where $n_1 = 10$ and $n_2 = 10$, $df = 18$. Referring to the t table (see page 160), we can see that with 18 degrees of freedom, a t statistic smaller than -2.88 or larger than $+2.88$ will occur only 1 time in 100 (with probability 0.01). So, to return finally to the initial example, it is most unlikely that we would observe two samples like this from the same normal distribution. We can reject the null hypothesis that the two kinds of dog chow do not differ in productivity. J.F.'s Superchow *is* super chow.

Summary Now let's review the logic of this analysis, since it is somewhat involved. We started out wondering whether Wilson's chow was significantly different from J.F.'s Superchow, in terms of puppy production. Put differently, we want to know if the population of litters produced when the bitch eats Wilson's chow is different from the population of litters produced when the bitch eats J.F.'s Superchow. We have two samples, each containing ten numbers, as a basis for our computations. We start out with a null hypothesis: H_0 : The two samples came from the same (normally distributed) population. When this null hypothesis is true, the *difference* between two sample means, M_1 minus M_2, will have a normal distribution with a zero mean and variance estimated by

$$\text{Est. } SD_{\text{diff}}^2 = \frac{\Sigma(X_1 - M_1)^2 + \Sigma(X_2 - M_2)^2}{n_1 + n_2 - 2} \times \left(\frac{1}{n_1} + \frac{1}{n_2}\right)$$

When the null hypothesis is true, the t statistic given by

$$t = \frac{M_1 - M_2}{\text{Est. } SD_{\text{diff}}}$$

will have a t distribution with $n_1 + n_2 - 2$ degrees of freedom. When the statistic is too extreme, we reject the null hypothesis.

TWO t TESTS COMPARED

A note on the similarities and differences between this t test and the t test for the mean of a single sample which was presented earlier.

In Chapters 18 and 19 you learned about the t test to evaluate a null hypothesis of the following sort: The (one) sample which I am considering came from a normal distribution with known mean and unknown variance. The t statistic is defined as

$$t = \frac{\text{Sample mean} - \text{Hypothesized population mean}}{\text{Est. } SD_{\text{samp dist of the mean}}}$$

The number of degrees of freedom, as shown in the t table, is $n - 1$, where n is the sample size.

In this chapter you learned about the t test to check a null hypothesis that both samples under consideration came from the same (normally distributed) population. In doing this, you computed the difference between the two sample means, M_1 minus M_2, and calculated a t statistic as follows:

$$t = \frac{M_1 - M_2}{\text{Est. } SD_{\text{samp dist of differences}}}$$

There is a close conceptual similarity, since $(M_1 - M_2)$ is an observed *difference*, and the numerator is sometimes written

$(M_1 - M_2) - 0$ or Observed difference − Mean of population of differences

The number of degrees of freedom in this situation is $(n_1 + n_2 - 2)$. This can also be written $(n_1 - 1) + (n_2 - 1)$, which increases the similarity with the previous case.

A formula which simplifies the actual calculation of the *t* statistic: In the dog chow example, as you probably observed, the sample means very conveniently came out as integers: $M_1 = 4$ and $M_2 = 6$. If the numbers had been 4.27 and 6.33, the computations would have been much less simple. In the real world, samples are rarely so well behaved. The formula presented above for computing the estimated SD^2 of the sampling distribution of differences, which is used in finding the *t* statistic, is easier to understand (honestly) than the following one. But the following one will be easier to use if you're computing a *t* statistic on a calculator, or with the tables in the appendix of this book.

The alternative formula for computing the *t* statistic for testing the significance of the difference between two sample means, M_1 and M_2 is:

$$t = \frac{M_1 - M_2}{\text{Est. } SD_{\text{diff}}} \quad \text{(this hasn't changed)}$$

where

$$\text{Est. } SD_{\text{diff}} = \sqrt{\frac{\Sigma(X_1^2) - (\Sigma X_1)^2/n_1 + \Sigma(X_2^2) - (\Sigma X_2)^2/n_2}{n_1 + n_2 - 2} \times \left(\frac{1}{n_1} + \frac{1}{n_2}\right)}$$

Here $\Sigma(X_1^2)$ is the sum of the squares of the numbers in the first sample: you square each number and add up the results. The similar looking $(\Sigma X_1)^2$ has a different meaning: you add up the numbers in the first sample (ΣX_1) and then you square that sum. The same relation holds for $\Sigma(X_2^2)$ and $(\Sigma X_2)^2$. Since this is just a different procedure for calculating the same standard deviation, the number of degrees of freedom, $df = n_1 + n_2 - 2$, is unchanged.

QUESTION AND ANSWER

Question: You say that the formula above is advantageous in many situations. Could you do a sample computation with it?

Answer: With pleasure. Here is a reanalysis of the dog chow example, using that formula. The data are repeated here for convenience:

Sample 1 (Wilson's chow): 2, 3, 5, 6, 2, 4, 4, 3, 4, 7 $M_1 = 4$
Sample 2 (J.F.'s chow): 7, 6, 5, 8, 4, 5, 6, 6, 7, 6 $M_2 = 6$

To compute the *t* statistic, we proceed as follows:

$$t = \frac{M_1 - M_2}{\text{Est. } SD_{\text{diff}}}$$

$$\text{Est. } SD_{\text{diff}} = \sqrt{\frac{\Sigma(X_1^2) - (\Sigma X_1)^2/n_1 + \Sigma(X_2^2) - (\Sigma X_2)^2/n_2}{n_1 + n_2 - 2} \times \left(\frac{1}{n_1} + \frac{1}{n_2}\right)}$$

We'll now proceed piece by piece. Since the first sample contains ten numbers, $n_1 = 10$, and the second contains ten numbers, $n_2 = 10$:

$\Sigma(X_1^2) = 2^2 + 3^2 + 5^2 + 6^2 + 2^2 + 4^2 + 4^2 + 3^2 + 4^2 + 7^2$

$\qquad = 4 + 9 + 25 + 36 + 4 + 16 + 16 + 9 + 16 + 49$

$\Sigma(X_1^2) = 184$

$\dfrac{(\Sigma X_1)^2}{n_1} = \dfrac{(2+3+5+6+2+4+4+3+4+7)^2}{10} = \dfrac{(40)^2}{10} = \dfrac{1600}{10}$

$\dfrac{(\Sigma X_1)^2}{n_1} = 160$

$\Sigma(X_2^2) = 7^2 + 6^2 + 5^2 + 8^2 + 4^2 + 5^2 + 6^2 + 6^2 + 7^2 + 6^2$

$\qquad = 49 + 36 + 25 + 64 + 16 + 25 + 36 + 36 + 49 + 36$

$\Sigma(X_2^2) = 372$

$\dfrac{(\Sigma X_2)^2}{n_2} = \dfrac{(7+6+5+8+4+5+6+6+7+6)^2}{10} = \dfrac{(60)^2}{10} = \dfrac{3600}{10}$

$\dfrac{(\Sigma X_2)^2}{n_2} = 360$

Turning to the denominator:

$n_1 + n_2 - 2 = 10 + 10 - 2$

$\qquad\qquad = 18$

Finally:

$\dfrac{1}{n_1} + \dfrac{1}{n_2} = \dfrac{1}{10} + \dfrac{1}{10}$

$\qquad\qquad = \dfrac{2}{10}$

Putting the pieces together:

$\text{Est.}\,SD_{\text{diff}}^2 = \dfrac{184 - 160 + 372 - 360}{18} \times \left(\dfrac{2}{10}\right)$

$\qquad\qquad = \dfrac{36}{18} \times \dfrac{2}{10} = \dfrac{4}{10}$

$\qquad\qquad = 0.400$

Taking the square root of both sides:

$\text{Est.}\,SD_{\text{diff}} = \sqrt{0.40} = \sqrt{\dfrac{40}{100}} = \dfrac{\sqrt{40}}{\sqrt{100}} = \dfrac{\sqrt{40}}{10} = \dfrac{6.32}{10}$

$\qquad\qquad = 0.632$

And finally, since $M_1 = 4$ and $M_2 = 6$, we can compute t:

$t = \dfrac{M_1 - M_2}{\text{Est.}\,SD_{\text{diff}}} = \dfrac{4 - 6}{0.632} = \dfrac{-2}{0.632}$

$t = -3.16$

Question: In the null hypothesis for the *t* test for the difference between two means, you always said you assumed the samples came from a normal distribution. Why did you mention that? And how do you tell if the assumption is satisfied?

Answer: The theory which underlies the *t* test requires the assumption that the samples come from a normal distribution. In fact, the test works well for distributions which even roughly resemble the normal distribution. As long as the population is "bump"-shaped or mound-shaped, or has a rough similarity to the normal distribution, the *t* test works well. In the event that you have to work with a population which has a strikingly different distribution, it is sometimes possible to transform the data to reduce the problem. Many advanced statistics texts discuss the process.

Summary

In this chapter a very common statistical test was introduced: the *t* test for the difference between two sample means. Often we wish to compare processes which have produced two different (and uncorrelated) samples. For example, we frequently compare an experimental group with a control group, or a supposed improvement in a process with the original process. When two samples have been obtained, we test the null hypothesis that they resulted from independent random sampling from a common, normally distributed parent population, that is, that the two processes or groups are not different. In order to test this hypothesis, we compute the means of of the samples, M_1 and M_2. We then compute a *t* statistic as follows:

$$t = \frac{M_1 - M_2}{\text{Est.} SD_{\text{diff}}}$$

$$\text{Est.} SD_{\text{diff}} = \sqrt{\frac{\Sigma(X_1 - M_1)^2 + \Sigma(X_2 - M_2)^2}{n_1 + n_2 - 2} \times \left(\frac{1}{n_1} + \frac{1}{n_2}\right)}$$

where the first summation is over all the numbers in the first sample, and the second summation is over all the numbers in the second sample. n_1 is the size of the first sample, and n_2 is the size of the second. When the null hypothesis is true, and the two groups aren't different, this *t* statistic has a *t* distribution with $(n_1 + n_2 - 2)$ degrees of freedom. When a sufficiently large positive or negative value of *t* is observed, we conclude that the null hypothesis is false, and the two groups or treatments *are* significantly different. A different formula, that is somewhat harder to understand but often somewhat easier to use, was also presented.

Problems

1. Two groups of ten people participated in an experiment on the effects of "positive thinking." People in the first group spent several minutes thinking positively, and then performed a learning task. The errors made for the ten people were 2, 8, 4, 7, 5, 3, 5, 6, 3, and 7. People in the second group spent a few minutes reading dull magazine stories, and then did the same learning task. Their errors were 10, 5, 5, 6, 8, 6, 7, 10, 6, and 7. Perform a *t* test to help decide whether "thinking positively" had a significant effect on error rate. Use a significance level of 0.05.

2. In a laboratory test of the effect of ambient noise on problem solving, eight people learned a vocabulary list in a quiet environment, and eight others learned the list in a noisy environment. The number of times they had to hear the list before mastering it is given below:

   Quiet learning: 1, 2, 2, 1, 3, 1, 3, 2
   Noisy learning: 3, 3, 2, 3, 1, 4, 2, 3

   Check to see if the environment had a significant effect on the learning score. Use a 0.10 significance level.

3. Suppose you have a box of ball bearings whose diameters are normally distributed. Describe how you could experimentally approximate a sampling distribution for the difference between two means, M_1 minus M_2, like that in Figure 24.1.

†4. Given the following assumptions, derive the formula for Est. SD_{diff}^2 which appears earlier in the chapter:

i. Given two samples from the same population, you can estimate the variance of the population from which they are taken as follows:

$$\text{Est. } SD_{\text{pop}}^2 = \frac{\Sigma(X_1 - M_1)^2 + \Sigma(X_2 - M_2)^2}{n_1 + n_2 - 2}$$

ii. Suppose you have two populations, population A and population B, with variances SD_A^2 and SD_B^2, respectively. Then if X_A is a number sampled at random from population A, and X_B is a number sampled at random from population B, the difference score $D = X_A - X_B$ will have a variance $SD_A^2 + SD_B^2$. If both A and B populations are normally distributed, the difference score population will be, too.

iii. As you already know, the variance of the sampling distribution of the mean, for samples of size n from a population with variance SD_{pop}^2, is given by $SD_{\text{samp dist}}^2 = SD_{\text{pop}}^2/n$.

Two-Tailed Tests and One-Tailed Tests

In this book, we have looked at a variety of rejection regions. Starting with the ten-trial ESP experiment, with a significance level of 0.05, we decided to reject the null hypothesis if we observed 0/10, 1/10, 9/10, or 10/10 correct (Figure 25.1).

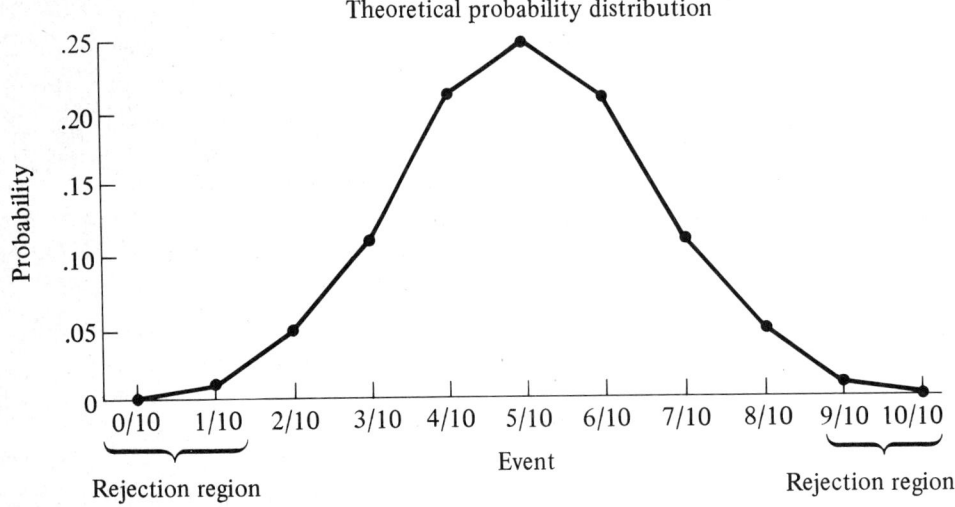

Figure 25.1 The rejection region for a 10-trial ESP experiment, when $\alpha = .05$.

In a variety of Z tests, again with a 0.05 significance level, we decided to reject the null hypothesis if Z were less than -1.96, or if Z were greater than $+1.96$ (Figure 25.2).

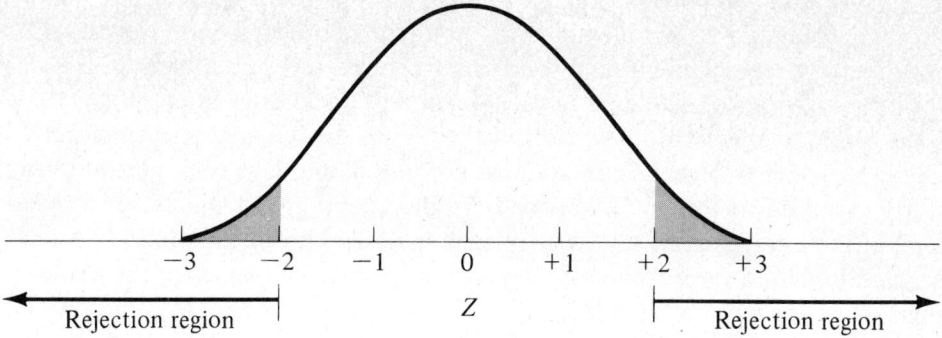

Figure 25.2 The rejection region for the Z test is composed of two infinite intervals, shown by arrows.

When we picked a rejection region for the *t* statistic, it had a form similar to that in Figure 25.2. All of these rejection regions contain an equal interval on each tail of the probability distribution. These are *two-tailed* rejection regions. Using them implies the assumption that we are just as likely to reject the null hypothesis if we see an extreme observation at one end as if we see it at the other.

There are other possibilities. For example, near the start of the book we recognized that the following would be a plausible decision rule: Decide the person has ESP if you observe 8/10 or 9/10 or 10/10 correct guesses; otherwise decide he was guessing randomly. The rejection region for this decision rule is not two tailed: it is *one tailed*. And there might be situations in which you would prefer a one-tailed rejection region. For example, if it is *absolutely inconceivable* for you that a person might consistently and regularly do worse than chance in an ESP experiment, you might prefer a one-tailed rejection region. If you were to announce *before conducting the experiment* that you would reject the null hypothesis only if the subject guessed 8/10, 9/10, or 10/10 correct, then the probability of a Type One error would be $0.044 + 0.010 + 0.001 = 0.55$ (see page 40). Now if 1000 experimenters acted this way, and each ran an ESP experiment like the one mentioned in Chapter 1, and if all the 1000 subjects had no ESP, then we would expect about 55 of the 1000 people, or a proportion of 0.055, to falsely reject the null hypothesis.

Why aren't rejection regions like this commonly used? The problem is that out of the 1000 experiments, we would expect to see, by pure chance, the result 0/10 correct about once. We would expect to see, by pure chance, the result 1/10 about 10 times, and to see the result 2/10 about 44 times. Now some of the people who observed 0/10 or 1/10 will probably start to scratch their heads. "Those are very unlikely events. It just might be, mightn't it, that a person with ESP would sort of get his signals crossed, and 'perceive' an X where there was an O, and vice versa. Then, instead of doing amazingly well, he would do amazingly badly. So perhaps we have seen a case of strong negative extrasensory perception." This might cause some of the people who observed 0/10 or 1/10 to decide to change their decision rules and rejection regions. If they *all* did, there would be 11 more experimenters who would (falsely) reject the null hypothesis. All 55 of the people who saw an extremely high success score (8/10 or higher) would reject the null hypothesis. And another 11, who saw an extremely low success score (0/10 or 1/10), would reject the null hypothesis too. So a total of 66 people would falsely reject the null hypothesis. And the probability of a Type One error would be 0.066, rather than the 0.055 we originally expected.

A similar danger exists whenever people report using a one-tailed test. If they had seen a very extreme result in the opposite side of the distribution from their rejection region, they *might* have changed their mind and decided to use a two-tailed analysis. This will have the net effect of increasing the number of Type One errors people make, without the people becoming aware of it.

Since this is a somewhat abstract argument, and sometimes appears to be an attack on the morals and character of people who use one-tailed tests, let's look at another example. I am interested in studying the effect of special study carrels in the library on the grades of randomly chosen students. I propose to select two groups of students at random and give each person in the first group a private study carrel in the library. The people in the second group will get no special treatment; I plan to look at their grades after a year and see if the carrel group is doing significantly better as a result of their improved study conditions. I decide to analyze the grades of the two groups using a t test for uncorrelated means, discussed in Chapter 24.[1] My scientific colleagues are interested only in learning about results that are significant at the 0.05 level. Since the experiment is somewhat time consuming, I want to maximize the chance that I find an effect if there is really one. So I decide to use a one-tailed t test. I decide that if the special-carrel group does sufficiently better than the control group — so that I get a t statistic of +1.70 or greater — I will reject the null hypothesis. Since I plan to run 15 subjects in each group, the number of degrees of freedom is $N_1 + N_2 - 2 = 28$. For 28 df, there is a probability of 0.10 of seeing a t value smaller than −1.70 or larger than +1.70. By using a one-tailed rejection region, I keep the chance of a Type One error down to half that level, or 0.05. (Remember that all t distributions, like the normal distribution, are symmetric.) The relevant row of the t table looks like this:

			Probability		
Degrees of freedom	0.50	0.10	0.05	0.02	0.01
28	.683	1.70	2.05	2.47	2.77

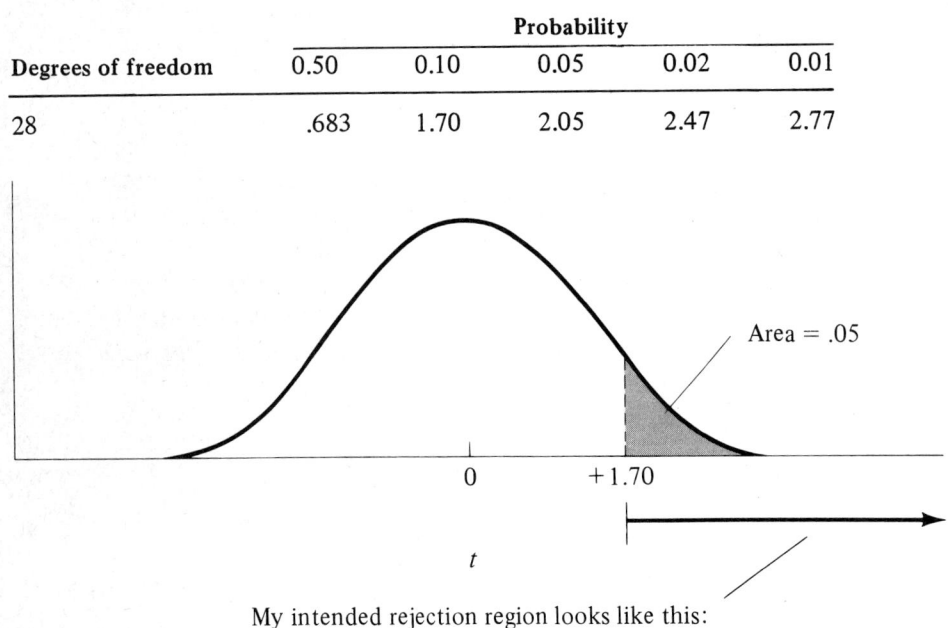

My intended rejection region looks like this:

Figure 25.3 The one-tailed rejection region consists of all t scores greater than +1.70. The probability of such a t score, when H_0 is true, is .05.

Now imagine that a year goes by, and the experiment proceeds as planned. I gather the data, compute the means, and finally compute the t statistic. To my great surprise, I find that the students with the special carrels have done *worse* than the control group, and *my t statistic has a value of* −2.34. This result, though most unexpected, is all the more remarkable for that reason. Why should having a private study area hurt people's grades? After interviewing the students extensively, two of them confess to me that they couldn't help falling asleep in the carrels.

[1] A suboptimal design, but that's beside the point.

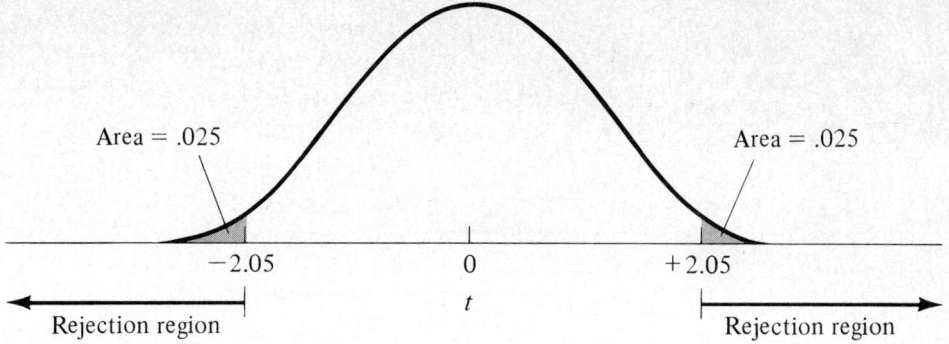

Figure 25.4 This two-tailed rejection region contains t scores greater than +2.05 and t scores less than −2.05. Again, $\alpha = .05$.

Without the social pressure of other students around them, they kept taking naps every time they remained in the library. It looks as if I've stumbled on something very interesting.

What do I do? My initial plan was to reject the null hypothesis only if I saw a t score greater than +1.70. The result, $t = -2.34$, is not in my rejection region. So I really shouldn't reject the null hypothesis. Still, I've done a lot of work and discovered something very interesting. It's clear that I *might just as well* have used a traditional rejection region, as in Figure 25.4. If I had planned on this rejection region, my results would have been significant. I'm sure, after the interviews, that I've discovered something interesting and important. In fact, looking back at my notes, I'm not even positive that I originally intended a one-tailed test. So why not just go ahead and report the findings?

There's no reason not to do so, except that I'm kidding myself and my colleagues a little. If I had observed a t score of +1.72, I probably would have stuck to the initial plan and reported a result significant at the 0.05 level using a one-tailed test. If I had observed a t score of −1.72, I might well have concluded "No significant results," unless my interviews with the students had convinced me that it was *obvious* that if the carrels had any effect it would be detrimental, so that I really should have had a one-tailed rejection region like that in Figure 25.5. This would be rather callous self-deception, and I doubt that I'd have done it to me.

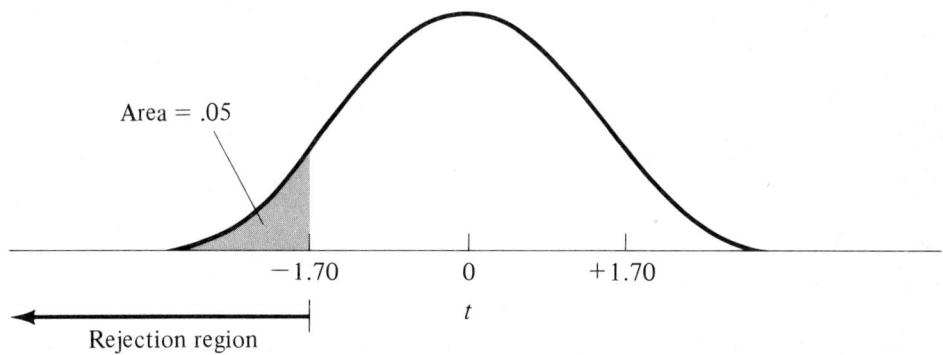

Figure 25.5 This is another one-tailed rejection region, with $\alpha = .05$; all t scores less than −1.70 are in the rejection region.

But it looks as if I *would* have reported "A significant effect at the 0.05 level" if I observed a t score greater than +1.70, in accordance with my original one-tailed analysis, *or* if I observed a

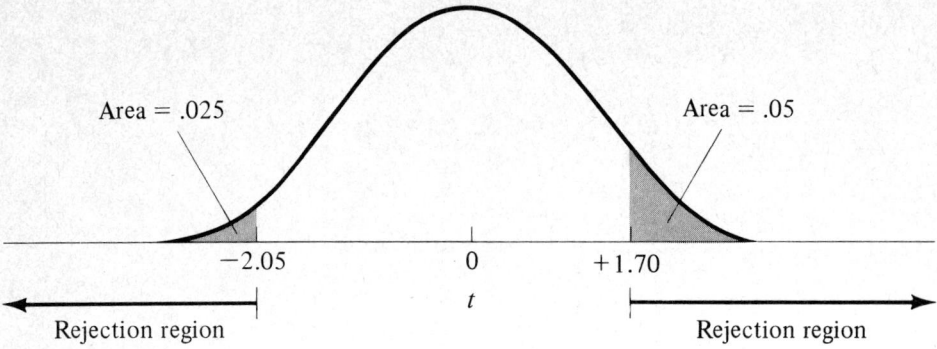

Figure 25.6 With a bit of selective forgetting, my effective rejection region might have looked like this.

t score less than -2.05, in accordance with a revised, two-tailed analysis. This effective composite rejection region, which I almost surely never sketched to myself, would look like Figure 25.6.

This is a very remarkable rejection region, for a test with a 0.05 significance level: the probability of an observation in the rejection region, if the null hypothesis is true, is 0.05 + 0.025 = 0.075. There were 7.5 chances in 100 of a Type One error. But because I switched intentions in the middle of the experiment, I may never have appreciated this. And even if I did, nobody else would ever know.

At this point, the problem with one-tailed rejection regions, and one-tailed tests, should be clear: If you should, by some remote chance, see a result way out in the opposite tail of the distribution, it's going to be very hard to totally ignore it. And this means, effectively, that your rejection region isn't really one tailed.

It is for this reason that all the analyses in this book have been two tailed.

Summary

All the rejection regions described in this book have been two-tailed. They have included extreme values in both possible directions. Some people advocate the use of one-tailed rejection regions under certain circumstances. Although there may be conditions in which such tests are advisable, there seem to be many more when they can lead to self-deception. A conservative position, with symmetric, two-tailed rejection regions, is recommended.

Reviewing for Final Exams

To help you prepare for a final exam on the material in this book this chapter is divided into three parts:

1. A sample final exam.
2. Solutions to the problems in the sample final.
3. Some questions and answers which may help prepare for a final.

Bonne chance!

SAMPLE FINAL:

1. Define each of the following in a few words:
   a. Type One error.
   b. Type Two error.
   c. Sample space.
   d. Sampling distribution of the mean.

   Note: State the null hypothesis for problems 2 through 5 below!

2. A study of Shakespeare's plays involved looking at each group of 100 consecutive words and counting the number of *nouns* in the group. This index, which we shall call N, was found to be very nearly normally distributed, with a mean of 17 and standard deviation of 6. An unsigned play written in a similar style is found, and a noun count is made for a sample of 50 100-word passages. The average of these 50 measurements is 19.7. Do you think this play was by Shakespeare? Support your conclusion with a statistical test.

3. In five previously unopened bottles labeled "100 Aspirin" we find 89, 93, 92, 95, and 91 pills. Perform a t test to see if this sample gives you statistically significant evidence that the average number of pills is less than 100. Use a 0.01 significance level.

4. An ESP experiment much like the one mentioned early in this book is conducted. On each trial the subject tries to predict whether an X or an O will appear inside an envelope. The experiment lasts for 64 trials. The subject guesses correctly 41 times out of 64. Do a sign test, using the normal approximation to the binomial, to see if this is significantly better than chance. Use $\alpha = 0.05$.

5. A drug study involved giving a placebo (sugar pill) to one group of people and a supposed hallucinogenic drug to a second group. Each person was asked half an hour later if he felt that his state of consciousness had changed. The results were:

	Change in consciousness	No change
Drug	38	12
Placebo	32	18

Analyze these data using a chi-square test, to see if the drug had any more effect than the placebo. Use $\alpha = 0.05$.

6. Estimate the correlation between the two variables represented in the following scatterplots:

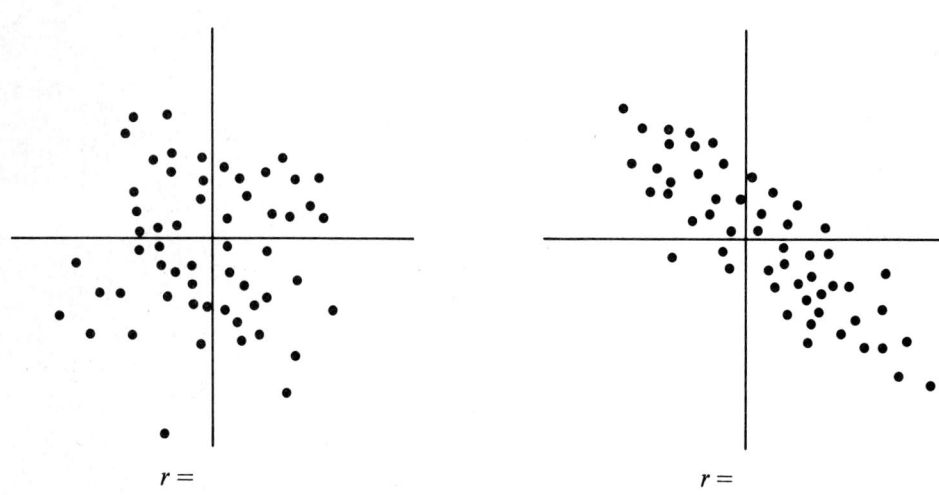

$r =$ $r =$

7. State the conditions under which the sampling distribution of the sample mean is a normal distribution.

8. Make up some data for two variables which are perfectly correlated. Prove this by calculating r.

	x	y
Case 1	_____	_____
Case 2	_____	_____
Case 3	_____	_____

9. For each of the following situations, state what statistical test would be appropriate, and what null hypothesis you would use:

   a. A local fisherman kept careful records of his halibut catch over the years prior to a big oil spill. He averaged 5000 pounds of halibut per month. After the spill, in a 6-month period he recorded the following poundages: 4200, 2100, 5300, 2600, 3400, and 1900. Had the situation deteriorated?

   b. A politician divides the country into four sections: East, West, South, and Midwest. He polls 200 people in each section, to see which of the following alternatives is favored: (1) keep present abortion laws; (2) liberalize abortion laws; (3) eliminate all legal regulation of abortion. After obtaining the data, he wants a test to find out whether people in different areas have significantly different opinions.

   c. Psychiatrists rate each of eight patients on "self-actualization" before group therapy and again after the therapy. What kind of test should be used to see if therapy has had a significant effect?

   d. The students in an English course decide to experiment on the professor. Each student has two papers to submit. Each will submit one handwritten and one typed paper. Each tosses a coin to decide which way to do the first paper. Out of 30 students, 22 get better grades on the typewritten paper, and 8 get better grades on the handwritten paper.

10. Sketch, very roughly, and label the Z distribution and the t distribution with 3 degrees of freedom on the same graph.

11. Suppose that you are taking samples from the standard normal distribution (Z distribution). What is the probability of observing a value between -1.96 and -2.58?

Solutions to Sample Final

1. a. Type One error: Rejecting the null hypothesis when it is true; saying there is an effect when there is none.

   b. Type Two error: Failing to reject the null hypothesis when you should; failing to say there's an effect when there is one.

   c. Sample space: The set of all possible outcomes of a conceptual experiment.

   d. Sampling distribution of the mean: The probability distribution which describes the likelihood of every possible value of the sample mean, based on samples of a given size, from the parent population under consideration.

2. This is a situation for a Z test for the sample mean. We are asking whether a sample of size 50 was taken from a known normal distribution with mean = 17 and standard deviation = 6.
   a. H_0: The 50 measurements were taken from a normal distribution with mean 17 and standard deviation 6. That is, the play was by Shakespeare (or at least it had the same N count).
   b. $\alpha = 0.05$. There is 0.05 chance that $|Z| > 1.96$,

   so

   c. Rejection region = $Z > 1.96$

   or

   $Z < -1.96$

d. Compute the test statistic Z. First we must find out the mean and standard deviation of the sampling distribution of the mean for samples of size 50 from the above distribution. The mean will be the same as the mean of the parent population:

$$M_M = M_{pop} = 17$$

The variance of the sampling distribution of the mean will be 1/50 times the variance of the parent population:

$$SD_M^2 = \frac{SD_{pop}^2}{n} = \frac{SD_{pop}^2}{50} = \frac{6 \times 6}{50} = \frac{36}{50}$$

$$= 0.72$$

Therefore,

$$SD_M = \sqrt{SD_M^2} = \sqrt{0.72}$$

$$\cong 0.85$$

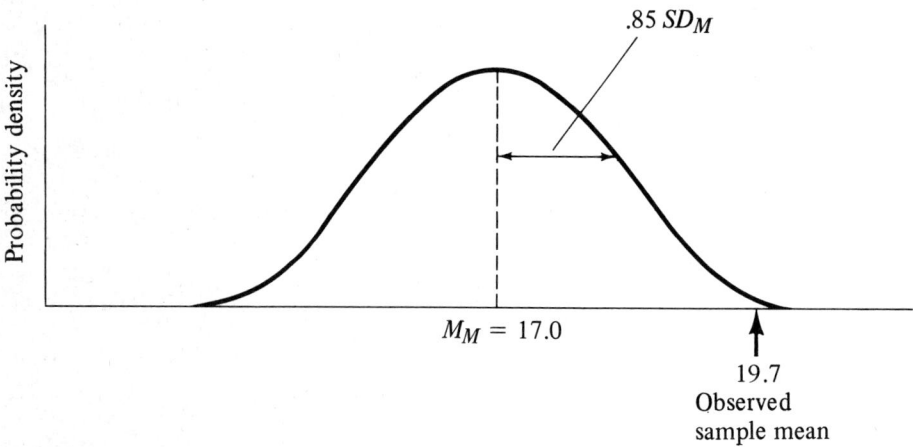

Sampling distribution of the mean

Z score:

$$Z = \frac{M - M_M}{SD_M} = \frac{19.7 - 17.0}{0.85}$$

$$= \frac{2.7}{0.85}$$

$$\cong 3.2$$

e. Observed Z score is in rejection region, therefore, reject H_0.
Decision: Play is *not* by Shakespeare.

3. The sample is 89, 93, 92, 95, 91. We wish to decide whether it could plausibly have come from a normal distribution with mean 100 and unknown variance. We run a *t* test.

a. H_0: The observed numbers, listed above, came from a normal distribution with mean 100. In general, the average number of aspirin in a bottle is 100.
b. Significance level = 0.01; $df = n - 1 = 5 - 1 = 4$

c. Critical value for t, with $df = 4$ and $\alpha = 0.01$, is 4.6 (from table): Rejection region is $t > 4.6$ or $t < -4.6$.

d. The mean of the sample is 92. $M = 92$. Estimate the population variance on the basis of the sample:

$$\text{Est. } SD^2_{pop} = \frac{\Sigma(X_i - M)^2}{n-1} = \frac{(89+92)^2 + (93-92)^2 + (92-92)^2 + (95-92)^2 + (91-92)^2}{n-1}$$

$$= \frac{9+1+0+9+1}{5-1} = \frac{20}{4}$$

$$= 5$$

Estimate the variance of the sampling distribution of the mean:

$$\text{Est. } SD^2_M = \frac{\text{Est. } SD^2_{pop}}{n} = \frac{5}{5} = 1$$

$$\text{Est. } SD_M = \sqrt{\text{Est. } SD^2_M} = \sqrt{1} = 1$$

The mean of the sampling distribution of the mean, $M_M = M_{pop} = 100$. Compute the t statistic:

$$t = \frac{M - M_M}{\text{Est. } SD_M} = \frac{92 - 100}{1} = -8$$

e. We observed $t = -8$; therefore, reject H_0. It is most unlikely that these bottles came from a population with $M = 100$. The manufacturer is cheating.

4. H_0 : The probability that the subject will guess correctly on any of the 64 trials is just the same as the probability that he will guess wrong: 1/2. The subject has no ESP. Possible events: 0/64, 1/64, 2/64, etc., up to 64/64.

Study H_0 : The shape of the distribution of sign-test probabilities can be closely approximated by a normal distribution with the same mean and variance. The mean of the binomial distribution (the sign-test distribution) with $N = 64$ and both outcomes equally likely is $M_{bin} = N/2 = 64/2 = 32$. The variance is $SD^2_{bin} = N/4 = 64/4 = 16$. So $SD_{bin} = \sqrt{SD^2_{bin}} = \sqrt{16} = 4$. The normal distribution which best approximates the exact probability distribution has the same mean, 32, and standard deviation, 4. Compute a Z score for the observed result, 41.

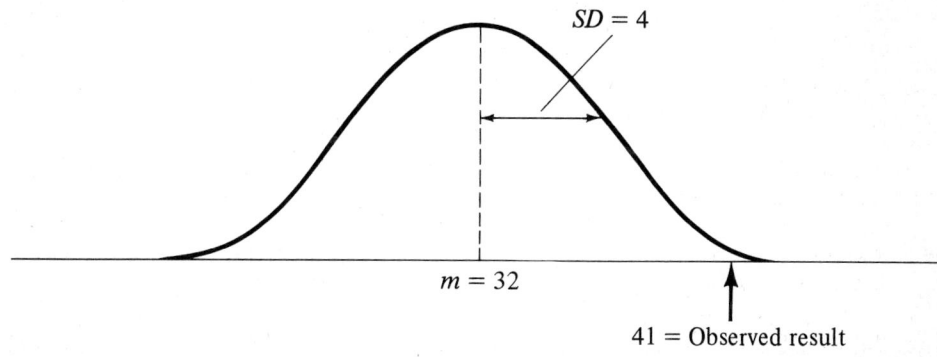

Here is a normal distribution with almost exactly the same shape as the binomial with the same mean and standard deviation.

$$Z = \frac{X_i - M}{SD} = \frac{41 - 32}{4} = \frac{9}{4} = 2.25$$

Significance level $\alpha = 0.05$. The rejection region, in the normal distribution with $\alpha = 0.05$, is the set of all Z scores greater than $+1.96$ or less than -1.96. We observed $Z = 2.25$. This is in the rejection region, so we reject H_0: The person has ESP.

5.

	Change	No Change	
Drug	38	12	50
Placebo	32	18	50
	70	30	

Observed

	Change	No Change	
Drug	35	15	50
Placebo	35	15	50
	70	30	

Expected

H_0: There is no relationship between pill and state of consciousness.
Compute χ^2:

$$\chi^2 = \sum_{\text{all cells}} \frac{(E - O)^2}{E}$$

$$= \frac{(35 - 38)^2}{35} + \frac{(15 - 12)^2}{15} + \frac{(35 - 32)^2}{35} + \frac{(15 - 18)^2}{15}$$

$$= \frac{9}{35} + \frac{9}{15} + \frac{9}{35} + \frac{9}{15}$$

$$= \frac{18}{35} + \frac{18}{15}$$

$$\cong 0.5 + 1.2$$

$$\cong 1.7$$

Degrees of freedom = $(2 - 1) \times (2 - 1) = 1$

$\alpha = 0.05$

Critical value of $\chi^2 = 3.84$ when $\alpha = 0.05$, $df = 1$. Observed χ^2, 1.7, is less than critical value. So you do not reject H_0: Drug and placebo have not been shown to have a different effect on state of consciousness.

6. In the first scatterplot, the correlation is about −0.10. You should have guessed between +0.10 and −0.30. In the second scatterplot, the correlation is about −0.85. Your answer should have been between −0.65 and −0.95.

7. The sampling distribution of sample means from a normal distribution will always be a normal distribution, regardless of sample size. The sampling distribution of the mean for large samples (more than 30 numbers in a sample) will also be a normal distribution, regardless of the parent population.

8. There are two problems here: one is to make up two data lists which are perfectly correlated, and the second is to calculate r. Here are some examples of perfect correlation. In every case the scatterplot is a straight line:

X	Y		X	Y		X	Y
1	1		5	1		6	5
2	2		10	2		12	10
3	3		15	3		18	15
(a)			(b)			(c)	

Consider table (a). The scatterplot looks like this:

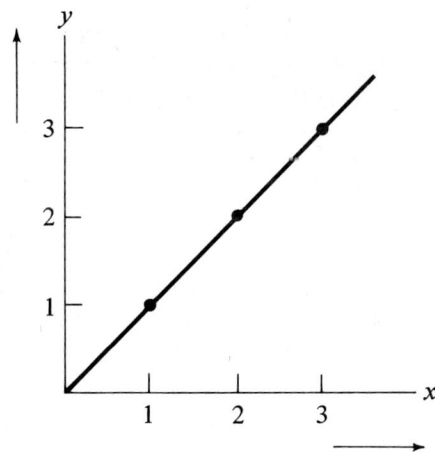

Computing r for the above scatterplot: X list = (1, 2, 3). So

$$M_X = \frac{1 + 2 + 3}{3} = \frac{6}{3} = 2$$

$$SD_X^2 = \frac{\Sigma(X_i - M_X)^2}{n}$$

$$= \frac{(1 - 2)^2 + (2 - 2)^2 + (3 - 2)^2}{3} = \frac{2}{3}$$

$SD_X = \sqrt{2/3}$

The Z scores are

$$Z_{X_i} = \frac{X_i - M_X}{SD_X}$$

or

X	Z_X
1	$-1/\sqrt{2/3}$
2	0
3	$+1/\sqrt{2/3}$

By an exactly similar process you can show that the Z_Y scores are:

Y	Z_Y
1	$-1/\sqrt{2/3}$
2	0
3	$+1/\sqrt{2/3}$

Putting the entries of the two tables together,

$$r = \frac{\Sigma Z_{X_i} \times Z_{Y_i}}{n} = \frac{\frac{-1}{\sqrt{2/3}} \times \frac{-1}{\sqrt{2/3}} + 0 \times 0 + \frac{1}{\sqrt{2/3}} \times \frac{1}{\sqrt{2/3}}}{3}$$

$$= \frac{1/(2/3) + 1/(2/3)}{3} = \frac{3/2 + 3/2}{3}$$

$$= 1$$

9. a. We wish to know whether the data listed could have come from a (presumably normal) distribution with a mean of 5000 and unknown variance. This is a situation for a t test.
H_0 : The data came from a normal distribution with mean 5000. The fishing is no worse.

b. The data table would look like this:

	Region			
	East	West	South	Midwest
Opinion a				
Opinion b				
Opinion c				

This is a situation for a chi-square test. The null hypothesis would be H_0 : There is no relationship between geographical origin and opinion on abortion. [*Note*: The number of degrees of freedom in this case would be $df = (3 - 1) \times (4 - 1) = 6$. Once six of the numbers in the table had been filled in, and the marginal totals specified, the remaining numbers could be deduced. Only six numbers are "free to vary."]

c. One would end up with eight difference scores. This would be a good place for a t test, or possibly a sign test. The t test would be more powerful, however. For the t test, the null hypothesis would be H_0 : The set of eight difference scores was taken from a normal distribution with mean zero. That is, the treatment had no effect on the measure.

d. This is a good situation for a sign test. There are 30 students, or trials. Each one will yield a T if the student does better on the typed paper, or an H if he does better on the handwritten paper. Because of the large number of cases, one would probably use the normal approximation to the binomial in working through the test. H_0 : The probability that a student will get a T is the same as the probability that he will get an H: 1/2. Typing doesn't affect grades.

10.

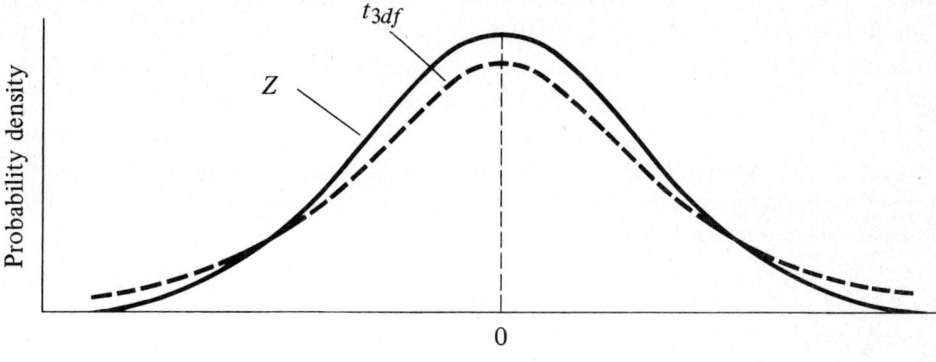

Value of t or Z statistic →

11. Since the probability of a Z score more extreme than 1.96 is 0.05, the probability of a Z score less than −1.96 is half of that, or 0.025. Similarly, since the probability of a score more extreme than 2.58 is 0.01, the probability of a score less than −2.58 is half of 0.01, or 0.005. Therefore the probability of an observation between −1.96 and −2.58 must be the difference between 0.025 and 0.005, or 0.020.

QUESTION AND ANSWER

These may help you to prepare for a final exam.

Question: For each of the following problems state a null hypothesis and tell what statistical test would be appropriate:

a. Suppose a used car dealer claims that you will get an average of 32 miles per gallon with a carefully maintained Volkswagon he wants to sell you. Assume that gasoline mileage is normally distributed with a standard deviation of 5. If you get 20 miles per gallon on your first tank full of gas, do you decide that he is a liar?

b. Two octogenarians have been checkers opponents for years. Over a long period they have been very evenly matched. Each has won half the time. Last September one of them read a book called *How to Play Better Checkers*. Since then he has lost 47 games while winning only 23. Have we a significant basis for concluding that the book affected his game?

c. Tryon B rats have been studied for decades at Berkeley. It is known that the trials to criterion or speed of learning for these rats on a black-white choice learning problem is normally distributed with mean 85 and standard deviation 21. We feed a special high protein diet to 20 rats from the time of weaning onward and train them on the standard tasks. We observe an average of 73 trials to criterion. Do we conclude that the diet affects learning?

d. Of twelve people who have played basketball for Ohio State recently, nine have had a better points-per-game average on the road than at home. Is this grounds for concluding that they do significantly better on the road?

e. The Miller radicalism test was standardized with college students several years ago. At that time it was found that the average score was 54 and the standard deviation 9. We tested 18 randomly selected students and observed that the average score was 59. Do we conclude that there has been a shift to the left?

f. A seat belt manufacturer specifies that his belts will, on the average, stand a 6000-pound strain before breaking. Eight are tested and we observe the following breaking points: 5600, 5300, 6100, 5300, 5700, 5400, 5400, 5900. Is the manufacturer honest?

Answer:

a. Run a Z test for a single observation from a known normal distribution. H_0 : The observation, 20, came from a normal distribution with mean 32 and standard deviation 5.

b. This is a situation for a sign test. With 70 trials, or games, use the normal approximation instead of calculating exact probabilities. H_0 : After reading the book, the probability of winning any single game was 1/2 for our hero.

c. This is an occasion for a Z test for a sample mean. You must first find the mean and standard deviation of the sampling distribution of the mean for samples of size 20 from the specified normal distribution. H_0 : The sample was taken from a normal distribution with mean 85 and standard deviation 21.

d. Sign test. H_0 : For any single player, the probability of doing better at home is the same as the probability of doing better on the road: 1/2.

e. This is an occasion for the Z test or the sample mean, if we're willing to assume that the test scores are normally distributed. The null hypothesis is that the 18 scores we observed were taken from the normal distribution with mean 54 and standard deviation 9. To solve this problem you have to find the mean and the standard deviation of the sampling distribution of the mean.

f. This is an occasion for a t test for the sample mean. You must estimate the population variance on the basis of the numbers in the sample. The null hypothesis is that these eight numbers came from a normal distribution with mean 6000.

Question: You have said that the t test is based on the assumption that the parent population from which the sample was taken is a normal distribution. How important is it that this assumption be exactly accurate?

Answer: As long as the distribution looks approximately like a bell curve, the t test will be quite accurate. If the distribution from which the samples were taken is dramatically different from a normal distribution in shape, then the t test is not accurate and you will make more errors than you expect.

Question: What is "null" about the null hypothesis in a t test?

Answer: The null hypothesis for the t test as presented here is always with the following form: The observed sample was taken from a normal distribution with mean [a number is inserted here]. Put into different words the null hypothesis could be written: The sample was taken from a population which is no different from the specified population. There is no significant difference between the numbers in the sample and the numbers you would expect if the sample had come from the distribution in question.

Question: In working with probability density functions, you repeatedly mentioned "the area over the interval." Did you mean the area divided by the interval?

Answer: No! This phrase seems to have generated some confusion, and more precisely one could say "the area between the horizontal axis and the curve bounded by the two sides of the interval."

Question: I'm not sure about how to use the correlation table. Suppose I compute a correlation of 2.31 based on the sample of 20 subjects and wish to set a significance level of 0.05. Then what do I do?

Answer: First of all, you should be very surprised on observing a correlation of 2.31. The largest possible correlation is 1.00, and the smallest possible correlation is −1.00. Any value outside of those two bounds must be the result of an error in calculation. Once you've taken care of that error, you can look at the correlation table in Chapter 21. Your sample contained 20 pairs.

Reading across the row corresponding to 20 pairs under the column labeled 0.05, you see the value .444. This is the critical value for the correlation. Any value greater than .444 would happen very rarely in a situation where you have 20 pairs of numbers. Indeed, from the table, we know that the chance of observing a correlation which is greater than +.444 or smaller than −.444 is just 0.05, or 1 in 20, when in fact the two variables are not correlated.

Question: Is it really true that the sampling distribution of the mean for large samples (more than 30 numbers in each sample) is a normal distribution no matter what the shape of the parent population is?

Answer: Except for some special cases of interest only to mathematicians, this is really true, amazing as it is. For any distribution you care to write down, the sampling distribution of the mean for large samples is very close to a normal distribution.

Question: What is it that a Z score tells you?

Answer: A Z score always tells you the location of one particular observation in a known distribution. The Z score tells you how many standard deviations away from the mean your observation lies. Sometimes, of course, the observation itself is a sample mean, and the distribution is the sampling distribution of sample means of that size.

Question: Would you say again what is meant by the *power* of a test?

Answer: The power of a test is defined as 1 minus the probability of a Type Two error. If the probability of a Type Two error is given the name β, then the power of a test is defined as Power = $1 - \beta$. The power of a test tells you how likely you are to find a significant difference if there in fact is one.

Appendix

Z TABLES: AREAS OF THE NORMAL PROBABILITY CURVE

Z	0.00	0.01	0.02	0.03	0.04	0.05	0.06	0.07	0.08	0.09
0.0	.0000	.0040	.0080	.0120	.0160	.0199	.0239	.0279	.0319	.0359
0.1	.0398	.0438	.0478	.0517	.0557	.0596	.0636	.0675	.0714	.0753
0.2	.0793	.0832	.0871	.0910	.0948	.0987	.1026	.1064	.1103	.1141
0.3	.1179	.1217	.1255	.1293	.1331	.1368	.1406	.1443	.1480	.1517
0.4	.1554	.1591	.1628	.1664	.1700	.1736	.1772	.1808	.1844	.1879
0.5	.1915	.1950	.1985	.2019	.2054	.2088	.2123	.2157	.2190	.2224
0.6	.2257	.2291	.2324	.2357	.2389	.2422	.2454	.2486	.2517	.2549
0.7	.2580	.2611	.2642	.2673	.2704	.2734	.2764	.2794	.2823	.2852
0.8	.2881	.2910	.2939	.2967	.2995	.3023	.3051	.3078	.3106	.3133
0.9	.3159	.3186	.3212	.3238	.3264	.3289	.3315	.3340	.3365	.3389
1.0	.3413	.3438	.3461	.3485	.3508	.3531	.3554	.3577	.3599	.3621
1.1	.3643	.3665	.3686	.3708	.3729	.3749	.3770	.3790	.3810	.3830
1.2	.3849	.3869	.3888	.3907	.3925	.3944	.3962	.3980	.3997	.4015
1.3	.4032	.4049	.4066	.4082	.4099	.4115	.4131	.4147	.4162	.4177
1.4	.4192	.4207	.4222	.4236	.4251	.4265	.4279	.4292	.4306	.4319
1.5	.4332	.4345	.4357	.4370	.4382	.4394	.4406	.4418	.4429	.4441
1.6	.4452	.4463	.4474	.4484	.4495	.4505	.4515	.4525	.4535	.4545
1.7	.4554	.4564	.4573	.4582	.4591	.4599	.4608	.4616	.4625	.4633
1.8	.4641	.4649	.4656	.4664	.4671	.4678	.4686	.4693	.4699	.4706
1.9	.4713	.4719	.4726	.4732	.4738	.4744	.4750	.4756	.4761	.4767
2.0	.4773	.4778	.4783	.4788	.4793	.4798	.4803	.4808	.4812	.4817
2.1	.4821	.4826	.4830	.4834	.4838	.4842	.4846	.4850	.4854	.4857
2.2	.4861	.4864	.4868	.4871	.4875	.4878	.4881	.4884	.4887	.4890
2.3	.4893	.4896	.4898	.4901	.4904	.4906	.4909	.4911	.4913	.4916
2.4	.4918	.4920	.4922	.4925	.4927	.4929	.4931	.4932	.4934	.4936
2.5	.4938	.4940	.4941	.4943	.4945	.4946	.4948	.4949	.4951	.4952
2.6	.4953	.4955	.4956	.4957	.4959	.4960	.4961	.4962	.4963	.4964
2.7	.4965	.4966	.4967	.4968	.4969	.4970	.4971	.4972	.4973	.4974
2.8	.4974	.4975	.4976	.4977	.4977	.4978	.4979	.4979	.4980	.4981
2.9	.4981	.4982	.4983	.4983	.4984	.4984	.4985	.4985	.4986	.4986
3.0	.4987	.4987	.4987	.4988	.4988	.4989	.4989	.4989	.4989	.4990
3.1	.4990	.4991	.4991	.4991	.4992	.4992	.4992	.4992	.4993	.4993
3.2	.4993	.4993	.4994	.4994	.4994	.4994	.4994	.4995	.4995	.4995
3.3	.4995	.4995	.4996	.4996	.4996	.4996	.4996	.4996	.4996	.4997
3.4	.4997	.4997	.4997	.4997	.4997	.4997	.4997	.4997	.4997	.4998
3.5	.4998	.4998	.4998	.4998	.4998	.4998	.4998	.4998	.4998	.4998
3.6	.4998	.4998	.4999	.4999	.4999	.4999	.4999	.4999	.4999	.4999
3.7	.4999	.4999	.4999	.4999	.4999	.4999	.4999	.4999	.4999	.4999
3.8	.4999	.4999	.4999	.4999	.4999	.4999	.4999	.4999	.4999	.5000
3.9	.5000	.5000	.5000	.5000	.5000	.5000	.5000	.5000	.5000	.5000
Z	0.00	0.01	0.02	0.03	0.04	0.05	0.06	0.07	0.08	0.09

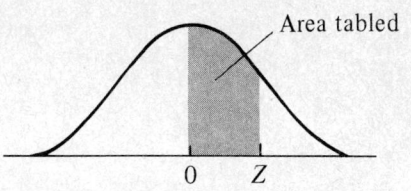

Example: The area corresponding to a Z score of 1.96 is .4750. Thus the probability of a Z score between 0 and 1.96 is .4750.

From Table II of Ronald A. Fisher and Frank Yates, *Statistical Tables for Biological, Agricultural and Medical Research*, 6th ed., published by Oliver and Boyd, Edinburgh, 1963, p. 45, and by permission of the authors and publishers.

Critical values of Z corresponding to three common significance levels (two-tailed test)

Significance level	0.10	0.05	0.01
Critical value	1.645	1.96	2.575

TABLE OF CRITICAL VALUES FOR THE t STATISTIC

Sample size for a test of a hypothesis about the mean of a single sample	Degrees of freedom	Probability				
		0.50	0.10	0.05	0.02	0.01
2	1	1.000	6.34	12.71	31.82	63.66
3	2	0.816	2.92	4.30	6.96	9.92
4	3	.765	2.35	3.18	4.54	5.84
5	4	.741	2.13	2.78	3.75	4.60
6	5	.727	2.02	2.57	3.36	4.03
7	6	.718	1.94	2.45	3.14	3.71
8	7	.711	1.90	2.36	3.00	3.50
9	8	.706	1.86	2.31	2.90	3.36
10	9	.703	1.83	2.26	2.82	3.25
11	10	.700	1.81	2.23	2.76	3.17
12	11	.697	1.80	2.20	2.72	3.11
13	12	.695	1.78	2.18	2.68	3.06
14	13	.694	1.77	2.16	2.65	3.01
15	14	.692	1.76	2.14	2.62	2.98
16	15	.691	1.75	2.13	2.60	2.95
17	16	.690	1.75	2.12	2.58	2.92
18	17	.689	1.74	2.11	2.57	2.90
19	18	.688	1.73	2.10	2.55	2.88
20	19	.688	1.73	2.09	2.54	2.86
21	20	.687	1.72	2.09	2.53	2.84
22	21	.686	1.72	2.08	2.52	2.83
23	22	.686	1.72	2.07	2.51	2.82
24	23	.685	1.71	2.07	2.50	2.81
25	24	.685	1.71	2.06	2.49	2.80
26	25	.684	1.71	2.06	2.48	2.79
27	26	.684	1.71	2.06	2.48	2.78
28	27	.684	1.70	2.05	2.47	2.77
29	28	.683	1.70	2.05	2.47	2.76
30	29	.683	1.70	2.04	2.46	2.76
31	30	.683	1.70	2.04	2.46	2.75
36	35	.682	1.69	2.03	2.44	2.72
41	40	.681	1.68	2.02	2.42	2.71
46	45	.680	1.68	2.02	2.41	2.69
51	50	.679	1.68	2.01	2.40	2.68
61	60	.678	1.67	2.00	2.39	2.66
71	70	.678	1.67	2.00	2.38	2.65
81	80	.677	1.66	1.99	2.38	2.64
91	90	.677	1.66	1.99	2.37	2.63
101	100	.677	1.66	1.98	2.36	2.63
126	125	.676	1.66	1.98	2.36	2.62
151	150	.676	1.66	1.98	2.35	2.61
201	200	.675	1.65	1.97	2.35	2.60
301	300	.675	1.65	1.97	2.34	2.59
401	400	.675	1.65	1.97	2.34	2.59
501	500	.674	1.65	1.96	2.33	2.59
1001	1000	.674	1.65	1.96	2.33	2.58
Infinity	∞	.674	1.64	1.96	2.33	2.58

Example
If a sample of size 8 is taken from a normal distribution with known mean and unknown variance, and a t statistic is computed, there is a 0.05 probability that the t score would be greater than $+2.36$ or smaller than -2.36.

From Table III of Ronald A. Fisher and Frank Yates, *Statistical Tables for Biological, Agricultural and Medical Research*, 6th ed., published by Oliver and Boyd, Edinburgh, 1963, p. 46, and by permission of authors and publishers.

r TABLE: CRITICAL VALUES OF THE CORRELATION COEFFICIENT

Number of pairs (number of points in scatterplot)	Level of significance for two-tailed test			
	0.10	0.05	0.02	0.01
3	.988	.997	.9995	.9999
4	.900	.950	.980	.990
5	.805	.878	.934	.959
6	.729	.811	.882	.917
7	.669	.754	.833	.874
8	.622	.707	.789	.834
9	.582	.666	.750	.798
10	.549	.632	.716	.765
11	.521	.602	.685	.735
12	.497	.576	.658	.708
13	.476	.553	.634	.684
14	.458	.532	.612	.661
15	.441	.514	.592	.641
16	.426	.497	.574	.623
17	.412	.482	.558	.606
18	.400	.468	.542	.590
19	.389	.456	.528	.575
20	.378	.444	.516	.561
21	.369	.433	.503	.549
22	.360	.423	.492	.537
23	.352	.413	.482	.526
24	.344	.404	.472	.515
25	.337	.396	.462	.505
26	.330	.388	.453	.496
27	.323	.381	.445	.487
28	.317	.374	.437	.479
29	.311	.367	.430	.471
30	.306	.361	.423	.463
31	.301	.355	.416	.456
32	.296	.349	.409	.449
37	.275	.325	.381	.418
42	.257	.304	.358	.393
47	.243	.288	.338	.372
52	.231	.273	.322	.354
62	.211	.250	.295	.325
72	.195	.232	.274	.303
82	.183	.217	.256	.283
92	.173	.205	.242	.267
102	.164	.195	.230	.254

Example
If two variables are really unrelated, and you observe 17 pairs of scores, there is a probability of 0.02 that you will observe a correlation greater than +0.558 or less than −0.558.

From Table VII of Ronald A. Fisher and Frank Yates, *Statistical Tables for Biological, Agricultural and Medical Research,* 6th ed., published by Oliver and Boyd, Edinburgh, 1963, p. 63, and by permission of the authors and publishers.

TABLE OF CRITICAL VALUES FOR THE χ^2 STATISTIC

Degrees of freedom	Significance level (α)								
	$P = 0.90$	0.70	0.50	0.30	0.20	0.10	0.05	0.02	0.01
1	0.02	0.15	0.45	1.07	1.64	2.71	3.84	5.41	6.63
2	0.21	0.71	1.39	2.41	3.22	4.60	5.99	7.82	9.21
3	0.58	1.42	2.37	3.66	4.64	6.25	7.81	9.84	11.34
4	1.06	2.19	3.36	4.88	5.99	7.78	9.49	11.67	13.28
5	1.61	3.00	4.35	6.06	7.29	9.24	11.07	13.39	15.09
6	2.20	3.83	5.35	7.23	8.56	10.64	12.59	15.03	16.81
7	2.83	4.67	6.35	8.38	9.80	12.02	14.07	16.62	18.47
8	3.49	5.53	7.34	9.52	11.03	13.36	15.51	18.17	20.09
9	4.17	6.39	8.34	10.66	12.24	14.68	16.92	19.68	21.67
10	4.86	7.27	9.34	11.78	13.44	15.99	18.31	21.16	23.21
11	5.58	8.15	10.34	12.90	14.63	17.27	19.67	22.62	24.72
12	6.30	9.03	11.34	14.01	15.81	18.55	21.03	24.05	26.22
13	7.04	9.93	12.34	15.12	16.98	19.81	22.36	25.47	27.69
14	7.79	10.82	13.34	16.22	18.15	21.06	23.68	26.87	29.14
15	8.55	11.72	14.34	17.32	19.31	22.31	25.00	28.26	30.58
16	9.31	12.62	15.34	18.42	20.46	23.54	26.30	29.63	32.00
17	10.08	13.53	16.34	19.51	21.61	24.77	27.59	30.99	33.41
18	10.86	14.44	17.34	20.60	22.76	25.99	28.87	32.35	34.80
19	11.65	15.35	18.34	21.69	23.90	27.20	30.14	33.69	36.19
20	12.44	16.27	19.34	22.77	25.04	28.41	31.41	35.02	37.57
21	13.24	17.18	20.34	23.86	26.17	29.61	32.67	36.34	38.93
22	14.04	18.10	21.34	24.94	27.30	30.81	33.92	37.66	40.29
23	14.85	19.02	22.34	26.02	28.43	32.01	35.17	38.97	41.64
24	15.66	19.94	23.34	27.10	29.55	33.20	36.41	40.27	42.98
25	16.47	20.87	24.34	28.17	30.67	34.38	37.65	41.57	44.31
26	17.29	21.79	25.34	29.25	31.79	35.56	38.88	42.86	45.64
27	18.11	22.72	26.34	30.32	32.91	36.74	40.11	44.14	46.96
28	18.94	23.65	27.34	31.39	34.03	37.92	41.34	45.42	48.28
29	19.77	24.58	28.24	32.46	35.14	39.09	42.56	46.69	49.59
30	20.60	25.51	29.34	33.53	36.25	40.26	43.77	47.96	50.89
Degrees of freedom	$P = 0.90$	0.70	0.50	0.30	0.20	0.10	0.05	0.02	0.01

Example There is 1 chance in 100 of seeing a χ^2 score of 6.63 or greater when the null hypothesis is true and you have 1 degree of freedom.

From Table IV of Ronald A. Fisher and Frank Yates, *Statistical Tables for Biological, Agricultural and Medical Research*, 6th ed., published by Oliver and Boyd, Edinburgh, 1963, p. 47, and by permission of the authors and publishers.

SQUARES, SQUARE ROOTS, AND RECIPROCALS

A note on the use of the following table of squares, square roots, and reciprocals of numbers from 1 to 1000

If you take about 10 minutes to learn how to get the most out of these tables, they will repay you handsomely for your efforts. If you are interested in finding the square or the square root of any number, N, between 1 and 1000, you simply find the appropriate line of the table. For example, to find the square of 46, 46^2, you find the number 46 in the first column of the table, and read off 2116 as the square of the number. Similarly, if you want to know the square root of 108, you can look in the second set of columns, and read off 10.3923. (For most purposes, 10.39 would give you plenty of accuracy.)

Square Roots

These tables will also help you find the square roots of numbers which do not lie between 1 and 1000. For example, you can read from the table that $\sqrt{46} = 6.7823$, and that $\sqrt{460} = 21.4476$. With this knowledge in hand, we can find lots of other square roots:

$$\sqrt{0.0046} = \sqrt{\frac{46}{10,000}} = \frac{\sqrt{46}}{\sqrt{10,000}} = \frac{\sqrt{46}}{100} = \frac{6.7823}{100} = 0.067823$$

$$\sqrt{0.046} = \sqrt{\frac{460}{10,000}} = \frac{\sqrt{460}}{\sqrt{10,000}} = \frac{\sqrt{460}}{100} = \frac{21.4476}{100} = 0.214476$$

$$\sqrt{0.46} = \sqrt{\frac{46}{100}} = \frac{\sqrt{46}}{\sqrt{100}} = \frac{\sqrt{46}}{10} = \frac{6.7823}{10} = 0.67823$$

$$\sqrt{4.6} = \sqrt{\frac{460}{100}} = \frac{\sqrt{460}}{\sqrt{100}} = \frac{\sqrt{460}}{10} = \frac{21.4476}{10} = 2.14476$$

$\sqrt{46} = 6.7823$ (from table)

$\sqrt{460} = 21.4476$ (from table)

$\sqrt{4,600} = \sqrt{46 \times 100} = \sqrt{46} \times \sqrt{100} = \sqrt{46} \times 10 = 6.7823 \times 10 = 67.823$

$\sqrt{46,000} = \sqrt{460 \times 100} = \sqrt{460} \times \sqrt{100} =$ etc.

$\sqrt{460,000} = \sqrt{46 \times 10,000} = \sqrt{46} \times \sqrt{10,000} = \sqrt{46} \times 100 =$ etc.

$\sqrt{4,600,000} = \sqrt{460 \times 10,000} = \sqrt{460} \times \sqrt{10,000} =$ etc.

Here are a few problems to check your skill at finding square roots with this table. (Answers follow.)

a. $\sqrt{85} =$

b. $\sqrt{817} =$

c. $\sqrt{6.72} =$

d. $\sqrt{14200} =$

e. $\sqrt{5800} =$

f. $\sqrt{1.68} =$

g. $\sqrt{.0625} =$

h. $\sqrt{.00071} =$

Using the Table of Reciprocals to Make Division Easier:

Suppose you want to find 300/581. By rewriting this as

300/581 = 300 x 1/581,

and then finding in the table that

1/581 = 0.00172117

you can change your division problem into a multiplication, which is easier:

$$\frac{300}{581} = 300 \times \frac{1}{581} = 300 \times 0.00172117$$

Any time you are doing a calculation involving division by a number between 1 and 1000, you can use the table in this simple, helpful way. This is especially sensible if you need to divide several different numbers by the same constant, as in finding a lot of Z scores.

Answers: a. 9.2195; b. 28.5832; c. 2.59230; d. 119.164; e. 76.158; f. 1.29615; g. 0.25; h. 0.0266458.

SQUARES, SQUARE ROOTS, AND RECIPROCALS

TABLE OF SQUARES, SQUARE ROOTS, AND RECIPROCALS OF NUMBERS FROM 1 TO 1000

N	N^2	$\sqrt{N}$	$1/N$	N	N^2	$\sqrt{N}$	$1/N$	N	N^2	$\sqrt{N}$	$1/N$
1	1	1.0000	1.000000	61	3721	7.8102	.016393	121	14641	11.0000	.00826446
2	4	1.4142	.500000	62	3844	7.8740	.016129	122	14884	11.0454	.00819672
3	9	1.7321	.333333	63	3969	7.9373	.015873	123	15129	11.0905	.00813008
4	16	2.0000	.250000	64	4096	8.0000	.015625	124	15376	11.1355	.00800452
5	25	2.2361	.200000	65	4225	8.0623	.015385	125	15625	11.1803	.00800000
6	36	2.4495	.166667	66	4356	8.1240	.015152	126	15876	11.2250	.00793651
7	49	2.6458	.142857	67	4489	8.1854	.014925	127	16129	11.2694	.00787402
8	64	2.8284	.125000	68	4624	8.2462	.014706	128	16384	11.3137	.00781250
9	81	3.0000	.111111	69	4761	8.3066	.014493	129	16641	11.3578	.00775194
10	100	3.1623	.100000	70	4900	8.3666	.014286	130	16900	11.4018	.00769231
11	121	3.3166	.090909	71	5041	8.4261	.014085	131	17161	11.4455	.00763359
12	144	3.4641	.083333	72	5184	8.4853	.013889	132	17424	11.4891	.00757576
13	169	3.6056	.076923	73	5329	8.5440	.013699	133	17689	11.5326	.00751880
14	196	3.7417	.071429	74	5476	8.6023	.013514	134	17956	11.5758	.00746269
15	225	3.8730	.066667	75	5625	8.6603	.013333	135	18496	11.6190	.00740741
16	256	4.0000	.062500	76	5776	8.7178	.013158	136	18496	11.6619	.00735294
17	289	4.1231	.058824	77	5929	8.7750	.012987	137	18769	11.7047	.00729927
18	324	4.2426	.055556	78	6084	8.8318	.012821	138	19044	11.7473	.00724638
19	361	4.3589	.052632	79	6241	8.8882	.012658	139	19321	11.7898	.00719424
20	400	4.4721	.050000	80	6400	8.9443	.012500	140	19600	11.8322	.00714286
21	441	4.5826	.047619	81	6561	9.0000	.012346	141	19881	11.8743	.00709220
22	484	4.6904	.045455	82	6724	9.0554	.012195	142	20164	11.9164	.00704225
23	529	4.7958	.043478	83	6889	9.1104	.012048	143	20449	11.9583	.00699301
24	576	4.8990	.041667	84	7056	9.1652	.011905	144	20736	12.0000	.00694444
25	625	5.0000	.040000	85	7225	9.2195	.011765	145	21025	12.0416	.00689655
26	676	5.0990	.038462	86	7396	9.2736	.011628	146	21316	12.0830	.00684932
27	729	5.1962	.037037	87	7569	9.3274	.011494	147	21609	12.1244	.00684272
28	784	5.2915	.035714	88	7744	9.3808	.011364	148	21904	12.1655	.00675676
29	841	5.3852	.034483	89	7921	9.4340	.011236	149	22201	12.2066	.00671141
30	900	5.4772	.033333	90	8100	9.4868	.011111	150	22500	12.2474	.00666667
31	961	5.5678	.032258	91	8281	9.5394	.010989	151	22801	12.2882	.00662252
32	1024	5.6569	.031250	92	8464	9.5917	.010870	152	23104	12.3288	.00657895
33	1089	5.7446	.030303	93	8649	9.6437	.010753	153	23409	12.3693	.00653595
34	1156	5.8310	.029412	94	8836	9.6954	.010638	154	23716	12.4097	.00649351
35	1225	5.9161	.028571	95	9025	9.7468	.010526	155	24025	12.4499	.00645161
36	1296	6.0000	.027778	96	9216	9.7980	.010417	156	24336	12.4900	.00641026
37	1369	6.0828	.027027	97	9409	9.8489	.010309	157	24649	12.5300	.00636943
38	1444	6.1644	.026316	98	9604	9.8995	.010204	158	24964	12.5698	.00632911
39	1521	6.2450	.025641	99	9801	9.9499	.010101	159	25281	12.6095	.00628931
40	1600	6.3246	.025000	100	10000	10.0000	.010000	160	25600	12.6491	.00625000
41	1681	6.4031	.024390	101	10201	10.0499	.00990099	161	25921	12.6886	.00621118
42	1764	6.4807	.023810	102	10404	10.0995	.00980392	162	26244	12.7279	.00617284
43	1849	6.5574	.023256	103	10609	10.1489	.00970874	163	26569	12.7671	.00613497
44	1936	6.6332	.022727	104	10816	10.1980	.00961538	164	26896	12.8062	.00609756
45	2025	6.7082	.022222	105	11025	10.2470	.00952381	165	27225	12.8452	.00606061
46	2116	6.7823	.021739	106	11236	10.2956	.00943396	166	27556	12.8841	.00602410
47	2209	6.8557	.021277	107	11449	10.3441	.00934579	167	27889	12.9228	.00598802
48	2304	6.9282	.020833	108	11664	10.3923	.00925926	168	28224	12.9615	.00595238
49	2401	7.0000	.020408	109	11881	10.4403	.00917431	169	28561	13.0000	.00591716
50	2500	7.0711	.020000	110	12100	10.4881	.00909091	170	28900	13.0384	.00588235
51	2601	7.1414	.019608	111	12321	10.5357	.00900901	171	29241	13.0767	.00584795
52	2704	7.2111	.019231	112	12544	10.5830	.00892857	172	29584	13.1149	.00581395
53	2809	7.2801	.018868	113	12769	10.6301	.00884956	173	29929	13.1529	.00578035
54	2916	7.3485	.018519	114	12996	10.6771	.00877193	174	30276	13.1909	.00574713
55	3025	7.4162	.018182	115	13225	10.7238	.00869565	175	30625	13.2288	.00571429
56	3136	7.4833	.017857	116	13456	10.7703	.00862069	176	30976	13.2665	.00568182
57	3249	7.5498	.017544	117	13689	10.8167	.00854701	177	31329	13.3041	.00564972
58	3364	7.6158	.017241	118	13924	10.8628	.00847458	178	31684	13.3417	.00561798
59	3481	7.6811	.016949	119	14161	10.9087	.00840336	179	32041	13.3791	.00558659
60	3600	7.7460	.016668	120	14400	10.9545	.00833333	180	32400	13.4164	.00555556

From Samuel M. Selby (Ed.), *Standard Mathematical Tables*, 16th ed., The Chemical Rubber Company, Cleveland, pp. 70–90.

SQUARES, SQUARE ROOTS, AND RECIPROCALS (Continued)

N	N^2	$\sqrt{N}$	$1/N$	N	N^2	$\sqrt{N}$	$1/N$	N	N^2	$\sqrt{N}$	$1/N$
181	32761	13.4536	.00552486	241	58081	15.5242	.00414938	301	90601	17.3494	.00332226
182	33124	13.4907	.00549451	242	58564	15.5563	.00413223	302	91204	17.3781	.00331126
183	33489	13.5277	.00546448	243	59049	15.5885	.00411523	303	91809	17.4069	.00330033
184	33856	13.5647	.00543478	244	59536	15.6205	.00409836	304	92416	17.4356	.00328047
185	34225	13.6015	.00540541	245	60025	15.6525	.00408163	305	93025	17.4642	.00328947
186	34596	13.6382	.00537634	246	60516	15.6844	.00406504	306	93636	17.4929	.00326797
187	34969	13.6748	.00534759	247	61009	15.7162	.00404858	307	94249	17.5214	.00325733
188	35344	13.7113	.00531915	248	61504	15.7480	.00403226	308	94864	17.5499	.00321675
189	35721	13.7477	.00529101	249	62001	15.7997	.00401606	309	95481	17.5784	.00323625
190	36100	13.7840	.00526316	250	62500	15.8114	.00400000	310	96100	17.6068	.00322581
191	36481	13.8203	.00523560	251	63001	15.8430	.00398406	311	96721	17.6352	.00321543
192	36864	13.8564	.00520833	252	63504	15.8745	.00396825	312	97344	17.6635	.00320513
193	37249	13.8924	.00518135	253	64009	15.9060	.00395257	313	97969	17.6918	.00319489
194	37636	13.9284	.00515464	254	64514	15.9374	.00393701	314	98596	17.7200	.00318471
195	38025	13.9642	.00512821	255	65025	15.9687	.00392157	315	99225	17.7482	.00317460
196	38416	14.0000	.00510204	256	65536	16.0000	.00390625	316	99856	17.7764	.00316456
197	38809	14.0357	.00507614	257	66049	16.0312	.00389105	317	100489	17.8045	.00315457
198	39204	14.0712	.00505051	258	66564	16.0624	.00387597	318	101124	17.8326	.00314465
199	39601	14.1067	.00152513	259	67081	16.0935	.00386100	319	101761	17.8606	.00313480
200	40000	14.1421	.00500000	260	67600	16.1245	.00384615	320	102400	17.8885	.00312500
201	40401	14.1774	.00497512	261	68121	16.1555	.00383142	321	103041	17.9165	.00311526
202	40804	14.2127	.00495050	262	68644	16.1864	.00381679	322	103684	17.9444	.00310559
203	41209	14.2478	.00492611	263	69169	16.2173	.00380228	323	104327	17.9722	.00309598
204	41616	14.2829	.00490196	264	69696	16.2481	.00378788	324	104976	18.0000	.00308642
205	42025	14.3178	.00487805	265	70225	16.2788	.00377358	325	105625	18.0278	.00307692
206	42436	14.3527	.00485437	266	70756	16.3095	.00375940	326	106276	18.0555	.00306748
207	42849	14.3875	.00483092	267	71289	16.3401	.00374532	327	106929	18.0831	.00305810
208	43264	14.4222	.00480769	268	71824	16.3707	.00373134	328	107584	18.1108	.00304878
209	43681	14.4568	.00478469	269	72361	16.4012	.00371747	329	108241	18.1384	.00303951
210	44100	14.4914	.00476190	270	72900	16.4317	.00370370	330	108900	18.1659	.00303030
211	44521	14.5258	.00473934	271	73441	16.4621	.00369004	331	109561	18.1934	.00302115
212	44944	14.5602	.00471698	272	73984	16.4924	.00367647	332	110224	18.2209	.00301205
213	45369	14.5945	.00469484	273	74529	16.5227	.00366300	333	110889	18.2483	.00300300
214	45796	14.6287	.00467290	274	75076	16.5529	.00364964	334	111556	18.2757	.00299401
215	46225	14.6629	.00465116	275	75625	16.5831	.00363636	335	112225	18.3030	.00298907
216	46656	14.6969	.00462963	276	76176	16.6132	.00362319	336	112896	18.3303	.00297619
217	47089	14.7309	.00460829	277	76729	16.6433	.00361011	337	113569	18.3576	.00296736
218	47524	14.7648	.00458716	278	77284	16.6733	.00359712	338	114244	18.3848	.00295858
219	47961	14.7986	.00456621	279	77841	16.7033	.00358423	339	114921	18.4120	.00294985
220	48400	14.8324	.00454545	280	78400	16.3732	.00357143	340	115600	18.4391	.00294118
221	48841	14.8661	.00452489	281	78961	16.7631	.00355872	341	116281	18.4662	.00293255
222	49284	14.8997	.00450450	282	79524	16.7929	.00354610	342	116964	18.4932	.00292398
223	49729	14.9332	.00448430	283	80089	16.8226	.00353357	343	117649	18.5203	.00291545
224	50176	14.9666	.00446429	284	80656	16.8523	.00352113	344	118336	18.5472	.00290698
225	50625	15.0000	.00444444	285	81225	16.8819	.00350877	345	119025	18.5742	.00289855
226	51076	15.0333	.00442478	286	81796	16.9115	.00349650	346	119716	18.6011	.00289017
227	51529	15.0665	.00440529	287	82369	16.9411	.00348432	347	120409	18.6279	.00288184
228	51984	15.0997	.00438596	288	82944	16.9706	.00347222	348	121104	18.6548	.00387356
229	52441	15.1327	.00436681	289	83521	17.0000	.00346021	349	121801	18.6815	.00286533
230	52900	15.1658	.00434783	290	84100	17.0294	.00344828	350	122500	18.7083	.00285714
231	53361	15.1987	.00432900	291	84681	17.0587	.00343643	351	123201	18.7350	.00284900
232	53824	15.2315	.00431034	292	85264	17.0880	.00342466	352	123904	18.7617	.00284091
233	54289	15.2643	.00429185	293	85849	17.1172	.00341297	353	124609	18.7883	.00283286
234	54756	15.2971	.00427350	294	86436	17.1464	.00340136	354	125316	18.8149	.00282486
235	55225	15.3297	.00425532	295	87025	17.1756	.00338983	355	126025	18.8414	.00281690
236	55696	15.3623	.00423739	296	87616	17.2047	.00337838	356	126736	18.8680	.00280899
237	56169	15.3948	.00421941	297	88209	17.2337	.00336700	357	127449	18.8944	.00280112
238	56644	15.4272	.00420168	298	88804	17.2627	.00335570	358	128164	18.9209	.00279330
239	57121	15.4596	.00418410	299	89401	17.2916	.00334448	359	128881	18.9473	.00278552
240	57600	15.4919	.00416667	300	90000	17.3205	.00333333	360	129600	18.9737	.00277778

SQUARES, SQUARE ROOTS, AND RECIPROCALS (Continued)

N	N²	√N	1/N	N	N²	√N	1/N	N	N²	√N	1/N
361	130321	19.0000	.00277008	421	177241	20.5183	.00237530	481	231361	21.9317	.00207900
362	131044	19.0263	.00276243	422	178084	20.5426	.00236967	482	232324	21.9545	.00207469
363	131769	19.0526	.00275482	423	178929	20.5670	.00236407	483	233289	21.9773	.00207039
364	132496	19.0788	.00274725	424	179776	20.5913	.00235849	484	234256	22.0000	.00206612
365	133225	19.1050	.00273973	425	180625	20.6155	.00235294	485	235225	22.0227	.00206186
366	133956	19.1311	.00273224	426	181476	20.6398	.00234742	486	236196	22.0454	.00205761
367	134689	19.1572	.00272480	427	182329	20.6640	.00234192	487	237169	22.0681	.00205339
368	135424	19.1833	.00271739	428	183184	20.6882	.00233645	488	238144	22.0907	.00204918
369	136161	19.2094	.00271003	429	184041	20.7123	.00233100	489	239121	22.1133	.00204499
370	136900	19.2354	.00270270	430	184900	20.7364	.00232558	490	240100	22.1359	.00204082
371	137641	19.2614	.00269542	431	185761	20.7605	.00232019	491	241081	22.1585	.00203666
372	138384	19.2873	.00268817	432	186624	20.7846	.00231481	492	242064	22.1811	.00203252
373	139129	19.3132	.00268097	433	187489	20.8087	.00230947	493	243049	22.2036	.00202840
374	139876	19.3391	.00267380	434	188356	20.8327	.00230415	494	244036	22.2261	.00202429
375	140625	19.3649	.00266667	435	189225	20.8567	.00229885	495	245025	22.2486	.00202020
376	141376	19.3907	.00265957	436	190096	20.8806	.00229358	496	246016	22.2711	.00201613
377	142129	19.4165	.00265252	437	190969	20.9045	.00228833	497	247009	22.2935	.00201207
378	142884	19.4422	.00264550	438	191844	20.9284	.00228311	498	248004	22.3159	.00200803
379	143641	19.4679	.00263852	439	192721	20.9523	.00227790	499	249001	22.3383	.00200401
380	144400	19.4936	.00263158	440	193600	20.9762	.00227273	500	250000	22.3607	.00200000
381	145161	19.5192	.00262467	441	194481	21.0000	.00226757	501	251001	22.3830	.00199601
382	145724	19.5448	.00261780	442	195364	21.0238	.00226244	502	252004	22.4054	.00199203
383	146689	19.5704	.00261097	443	196249	21.0476	.00225734	503	253009	22.4277	.00198807
384	147456	19.5959	.00260417	444	197136	21.0713	.00225225	504	254016	22.4499	.00198413
385	148225	19.6214	.00259740	445	198025	21.0950	.00224719	505	255025	22.4722	.00198020
386	148996	19.6469	.00259067	446	198916	21.1187	.00224215	506	256036	22.4944	.00197628
387	149769	19.6723	.00258398	447	199809	21.1424	.00223714	507	257049	22.5167	.00197239
388	150544	19.6977	.00257732	448	200704	21.1660	.00223214	508	258064	22.5389	.00196850
389	151321	19.7231	.00257069	449	201601	21.1896	.00222717	509	259081	22.5610	.00196464
390	152100	19.7484	.00256410	450	202500	21.2132	.00222222	510	260100	22.5832	.00196078
391	152881	19.7737	.00255754	451	203401	21.2368	.00221729	511	261121	22.6053	.00195695
392	153664	19.7990	.00255102	452	204304	21.2603	.00221239	512	262144	22.6274	.00195312
393	154449	19.8242	.00254453	453	205209	21.2838	.00220751	513	263169	22.6495	.00194932
394	155236	19.8494	.00253807	454	206116	21.3073	.00220264	514	264196	22.6716	.00194553
395	156025	19.8746	.00253165	455	207025	21.3307	.00219870	515	265225	22.6936	.00194175
396	156816	19.8997	.00252525	456	207936	21.3542	.00219298	516	266256	22.7156	.00193798
397	157609	19.9249	.00251889	457	208849	21.3776	.00218818	517	267289	22.7376	.00193424
398	158404	19.9499	.00251256	458	209764	21.4009	.00218341	518	268324	22.7596	.00193050
399	159201	19.9750	.00250627	459	210681	21.4243	.00217865	519	269361	22.7816	.00192678
400	160000	20.0000	.00250000	460	211600	21.4476	.00217391	520	270400	22.8035	.00192308
401	160801	20.0250	.00249377	461	212521	21.4709	.00216920	521	271441	22.8254	.00191939
402	161604	20.0499	.00248756	462	213444	21.4942	.00216450	522	272484	22.8473	.00191571
403	162409	20.0749	.00248139	463	214369	21.5174	.00215983	523	273529	22.8692	.00191205
404	163216	20.0998	.00247525	464	215296	21.5407	.00215517	524	274576	22.8910	.00190840
405	164025	20.1246	.00246914	465	216225	21.5639	.00215054	525	275625	22.9129	.00190476
406	164836	20.1494	.00246305	466	217156	21.5870	.00214592	526	276676	22.9347	.00190114
407	165649	20.1742	.00245700	467	218089	21.6102	.00214133	527	277729	22.5965	.00189653
408	166464	20.1990	.00245098	468	219024	21.6333	.00213675	528	278784	22.9783	.00189394
409	167281	20.2237	.00244499	469	219961	21.6564	.00213220	529	279841	23.0000	.00189036
410	168100	20.2485	.00243902	470	220900	21.6795	.00212766	530	280900	23.0217	.00188679
411	168921	20.2731	.00243309	471	221841	21.7025	.00212314	531	281961	23.0434	.00188324
412	169744	20.2978	.00242718	472	222784	21.7256	.00211864	532	283024	23.0651	.00187970
413	170569	20.3224	.00242131	473	223729	21.7486	.00211416	533	284089	23.0868	.00187617
414	171396	20.3470	.00241546	474	224676	21.7715	.00210970	534	285156	23.1084	.00187266
415	172225	20.3715	.00240964	475	225625	21.7945	.00210526	535	286225	23.1301	.00186916
416	173056	20.3961	.00240385	476	226576	21.8174	.00210084	536	287296	23.1517	.00186567
417	173889	20.4206	.00239808	477	227529	21.8403	.00209644	537	288369	23.7033	.00186220
418	174724	20.4450	.00239234	478	228484	21.8632	.00209205	538	289444	23.1948	.00185874
419	175561	20.4695	.00238663	479	229441	21.8861	.00208768	539	290521	23.2164	.00185529
420	176400	20.4939	.00238095	480	230400	21.9089	.00208333	540	291600	23.2379	.00185185

SQUARES, SQUARE ROOTS, AND RECIPROCALS (Continued)

N	N^2	$\sqrt{N}$	$1/N$	N	N^2	$\sqrt{N}$	$1/N$	N	N^2	$\sqrt{N}$	$1/N$
541	292681	23.2594	.00184843	601	361201	24.5153	.00166389	661	436921	25.7099	.00151286
542	293764	23.2809	.00184502	602	362404	24.5357	.00166113	662	438344	25.7294	.00151057
543	294849	23.3024	.00184162	603	363609	24.5561	.00165837	663	439569	25.7488	.00150830
544	295936	23.3238	.00183824	604	364816	24.5764	.00165563	664	440896	25.7682	.00150602
545	297025	23.3452	.00183486	605	366025	24.5967	.00165289	665	442225	25.7876	.00150376
546	298116	23.3666	.00183150	606	367236	24.6171	.00165070	666	443556	25.8070	.00150150
547	299209	23.3880	.00182815	607	368449	24.6374	.00164745	667	444889	25.8263	.00149925
548	300304	23.4094	.00182482	608	369664	24.6577	.00164474	668	446224	25.8457	.00149701
549	301401	23.4307	.00182149	609	370881	24.6779	.00164204	669	447561	25.8650	.00149477
550	302500	23.4521	.00181818	610	372100	24.6982	.00163934	670	448900	25.8844	.00149254
551	303601	23.4734	.00181488	611	373321	24.7164	.00163666	671	450241	25.9037	.00149031
552	304704	23.4947	.00181159	612	374544	24.7386	.00163399	672	451584	25.9230	.00148810
553	305809	23.5160	.00180832	613	375769	24.7588	.00163132	673	452929	25.9422	.00148588
554	306916	23.5372	.00180505	614	376996	24.7790	.00162861	674	454276	25.9615	.00148368
555	308025	23.5584	.00180180	615	378225	24.7992	.00162602	675	455625	25.9808	.00148148
556	309136	23.5797	.00189856	616	379456	24.8193	.00162338	676	456976	26.0000	.00147929
557	310249	23.6008	.00179533	617	380689	24.8395	.00162075	677	458329	26.0192	.00147710
558	311364	23.6220	.00179211	618	381924	24.8596	.00161812	678	459684	26.0384	.00147493
559	312481	23.6432	.00178891	619	383161	24.8797	.00161551	679	461041	26.0576	.00147275
560	313600	23.6643	.00178571	620	384400	24.8998	.00161290	680	462400	26.0768	.00147059
561	314721	23.6854	.00178253	621	385641	24.9199	.00161031	681	463761	26.0960	.00146843
562	315844	23.7065	.00177936	622	386884	24.9299	.00160772	682	465124	26.1151	.00146628
563	316969	23.7276	.00177620	623	388129	24.9600	.00160514	683	466489	26.1343	.00146413
564	318096	23.7487	.00177305	624	389376	24.9800	.00160256	684	467856	26.1534	.00146199
565	319225	23.7697	.00176991	625	390625	25.0000	.00160000	685	469225	26.1725	.00145985
566	320356	23.7908	.00176678	626	391876	25.0200	.00159744	686	470596	26.1916	.00145773
567	321489	23.8118	.00176367	627	393129	25.0400	.00159490	687	471969	26.2107	.00145560
568	322624	23.8328	.00176056	628	394384	25.0599	.00159236	688	473344	26.2298	.00145349
569	323761	23.8537	.00175747	629	395641	25.0799	.00158983	689	474721	26.2488	.00145138
570	324900	23.8747	.00175439	630	396900	25.0998	.00158730	690	476100	26.2679	.00144928
571	326041	23.8956	.00175143	631	398161	25.1197	.00158479	691	477481	26.2869	.00144718
572	327184	23.9165	.00174825	632	399424	25.1396	.00158228	692	478864	26.3059	.00144509
573	328329	23.9374	.00174520	633	400689	25.1595	.00157978	693	480249	26.3249	.00144300
574	329476	23.9583	.00174216	634	401956	25.1794	.00157729	694	481636	26.3439	.00144092
575	330625	23.9792	.00173913	635	403225	25.1992	.00157480	695	483025	25.3629	.00143885
576	331776	24.0000	.00173611	636	404496	25.2190	.00157243	696	484416	26.3818	.00143678
577	332929	24.0208	.00173310	637	405769	25.2389	.00156986	697	485809	26.4008	.00143472
578	334084	24.0416	.00173010	638	407044	25.2587	.00156740	698	487204	26.4197	.00143266
579	335241	24.0624	.00172712	639	408321	25.2784	.00156495	699	488601	26.4386	.00143062
580	336400	24.0832	.00172414	640	409600	25.2982	.00156250	700	490000	26.4575	.00142857
581	337561	24.1039	.00172117	641	410881	25.3180	.00156006	701	491401	26.4764	.00142653
582	338724	24.1247	.00171821	642	412164	25.3377	.00155763	702	492804	26.4953	.00142450
583	339889	24.1454	.00171527	643	413449	25.3574	.00155521	703	494209	26.5141	.00142248
584	341056	24.1661	.00171233	644	414736	25.3772	.00155280	704	495616	26.5330	.00142045
585	342225	24.1868	.00170940	645	416025	25.3969	.00155039	705	497025	26.5518	.00141844
586	343396	24.2074	.00170648	646	417316	25.4165	.00154739	706	498436	26.5707	.00141643
587	344569	24.2281	.00170358	647	418609	25.4362	.00154560	707	499849	26.7895	.00141443
588	345744	24.2487	.00170068	648	419904	25.4558	.00154321	708	501264	26.6083	.00141243
589	346921	24.2693	.00169779	649	421201	25.4755	.00154083	709	502681	26.6271	.00141044
590	348100	24.2899	.00169492	650	422500	25.4951	.00153846	710	504100	26.6458	.00140845
591	349281	24.3105	.00169205	651	423801	25.5147	.00153610	711	505521	26.6646	.00140647
592	350464	24.3311	.00168919	652	425104	25.5343	.00153374	712	506944	26.6833	.00140449
593	351649	24.3516	.00168634	653	426409	25.5539	.00153139	713	508369	26.7021	.00140252
594	352836	24.3721	.00168350	654	427716	25.5734	.00152905	714	509796	26.7208	.00140056
595	354025	24.3926	.00168067	655	429025	25.5930	.00152672	715	511225	26.7395	.00139860
596	355216	24.4131	.00167785	656	430336	25.6125	.00152439	716	512656	26.7582	.00139665
597	356409	24.4336	.00167504	657	431649	25.6320	.00152207	717	514089	26.7769	.00139470
598	357604	24.4540	.00167224	658	432964	25.6515	.00151976	718	515524	26.7955	.00139276
599	358801	24.4745	.00166945	659	434281	25.6710	.00151745	719	516971	26.8142	.00139082
600	360000	24.4949	.00166667	660	435600	25.6905	.00151515	720	518400	26.8328	.00138889

SQUARES, SQUARE ROOTS, AND RECIPROCALS (Continued)

N	N^2	$\sqrt{N}$	$1/N$	N	N^2	$\sqrt{N}$	$1/N$	N	N^2	$\sqrt{N}$	$1/N$
721	519841	26.8514	.00138696	781	609961	27.9464	.00128041	841	707281	29.0000	.00118906
722	521284	26.8701	.00138504	782	611524	27.9643	.00127877	842	708964	29.0172	.00118765
723	522729	26.8887	.00138313	783	613089	27.9821	.00127714	843	710649	29.0345	.00118624
724	524176	26.9072	.00138122	784	614656	28.0000	.00127551	844	712336	29.0517	.00118483
725	525625	26.9258	.00137931	785	616225	28.0179	.00127389	845	714025	29.0689	.00118343
726	527076	26.9444	.00137741	786	617796	28.0357	.00127226	846	715716	29.0861	.00118203
727	528529	26.9629	.00137552	787	619369	28.0535	.00127065	847	717409	29.1033	.00118064
728	529984	26.9815	.00137363	788	620944	28.0713	.00126904	848	719104	29.1204	.00117925
729	531441	27.0000	.00137174	789	622521	28.0891	.00126743	849	720801	29.1376	.00117786
730	532900	27.0185	.00136986	790	624100	28.1069	.00126582	850	722500	29.1548	.00117647
731	534361	27.0370	.00136799	791	625681	28.1247	.00126422	851	724201	29.1719	.00117509
732	535824	27.0555	.00136612	792	627264	28.1425	.00126263	852	725904	29.1890	.00117371
733	537289	27.0740	.00136426	793	628849	28.1603	.00126103	853	727609	29.2062	.00117233
734	538756	27.0924	.00136240	794	630436	28.1780	.00125945	854	729316	29.2233	.00117096
735	540225	27.1109	.00131654	795	632025	28.1957	.00125786	855	731025	29.2404	.00116959
736	541696	27.1293	.00135870	796	633616	28.2135	.00125628	856	732736	29.2575	.00116822
737	543169	27.1477	.00135685	797	635209	28.2312	.00125471	857	734449	29.2746	.00116686
738	544644	27.1662	.00135501	798	636804	28.2489	.00125313	858	736164	29.2916	.00116550
739	546121	27.1846	.00135318	799	638401	28.2666	.00125156	859	737881	29.3087	.00116414
740	547600	27.2029	.00135135	800	640000	28.2843	.00125000	860	739600	29.3258	.00116279
741	549081	27.2213	.00134953	801	641601	28.3019	.00124844	861	741321	29.3428	.00116144
742	550564	27.2397	.00134771	802	643204	28.3196	.00124688	862	743044	29.3598	.00116009
743	552049	27.2580	.00134590	803	644809	28.3373	.00124533	863	744769	29.3769	.00115875
744	553536	27.2764	.00134409	804	646416	28.3549	.00124378	864	746496	29.3939	.00115741
745	555025	27.2947	.00134228	805	648025	28.3725	.00124224	865	748225	29.4109	.00115607
746	556516	27.3130	.00134048	806	649636	28.3901	.00124069	866	749956	29.4279	.00115473
747	558009	27.3313	.00133869	807	651249	28.4077	.00123916	867	751689	29.4449	.00115340
748	559504	27.3496	.00133690	808	652864	28.4253	.00123762	868	753424	29.4618	.00115207
749	561001	27.3679	.00133511	809	654481	28.4429	.00123609	869	755161	29.4788	.00115075
750	562500	27.3861	.00133333	810	656100	28.4605	.00123457	870	756900	29.4958	.00114943
751	564001	27.4044	.00133156	811	657721	28.4781	.00123305	871	758641	29.5127	.00114811
752	565504	27.4226	.00132979	812	659344	28.4956	.00123153	872	760384	29.5296	.00114679
753	567009	27.4408	.00132802	813	660969	28.5132	.00123001	873	762129	29.5466	.00114548
754	568516	27.4591	.00132626	814	662596	28.5307	.00122850	874	763876	29.5635	.00114416
755	570025	27.3773	.00132450	815	664225	28.5482	.00122699	875	765625	29.5804	.00114286
756	571536	27.4955	.00132275	816	665856	28.5657	.00122549	876	767376	29.5973	.00114155
757	573049	27.5136	.00132100	817	667489	28.5832	.00122399	877	769129	29.6142	.00114025
758	574564	27.5318	.00131926	818	669124	28.6007	.00122249	878	770884	29.6311	.00113895
759	576081	27.5500	.00131752	819	670761	28.6182	.00122100	879	772641	29.6479	.00113762
760	577600	27.5681	.00131579	820	672400	28.6356	.00121951	880	774400	29.6848	.00113636
761	579121	27.5862	.00131406	821	674041	28.6531	.00121803	881	776161	29.6816	.00113507
762	580644	27.6043	.00131234	822	675684	28.6705	.00121655	882	777924	29.6985	.00113379
763	582169	27.6225	.00131062	823	677329	28.6880	.00121507	883	779689	29.7153	.00117250
764	583696	27.6405	.00130890	824	678976	28.7054	.00121359	884	781456	29.7321	.00113122
765	585225	27.6586	.00130719	825	680625	28.7228	.00121212	885	783225	29.7489	.00112994
766	586756	27.6767	.00130548	826	682276	28.7402	.00121065	886	784996	29.7658	.00112867
767	588289	27.6948	.00130378	827	683929	28.7576	.00120919	887	786769	29.7825	.00112740
768	589824	27.7128	.00130208	828	685584	28.7750	.00120773	888	788544	29.7993	.00112613
769	591361	27.7308	.00130039	829	687241	28.7924	.00120627	889	790321	29.8161	.00112486
770	592900	27.7489	.00129870	830	688900	28.8097	.00120482	890	792100	29.8329	.00112360
771	594441	27.7669	.00129702	831	690561	28.8271	.00120337	891	793881	29.8496	.00112233
772	595984	27.7849	.00129534	832	692224	28.8444	.00120192	892	795664	29.8664	.00112108
773	597529	27.8029	.00129366	833	694889	28.8617	.00120048	893	797449	29.8831	.00111982
774	599076	27.8209	.00129199	834	695556	28.8791	.00119904	894	799236	29.8998	.00111857
775	600625	27.8388	.00129032	835	697225	28.8964	.00117960	895	801025	29.9166	.00111732
776	602176	27.8568	.00128866	836	698896	28.9137	.00119617	896	802816	29.9333	.00111607
777	603729	27.8747	.00128700	837	700569	28.9310	.00119474	897	804609	29.9500	.00111483
778	605284	27.8927	.00128535	838	702244	28.9482	.00119332	898	806404	29.9666	.00111359
779	606841	27.9106	.00128370	839	703921	28.9655	.00119190	899	808201	29.9833	.00111235
780	608400	27.9285	.00128205	840	705600	28.9828	.00119048	900	810000	30.0000	.00111111

SQUARES, SQUARE ROOTS, AND RECIPROCALS (*Continued*)

N	N^2	$\sqrt{N}$	$1/N$	N	N^2	$\sqrt{N}$	$1/N$	N	N^2	$\sqrt{N}$	$1/N$
901	811801	30.0167	.00110988	936	876096	30.5941	.00106838	971	942841	31.1609	.00102987
902	813604	30.0333	.00110865	937	877969	30.6105	.00106724	972	944784	31.1769	.00102881
903	815409	30.0500	.00110742	938	879844	30.6268	.00106610	973	946729	31.1929	.00102775
904	817216	30.0666	.00110619	939	881721	30.6431	.00106496	974	948676	31.2090	.00102669
905	819025	30.0832	.00110497	940	883600	30.6594	.00106383	975	950625	31.2250	.00102564
906	820836	30.0998	.00110375	941	885481	30.6757	.00106270	976	952576	31.2410	.00102459
907	822649	30.1164	.00110254	942	887364	30.6920	.00106157	977	954529	31.2570	.00102354
908	824464	30.1330	.00110132	943	889249	30.7083	.00106045	978	956484	31.2730	.00102249
909	826281	30.1496	.00110011	944	891136	30.7246	.00105932	979	958441	31.2890	.00102145
910	828100	30.1662	.00109890	945	893025	30.7409	.00105820	980	960400	31.3050	.00102041
911	829921	30.1828	.00109769	946	894916	30.7571	.00105708	981	962361	31.3209	.00101937
912	831744	30.1993	.00109649	947	896809	30.7734	.00105547	982	964324	31.3369	.00101833
913	833569	30.2159	.00109529	948	898704	30.7896	.00105485	983	966289	31.3528	.00101729
914	835396	30.2324	.00109409	949	900601	30.8058	.00105374	984	968256	31.3688	.00101626
915	837225	30.2490	.00109290	950	902500	30.8221	.00105263	985	970225	31.3847	.00101523
916	839056	30.2655	.00109170	951	904401	30.8383	.00105152	986	972196	31.4006	.00101420
917	840889	30.2820	.00109051	952	906304	30.8545	.00105042	987	974169	31.4166	.00101317
918	842724	30.2985	.00108932	953	908209	30.8707	.00104932	988	976144	31.4325	.00101215
919	844561	30.3150	.00108814	954	910116	30.8869	.00104822	989	978121	31.4484	.00101112
920	846400	30.3315	.00108696	955	912025	30.9031	.00104712	990	980100	31.4643	.00101010
921	848241	30.3480	.00108578	956	913936	30.9192	.00104603	991	982081	31.4802	.00100908
922	850084	30.3645	.00108460	957	915849	30.9354	.00104493	992	984064	31.4960	.00100806
923	851929	30.3809	.00108342	958	917764	30.9516	.00104384	993	986049	31.5119	.00100705
924	853776	30.3974	.00108225	959	919681	30.9677	.00104275	994	988036	31.5278	.00100604
925	855625	30.4138	.00108108	960	921600	30.9839	.00104167	995	990025	31.5436	.00100503
926	857476	30.4302	.00107991	961	923521	31.0000	.00104058	996	992016	31.5595	.00100402
927	859329	30.4467	.00107875	962	925444	31.0161	.00103950	997	994009	31.5753	.00103842
928	861184	30.4631	.00107759	963	927369	31.0322	.00103842	998	996004	31.5911	.00100200
929	863041	30.4795	.00107643	964	929296	31.0483	.00103734	999	998001	31.6070	.00100100
930	864900	30.4959	.00107527	965	931225	31.0644	.00103627	1000	1000000	31.6228	.00100000
931	866761	30.5123	.00107411	966	933156	31.0805	.00103520				
932	868624	30.5287	.00107296	967	935089	31.0966	.00103413				
933	870489	30.5450	.00107181	968	937024	31.1127	.00103306				
934	872356	30.5614	.00107066	969	938961	31.1288	.00103199				
935	874225	30.5778	.00106952	970	940900	31.1448	.00103093				

TABLE OF DECIMAL EQUIVALENTS OF VARIOUS FRACTIONS USED FOR SIGN TESTS

N	N/8 ($N/2^3$)	N/16 ($N/2^4$)	N/32 ($N/2^5$)	N/64 ($N/2^6$)	N/128 ($N/2^7$)	N/256 ($N/2^8$)	N/512 ($N/2^9$)	N/1024 ($N/2^{10}$)	N/2048 ($N/2^{11}$)
1	0.125	0.063	0.031	0.016	0.008	0.004	0.002	0.001	0.001
2	0.250	0.125	0.063	0.031	0.016	0.008	0.004	0.002	0.001
3	0.375	0.188	0.094	0.047	0.024	0.012	0.006	0.003	0.002
4	0.500	0.250	0.125	0.063	0.031	0.016	0.008	0.004	0.002
5	0.625	0.313	0.156	0.078	0.039	0.020	0.010	0.005	0.003
6	0.750	0.375	0.188	0.094	0.047	0.024	0.012	0.006	0.003
7	0.875	0.438	0.219	0.109	0.055	0.027	0.014	0.007	0.004
8		0.500	0.250	0.125	0.063	0.031	0.016	0.008	0.004
9		0.563	0.281	0.141	0.070	0.035	0.018	0.009	0.004
10		0.625	0.313	0.156	0.078	0.039	0.020	0.010	0.005
11		0.688	0.344	0.172	0.086	0.043	0.022	0.011	0.005
12		0.750	0.375	0.188	0.094	0.047	0.024	0.012	0.006
13		0.813	0.406	0.203	0.102	0.051	0.025	0.013	0.006
14		0.875	0.438	0.219	0.109	0.055	0.027	0.014	0.007
15		0.938	0.469	0.234	0.117	0.059	0.029	0.015	0.007
16			0.500	0.250	0.125	0.063	0.031	0.016	0.008
17			0.531	0.266	0.133	0.067	0.033	0.017	0.008
18			0.563	0.281	0.141	0.070	0.035	0.018	0.009
19			0.594	0.297	0.149	0.074	0.037	0.019	0.009
20			0.625	0.313	0.156	0.078	0.039	0.020	0.010
21			0.656	0.328	0.164	0.082	0.041	0.021	0.010
22			0.688	0.344	0.172	0.086	0.043	0.022	0.011
23			0.719	0.359	0.180	0.090	0.045	0.023	0.011
24			0.750	0.375	0.188	0.094	0.047	0.024	0.012
25			0.781	0.391	0.195	0.098	0.049	0.025	0.012
26			0.813	0.406	0.203	0.102	0.051	0.025	0.013
27			0.844	0.422	0.211	0.106	0.053	0.026	0.013
28			0.875	0.438	0.219	0.109	0.055	0.027	0.014
29			0.906	0.453	0.227	0.113	0.057	0.028	0.014
30			0.938	0.469	0.234	0.117	0.059	0.029	0.015
31			0.969	0.484	0.242	0.121	0.061	0.030	0.015
32				0.500	0.250	0.125	0.063	0.031	0.016
33				0.516	0.258	0.129	0.065	0.032	0.016
34				0.531	0.266	0.133	0.067	0.033	0.017
35				0.547	0.273	0.137	0.068	0.034	0.017
36				0.563	0.281	0.141	0.070	0.035	0.018
37				0.578	0.289	0.145	0.072	0.036	0.018
38				0.594	0.297	0.149	0.074	0.037	0.019
39				0.609	0.305	0.152	0.176	0.038	0.019
40				0.625	0.313	0.156	0.078	0.039	0.020
41				0.641	0.320	0.160	0.080	0.040	0.020
42				0.656	0.328	0.164	0.082	0.041	0.021
43				0.672	0.336	0.168	0.084	0.042	0.021
44				0.688	0.344	0.172	0.086	0.043	0.022
45				0.703	0.352	0.176	0.088	0.044	0.022
46				0.719	0.359	0.180	0.090	0.045	0.023
47				0.734	0.367	0.184	0.092	0.046	0.023
48				0.750	0.375	0.188	0.094	0.047	0.024
49				0.766	0.383	0.192	0.096	0.048	0.024
50				0.781	0.391	0.195	0.098	0.049	0.025

Example
The decimal equivalent of 15/32 is 0.469

Flash Cards

It has been stressed that a certain amount of rote memorization is essential for statistics. To help you with this memorization, a set of 57 flash cards follows. On each card is a question or a term for you to define. On the back of that card is an answer. I recommend that you cut out the cards and learn them very thoroughly as you work through the chapters. This will greatly simplify your job in learning statistics.

QUESTION 1

Define *statistics*.

QUESTION 5.

There is a "fundamental" question which we are often trying to answer in statistics: What is it?

QUESTION 2.

What is a Type One error?

QUESTION 6.

What is a decision rule?

QUESTION 3.

What is a Type Two error?

QUESTION 7.

Define *Significance level*.

QUESTION 4.

For any decision rule, there is a certain chance of a Type One error and a certain chance of a Type Two error. If I change to a new decision rule which *reduces* my chance of a Type One error, what will probably happen to the chance of a Type Two error?

QUESTION 8.

Define *Rejection region*.

ANSWER 5.

Is the observation the result of a real underlying process, or is it just the result of random variation?

ANSWER 1.

A set of tools to help you make good decisions on the basis of limited and variable data.

ANSWER 6.

A rule which specifies what your decision will be (reject or don't reject the null hypothesis) for every possible result of the experiment.

ANSWER 2.

A false alarm: Rejecting the null hypothesis when it is in fact true; saying there's an effect when there is none.

ANSWER 7.

The maximum acceptable risk of a Type One error.
The significance level associated with a decision rule is the risk of a Type One error when you use that decision rule. If there is no effect, and you use that rule, then the chance of deciding "There *is* an effect" is the significance level. The symbol α (alpha) is often used.

ANSWER 3.

A miss: Failing to reject the null hypothesis when you should; saying there's no effect when in fact there is one.

ANSWER 8.

The rejection region associated with a decision rule specifies those results which, if observed, will induce us to *reject* the null hypothesis. The *rejection region* is a set of results.

ANSWER 4.

It will probably increase. There is a trade-off between the two error types. As you reduce the chance of one kind of error, you generally increase your chance of the other kind.

QUESTION 9.

List the six steps in a sign test.

QUESTION 10.

If you toss a coin 17 times, how many different *outcomes* are possible?

If you toss it *n* times?

QUESTION 11.

Define *outcome*.

QUESTION 12.

Define *event*. Give an example.

QUESTION 13.

Define *sample space*.

QUESTION 14.

If a "fair" coin is tossed ten times, among the possible outcomes are HHHHHHHHHH, HTHTTHHHTT, HHHHHTTTTT, and HTHTHTHTHT. Which is the most likely? The least likely?

QUESTION 15.

How do you transform the numbers in a frequency distribution into a relative frequency distribution?

QUESTION 16.

What is a null hypothesis?

ANSWER 13.

A sample space is the set of all the possible outcomes of an experiment.

ANSWER 9.

1. List the possible results (events).
2. State a null hypothesis.
3. Study the null hypothesis: Find probabilities of various events when it is true.
4. Pick a significance level, α.
5. Pick rejection region.
6. Run an experiment and make a decision.

ANSWER 14.

They are all equally likely. The probability of any specific pattern of Hs and Ts is just the same as any other. All outcomes are equally likely in this situation.

ANSWER 10.

2^{17}

2^n

ANSWER 15.

Divide each frequency by the total number of observations.

ANSWER 11.

An outcome is one specific result of an experiment.

ANSWER 16.

A null hypothesis is a precise statement that you are observing only the results of chance variation, and that there is no effect.

ANSWER 12.

An event is a collection of outcomes in a sample space. The *event* "2 heads" in a three-coin-toss experiment includes the *outcomes* HHT, HTH, and THH.

QUESTION 17.

When do you run a sign test?

QUESTION 21.

What is the mode of a set of numbers?

QUESTION 18.

A sign test is useful for testing a null hypothesis of what sort?

QUESTION 22.

Define the *variance* of a population, SD^2

QUESTION 19.

What is the mean of a set of numbers, M?

QUESTION 23.

What is the standard deviation?

QUESTION 20.

What is the *median* of a set of numbers, Md?

QUESTION 24.

Define a *statistic*.

ANSWER 21.

The mode of a set of numbers is the number which occurs more often than any other number.

ANSWER 17.

When your experiment consists of a series of independent tries or trials or subjects, each of which falls into one of two separate categories; and when you are interested in finding out whether one category is more probable than the other.

ANSWER 22.

The variance of a population is the average squared deviation from the mean, averaged over all the numbers in the population:

$SD^2_{pop} = \Sigma(X_i - M_{pop})^2 / N$

ANSWER 18.

Each of the two classes of observation is equally likely. There is a 50 : 50 chance that any given observation will lie in either class.

For every _____ there is a 0.5 chance that you will observe _____ and an equal chance that you will observe _____.

ANSWER 23.

The standard deviation (SD) of a population is the square root of the variance:

$SD = \sqrt{\text{variance}} = \sqrt{SD^2}$.

ANSWER 19.

The mean is the average: You add up all the numbers and divide by n, the number of values:

$M = \Sigma X_i / n$.

ANSWER 24.

A statistic is a measurement of a sample: a number calculated on the basis of a sample. Examples include the sample mean, the number of elements in the sample, and the correlation between IQ and GPA for a sample of students.

ANSWER 20.

The median in a group of numbers is a number chosen so that as many numbers are smaller than the median as are bigger than the median.

QUESTION 25.

Define a *parameter*.

QUESTION 26.

What is the rule for estimating the variance of a population, on the basis of the numbers in a sample.

QUESTION 27.

State the rule of thumb.

QUESTION 28.

What is the standard score or Z score?

QUESTION 29.

How many numbers do you need to completely determine or specify a normal distribution?

QUESTION 30.

For a continuous probability distribution, what is the rule for computing the probability of an observation between a and b?

QUESTION 31.

What is the area under a probability density function?

QUESTION 32.

What is the sampling distribution of the mean, for a given sample size, n? How could you approximate it experimentally?

ANSWER 29.

Two: the mean and the variance.

ANSWER 25.

A parameter is a measurement of an entire population; for example, the population mean, or the true correlation between IQ and GPA for an entire population.

ANSWER 30.

The probability of an observation between a and b is the area under the curve bounded by vertical lines at a and b.

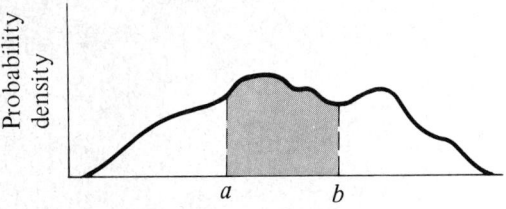

ANSWER 26.

$$\text{Est } SD^2_{pop} = \frac{\Sigma(X_i - M_{samp})^2}{n - 1}$$

where M_{samp} is the mean of the sample, n is the number of values in the sample, and the summation is over all the values, X_i, in the sample.

ANSWER 31.

One square unit: 1.00.

ANSWER 27.

For most populations you are likely to encounter, about 95 percent of the members of the population lie within 2 standard deviations of the mean.

ANSWER 32.

The sampling distribution of the mean specifies the probability of various possible values of the sample mean, M_{samp}, when samples of size n are taken from a population. You could approximate it by taking a sample of n numbers, finding the mean, recording it, and replacing the numbers; repeat this operation a great many times, and plot the resulting frequency distribution.

ANSWER 28.

A Z score tells you how many standard deviations there are between the mean and a particular observation.

$$Z = \frac{X_i - M}{SD}$$

where M is the mean and SD the standard deviation of the population in question.

QUESTION 33.

What is the unit normal distribution? What is another name for it?

QUESTION 34.

What is the addition rule for probabilities?

QUESTION 35.

What percentage of the observations from a normal distribution lie within 1 standard deviation of the mean? 2 SDs? 3 SDs?

QUESTION 36.

What is the variance of the sampling distribution of the mean, SD_M^2, for samples of size n?

QUESTION 37.

What is the mean of the sampling distribution of the mean, M_M?

QUESTION 38.

What does the central limit theorem say about the shape of the sampling distribution of the mean?

QUESTION 39.

When do you run a Z test?

QUESTION 40.

What are the five steps involved in a Z test?

ANSWER 37.

It is equal to the mean of the population from which the samples are drawn:

$M_M = M_{pop}$

ANSWER 38.

It says that the sampling distribution of the mean will be a normal distribution if the parent population is normal, or if the sample size is large (greater than 30).

ANSWER 39.

When you want to know if a sample came from a specified population with known mean and variance. In order to use a Z test, the sampling distribution of the mean must be a normal distribution.

ANSWER 40.

A. State the null hypothesis.
B. Set a significance level, α.
C. Determine a rejection region for the test statistic, Z.
D. Compute Z.
E. Make a decision.

ANSWER 33.

A normal distribution with mean 0 and variance 1. It is also called the Z distribution.

ANSWER 34.

If two events, A and B, are nonoverlapping, then the probability that either A or B occurs is the sum of the two probabilities:

$P(A \text{ or } B) = P(A) + P(B)$.

ANSWER 35.

68 percent; 95 percent; 99.7 percent

ANSWER 36.

$SD_M^2 = \dfrac{SD_{pop}^2}{n}$

where SD_{pop}^2 is the variance of the parent population from which the samples are drawn.

QUESTION 41.

What is a binomial distribution? A symmetric binomial distribution? Give an example.

QUESTION 42.

For the symmetric binomial distribution, and N trials, what are the mean and the variance?

QUESTION 43.

When do you run a sign test using the normal approximation to the binomial?

QUESTION 44.

When do you run a t test?

QUESTION 45.

Write the formula for the t statistic, in terms of M_M, the mean of the sampling distribution of the mean; Est SD_M, the estimated SD of the sampling distribution of the mean; and M_{samp}, the sample mean.

QUESTION 46.

In computing a t statistic, you find Est SD_M, the estimated SD of the sampling distribution of the mean, on the basis of the numbers in the sample. Indicate the two rules you use.

QUESTION 47.

For samples of size 1000, how will the t statistic be distributed?

QUESTION 48.

Suppose you are using a t test to decide whether a person's IQ score has been influenced by a training program. For 15 people you have pretest and posttest scores. State the null hypothesis.

ANSWER 45.

$$t = \frac{M_{samp} - M_M}{\text{Est. } SD_M}$$

Note the similarity to a Z score.

ANSWER 46.

1. $\text{Est. } SD_{pop}^2 = \dfrac{\Sigma(X_i - M_{samp})^2}{n-1}$

2. $\text{Est. } SD_M^2 = \dfrac{\text{Est. } SD_{pop}^2}{n}$

ANSWER 47.

Like the Z statistic: according to the unit normal distribution.

ANSWER 48.

H_0 : The sample of 15 difference scores (posttest minus pretest) came from a normal distribution with a zero mean.

ANSWER 41.

A binomial distribution is a probability distribution giving the probability of each of the $(n + 1)$ possible events when a simple dichotomous (two-alternative) experiment, like coin tossing, is carried out n times.

A symmetric binomial distribution is one in which each of the two alternatives is equally likely; for example, heads is as likely as tails.

The probability distribution giving the probabilities of 0/10, 1/10, etc., 10/10 correct in the ESP experiment is a symmetric binomial distribution.

ANSWER 42.

$M = N/2; \quad SD^2 = N/4.$

ANSWER 43.

In a standard sign-test situation where the number of trials is large. When you have at least 20 trials, the approximation is good.

ANSWER 44.

When you have a hypothesis about the mean of the parent population and know that the parent population is approximately a normal distribution, but you do not know the variance of the parent population.

QUESTION 49.

What is the rule for calculating a correlation coefficient?

QUESTION 50.

Sketch scatterplots for correlations of -1; 0; $+0.8$; and $+1$.

QUESTION 51.

If you know that two variables are highly correlated, what can you deduce about the causal relationship between them?

QUESTION 52.

When do you run an r test?

QUESTION 53.

What is the rule for computing a chi-square statistic?

QUESTION 54.

What is the special restriction on expected frequencies, for the chi-square test?

QUESTION 55.

How can you compute the expected frequency, E, in a contingency table like this?

QUESTION 56.

If a contingency table has R rows and C columns, how many degrees of freedom has the chi-square statistic when the null hypothesis is true?

ANSWER 53.

$$\chi^2 = \sum \frac{(E-O)^2}{E}$$

where the summation is over all the cells in a contingency table. For each cell, E is the expected frequency and O is the observed frequency.

ANSWER 54.

Each expected frequency, E, must be 5 or greater.

ANSWER 55.

$$E = \frac{R \times C}{T}$$

where R is the row total, and C is the column total, for the row and column of the cell in question. T is the total number of entries in the table.

ANSWER 56.

$(R - 1) \times (C - 1)$

ANSWER 49.

$$r = \frac{\Sigma Z_{X_i} \times Z_{Y_i}}{n}$$

where the summation is over all the n pairs; where the Z_{X_i} are computed on the basis of the numbers in the X list, and the Z_{Y_i} are computed on the basis of the numbers in the Y list.

ANSWER 50.

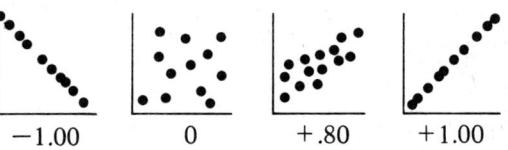

ANSWER 51.

Nothing.

ANSWER 52.

When you have computed a correlation on the basis of n pairs of numbers, and you wish to decide if there is any reasonable chance of observing so extreme a correlation coefficient if the two variables (sets of scores) in the population from which your sample of pairs is taken were totally unrelated.

QUESTION 57.

Define the power of a test. Which is usually more powerful, the *t* test or the sign test?

ANSWER 57.

Power is defined as $1 - \beta$, where β (beta) is the probability of a Type Two error. The power of a test is the chance of finding an effect (rejecting the null hypothesis) when it is really there. In general, the t test is more powerful than the sign test.

Index

Absolute value, 66
Addition rule for probabilities, 91–92
Average. *See* Mean
Average absolute deviation, 66–67
Average squared deviation, 67. *See also* Variance

Bimodal distribution, 62–63, 105
Binomial distribution, 130–134
    asymmetric, 134
    finding a similar normal distribution, 131–132
    formulas for mean and variance, 132
    probabilities for 20 trials, 130–131

Central limit theorem, 105–108, 110–117
    concise statement of theorem, 115–116
Central tendency, measures of, 61–62
Chi-square statistic, 196–197
    formula for expected frequencies, 201–202
    table of critical values, 198
Chi-square test, 194–210
    degrees of freedom, 203
    restriction on expected frequencies, 207
    for theory checking, 208–209
    worked problem, 205–207
Chlorine example, 128–130
Coin-toss tree, 31, 32
Contingency table, 195
Correlation, curvilinear, 173
Correlation coefficient, r, 173–176
    computational formula, 178–179
    sampling distribution, 185–186
    table of critical values, 186, 187
    test of significance, 185–188, 191
    when $n = 2$, 182–185
Counting rules, 43, 29–44
Critical value, 139

Darwin, Charles, 106
Decisions and errors, 2-by-2 table, 18
Decision rules, 17–18
    how to choose, 21–28
Difference scores
    sign test for, 51
    difference of zero, 57
    size ignored, 57
    t-test for, 165–166

Error, Type One. *See* Type One error
Error, Type Two. *See* Type Two error
ESP experiment, 5–49
Est. SD^2 pop. *See* Estimated variance
Estimated variance, 72
    why you divide by $n - 1$, 74, 75
    worked problem, 96, 99
Events, 43
    coin example, 30, 36
    dice example, 48–49
    mutually exclusive, 92
    probability of, 36

Five steps for statistical tests, 127
Flex point on a curve, 132, 135
Frequency, relative, 42, 44
Frequency distribution, 13
   graph of, 13
Frequency polygon, 14
Fundamental question in statistics, 16–17

Galton, Sir Francis, 106
GPA example, 78–80, 105–108, 111–113

H_0. See Null hypothesis

Independence, 56

Law of Yerkes and Dodson, 173

Marble box experiment, 12–14
   1000-trial results, 27–28
Marginal totals, 195
Marijuana example, 9–10
Mean, 62
Median, 62
Mode, 62
Models and their uses, 11–13
Mozart, 194–196
Mutually exclusive events, 92

N, for populations, 74
n, for samples, 74
Normal approximation to the binomial, 130–134, 137–141
   worked example, 138–139
Normal distribution, 102–108. See also Central limit theorem; Unit normal distribution
   graph with 100 squares, 141
   table, 143, 142–147
Normal test. See Z test
Null hypothesis, 7
   practice problems, 52, 227, 232–233
   standard null hypothesis for various tests, 191
   what if you can't reject H_0?, 55, 57

Outcomes, 43. See also Events

Parameter, 72
Power of a test, 7, 166
Probability density functions, 81–89
   rectangular, 82
   triangular, 84
   solved problem, 93, 97
Probability distributions, continuous vs. discrete, 80–82. See also Normal distribution; Binomial distribution
Probability tree, 31, 32, 48
   solved problem, 96, 100

r. See Correlation coefficient
r table, 186–187

Random sampling, 79, 88
   decision whether sample is random, 78ff.
Range, 65
Rectangular distribution, 82
Regression line, 181
Rejection region, 25
   one tailed vs two tailed, 220–224
   for Z test, 122
Relative frequencies, 42, 44
Rule of thumb, 73

Sample space, 43
Sampling, biased, 110–117
   random, 79, 88
   with replacement, 88
Sampling distribution of the mean, 80. See also Central limit theorem
   GPA example, 78
   how it changes with sample size, 113–115
   how to approximate, 90
   mean of, 107
   sample size 1, 106–107
   variance of, 110–111
Sampling distributions, 163–164
   of chi-square, 202
   of difference scores, 213
   marble box, 14
   of the mean. See also Sampling distribution of the mean
   of the proportion, 88
   of the r statistic, 185–186
   of t and Z, 162–164
   of the Z statistic, 126
Scatterplots, 170–173, 188–190
   with Z scores, 178
Schoenberg, 194–196
SD. See Standard deviation
SD^2. See Variance
Self-test for review, 51–55, 92–100
Sign test
   assumptions involved, 47
   defined, 55–56
   difference scores, 51
      difference of zero, 57
   null hypothesis for, 56, 191
   probability distribution. See Binomial distribution
   six steps, 27
   when to use, 56
   worked problems, 26–27, 51–54
Sigma notation, 63–64
Significance level, 24
Sinistralics example, 123–125
Six steps for sign test, 25, 27
Square of a number, 67
Square root of a number, 68
Standard deviation, 67–68, 72–73
Standard score, 73

Statistic
    compared with parameter, 72
    defined, 72
Statistical tests: Which one do you use where?, 191
Statistics, definition, 4
    descriptive, 61ff.
    inferential, 61
Summation, sigma notation, 63–64
Swallows and storms, 9–10

t distribution, 157
t statistic compared with Z statistic, 156–157
    comparison in a sampling experiment, 162–164
t table, 160, 240
t test for means of uncorrelated samples, 212–219
    compared to t test for a single mean, 215–216
    computational formula, 216
t test for sample mean, 151–169
    compared to sign test, 166
    difference scores, 165–166
    where to use it, 164–165
Tables
    chi-square, critical values, 242
    of decimal equivalents, 251
    r, critical values, 241
    squares, square roots, and reciprocals, 243–250
    t, critical values, 240
    Z, critical values, 239
Ten-coin model, probabilities, 40
Ten-sided cylinder example, 113–115

Trees, probability, 21, 32, 48
Two-tailed tests and one-tailed tests compared, 220–224
Type One and Type Two errors,
    a comparison of the costs, 129
    trade off between probabilities, 93, 97, 129
Type One error, 7
    probability of, 21, 26
Type Two error, 7

Unit normal distribution, 126, 141

Variability of a set of numbers, 64–68
Variance, estimated, 72
    why divide by $n-1$?, 74, 75
Variance of a set of numbers, 67
    formula, 71
Volvo example, 16

Wheel of fortune example, 81

Z distribution, 126, 141
Z score, 73
    solved problems, 95, 99
Z test
    compared to sign test, 120–122
    five steps, 127
    for the mean, example, 123–125
    for a single observation, 128–130
Z table, 143, 239. *See also* Normal distribution table